124 Springer Series in Solid-State Sciences

Edited by Peter Fulde

Springer

Berlin
Heidelberg
New York
Barcelona
Budapest
Hong Kong
London
Milan
Paris
Santa Clara
Singapore
Tokyo

Springer Series in Solid-State Sciences

Editors: M. Cardona P. Fulde K. von Klitzing H.-J. Queisser

Managing Editor: H. K.V. Lotsch

Volumes 1–89 are listed at the end of the book

90 **Earlier and Recent Aspects of Superconductivity**
Editors: J. G. Bednorz and K. A. Müller

91 **Electronic Properties of Conjugated Polymers III** Basic Models and Applications
Editors: H. Kuzmany, M. Mehring, and S. Roth

92 **Physics and Engineering Applications of Magnetism** Editors: Y. Ishikawa and N. Miura

93 **Quasicrystals** Editors: T. Fujiwara and T. Ogawa

94 **Electronic Conduction in Oxides**
By N. Tsuda, K. Nasu, A. Yanase, and K. Siratori

95 **Electronic Materials**
A New Era in Materials Science
Editors: J. R. Chelikowsky and A. Franciosi

96 **Electron Liquids** 2nd Edition By A. Isihara

97 **Localization and Confinement of Electrons in Semiconductors**
Editors: F. Kuchar, H. Heinrich, and G. Bauer

98 **Magnetism and the Electronic Structure of Crystals** By V.A. Gubanov, A.I. Liechtenstein, and A.V. Postnikov

99 **Electronic Properties of High-T_c Superconductors and Related Compounds**
Editors: H. Kuzmany, M. Mehring, and J. Fink

100 **Electron Correlations in Molecules and Solids** 3rd Edition By P. Fulde

101 **High Magnetic Fields in Semiconductor Physics III** Quantum Hall Effect, Transport and Optics By G. Landwehr

102 **Conjugated Conducting Polymers**
Editor: H. Kiess

103 **Molecular Dynamics Simulations**
Editor: F. Yonezawa

104 **Products of Random Matrices** in Statistical Physics By A. Crisanti, G. Paladin, and A. Vulpiani

105 **Self-Trapped Excitons**
2nd Edition By K. S. Song and R. T. Williams

106 **Physics of High-Temperature Superconductors**
Editors: S. Maekawa and M. Sato

107 **Electronic Properties of Polymers**
Orientation and Dimensionality of Conjugated Systems Editors: H. Kuzmany, M. Mehring, and S. Roth

108 **Site Symmetry in Crystals**
Theory and Applications 2nd Edition
By R. A. Evarestov and V. P. Smirnov

109 **Transport Phenomena in Mesoscopic Systems** Editors: H. Fukuyama and T. Ando

110 **Superlattices and Other Heterostructures**
Symmetry and Optical Phenomena 2nd Edition
By E. L. Ivchenko and G. E. Pikus

111 **Low-Dimensional Electronic Systems**
New Concepts
Editors: G. Bauer, F. Kuchar, and H. Heinrich

112 **Phonon Scattering in Condensed Matter VII**
Editors: M. Meissner and R. O. Pohl

113 **Electronic Properties of High-T_c Superconductors**
Editors: H. Kuzmany, M. Mehring, and J. Fink

114 **Interatomic Potential and Structural Stability**
Editors: K. Terakura and H. Akai

115 **Ultrafast Spectroscopy of Semiconductors and Semiconductor Nanostructures** By J. Shah

116 **Electron Spectrum of Gapless Semiconductors**
By J. M. Tsidilkovski

117 **Electronic Properties of Fullerenes**
Editors: H. Kuzmany, J. Fink, M. Mehring, and S. Roth

118 **Correlation Effects in Low-Dimensional Electron Systems**
Editors: A. Okiji and N. Kawakami

119 **Spectroscopy of Mott Insulators and Correlated Metals**
Editors: A. Fujimori and Y. Tokura

120 **Optical Properties of III–V Semiconductors**
The Influence of Multi-Valley Band Structures
By H. Kalt

121 **Elementary Processes in Excitations and Reactions on Solid Surfaces**
Editors: A. Okiji, H. Kasai, and K. Makoshi

122 **Theory of Magnetism**
By K. Yosida

123 **Quantum Kinetics in Transport and Optics of Semiconductors**
By H. Haug and A.-P. Jauho

124 **Relaxations of Excited States and Photo-Induced Structural Phase Transitions**
Editor: K. Nasu

K. Nasu (Ed.)

Relaxations of Excited States and Photo-Induced Phase Transitions

Proceedings of the
19th Taniguchi Symposium
Kashikojima, Japan, July 18–23, 1996

With 135 Figures

Springer

Professor Keiichiro Nasu
Photon Factory, National Laboratory for High Energy Physics
The Graduate University for Advanced Studies
1-1, Oho, Tsukuba, Ibaraki 305, Japan

Series Editors:

Professor Dr., Dres. h. c. Manuel Cardona
Professor Dr., Dres. h. c. Peter Fulde*
Professor Dr., Dres. h. c. Klaus von Klitzing
Professor Dr., Dres. h. c. Hans-Joachim Queisser

Max-Planck-Institut für Festkörperforschung, Heisenbergstrasse 1, D-70569 Stuttgart, Germany
* Max-Planck-Institut für Physik komplexer Systeme, Bayreuther Strasse 40, Haus 16
D-01187 Dresden, Germany

Managing Editor:

Dr.-Ing. Helmut K.V. Lotsch

Springer-Verlag, Tiergartenstrasse 17, D-69121 Heidelberg, Germany

Library of Congress Cataloging-in-Publication Data applied for

Die Deutsche Bibliothek - CIP-Einheitsaufnahme

Relaxations of excited states and photo induced structural phase transitions : proceedings of the 19th Taniguchi Symposium, Kashikojima, Japan, July 18 - 23, 1996 / K. Nasu (ed.). - Berlin ; Heidelberg ; New York ; Barcelona ; Budapest ; Hong Kong ; London ; Milan ; Paris ; Santa Clara ; Singapore ; Tokyo : Springer, 1997
(Springer series in solid-state sciences ; 124)
ISBN-13:978-3-642-64516-7 e-ISBN-13:978-3-642-60702-8
DOI:10.1007/978-3-642-60702-8

NE: Nasu, Keiichiro [Hrsg.]; Taniguchi International Symposium <19, 1996, Kashikojima>; GT

ISSN 0171-1873
ISBN-13:978-3-642-64516-7 Springer-Verlag Berlin Heidelberg New York

Softcover reprint of the hardcover 1st edition 1997

Typesetting:Camera-ready copy from the authors/editor
Cover design: *design & production* GmbH, Heidelberg

SPIN: 10566804 54/3144 – 5 4 3 2 1 0 – Printed on acid-free paper

Foreword

Message from The Taniguchi Foundation

Dr. Kanamori, Distinguished Guests and Friends:

The Taniguchi Foundation wishes to welcome the participants of the nineteenth International Symposium on the Theory of Condensed Matter, who have come from within this country and from different parts of the world.

The concept of the symposium is unique in that participants, both Japanese and from abroad, are limited in number to small discussion groups, and live together, although for a short period, as a close-knit community. We feel that this kind of environment will assist towards the strengthening of understanding and the fostering of friendship among the attendees.

It is easy to talk about, but difficult to realize, the ideal of international friendship and understanding in a world which is steadily growing smaller.

So far, the Foundation has invited a total of 149 participants in this division from 24 foreign countries and 299 participants from Japan. And we are all friends.

We hope and trust that even after they have reached the heights of academic fame during the coming decades, the participants will continue to join forces and help to forge closer bonds of friendship and cooperation that will make major contributions not only to academia, but also towards world peace and the welfare of mankind.

We hope that all the participants will return home with warm memories of both this symposium and the pleasant times that we have shared.

Thank you.

Kyoto Japan,
July 23, 1996

The Taniguchi Foundation

Participants of the 19th Taniguchi International Symposium

July 18-23, 1996 at Kashikojima, Mie-prefecture, Japan

Preface

This volume contains 28 papers presented at the Nineteenth Taniguchi Symposium on the Theory of Condensed Matter, which was held during 18-23 July 1996 at Shima-Kanko Hotel in Kashikojima, Mie-prefecture, Japan. The topics of the symposium was "Relaxations of Excited States and Photo-Induced Structural Phase Transitions".

As is already well known, an electron in an insulating solid induces a local lattice distortion around itself when it is excited by a photon. This phenomenon, called "lattice relaxation" of an optical excitation, has been studied in various kinds of insulating solids for the past fourty years. According to the original concept of the lattice relaxation process, however, it is tacitly assumed to be a microscopic phenomenon, in the sense that only a few atoms and electrons are involved.

In recent years, on the other hand, there have appeared various unconventionally photo-active materials, in which the relaxations of optically excited states result in collective motions or drastic structural changes, involving a large number of atoms and electrons. In some cases, this results in a semi-macroscopic excited domain with a new lattice structure quite different from the original material. This situation can be called "photo-induced structural phase transition".

The aim of this symposium was to discuss this new problem from the view points of both material science and solid-state spectroscopy. Phenomena within the scope of this conference are as follows.

- Optical excitations, lattice relaxations and photo-induced phenomena in CH_x, polydiacetylene, M-X chains, $BaBaO_3$, La_2CuO_4 and other perovskites.
- Optical properties of the C_{60} crystal and photo-induced polymerization.
- Self-trapped excitons, polarons, solitons and other nonlinear photo-excitations various insulators.
- Neutral-ionic transitions of organic charge-transfer salts
- Photo-induced structural changes in semiconductors and amorphous materials.

In order to avoid hurried and superficial discussions, the total number of participants in the symposium was limited to 28 (10 foreign guests) and a sufficiently long time was allocated for each talk in an informal and relaxed atmosphere, so that we can attain a deep mutual understanding. This is the original intention of the Taniguchi Symposium.

The present symposium was fully supported by the Taniguchi Foundation, which has also supported the other 18 symposia of this type. The basic idea of the symposia was developed by Mr. Toyosaburo Taniguchi, one of the most out-standing entrepreneurs in Japan, who regrettably passed away two years ago. On behalf of all the participants of this symposium, I would like to cordially thank the Taniguchi Foundation for its support to make this symposium possible.

Tsukuba Japan,
September 12, 1996

Keiichiro Nasu

Contents

Part I

General Theory for Photo-Induced Phase Transitions

Multistabilities of the Ground States, Proliferations of Excitons and Photo-Induced Structural Phase Transitions

K. Nasu

**Photon Factory, National Laboratory for High Energy Physics,
The Graduate University for Advanced Studies,
1-1, Oho, Tsukuba, Ibaraki, 305, Japan**

Abstract An electron in an isulating solid induces a local lattice distortion around itself, when it is excited by a photon. This phenomenon called "lattice relaxation" of an optical excitation has been studied in various kinds of insulating solids. However, according to the conventional concept of this relaxation, it is tacitly assumed to be a microscopic phenomenon in the sense that only few atoms and electrons are involved. In the recent years, on the other hand, there have been appeared various new unconventionally photo-active solids, where the relaxation of optical excited states results in a semi-macroscopic excited domain with a new structural order different from the original one. This situation can be called "photo-induced structural phase transition". In this paper, we review our recent theoretical studies on this problem.

1. General Concept of the Photoinduced Phase Transition

As is already well known, an electron in an isulating solid induces a local lattice distortion around itself, when it is excited by a photon. This phenomenon called "lattice relaxation" of an optical excitation has been studied in various kinds of insulating solids for these fourty years. According to the original concept of this lattice relaxation, however, it is tacitly assumed to be a microscopic phenomenon, in which only few atoms and electrons are involved.

In the recent years, on the other hand, there have been appeared many unconventionally photo-active solids, where the relaxation of optical excited states results in various collective motions involving a large number of atoms and electrons. In some cases, it results in a semi-macroscopic excited domain with new structural and electronic orders different from the original ones. This situation can be called "photo-induced structural phase transition", and the purpose of this paper is to review our recent theoretical studies on this phenomenon.

This phenomenon is closely related with the structural multi-stability intrinsic to each solid. If the ground state of a solid is degenerate or pseudo-degenerate, being composed of several true and false ground states with each structural and electronic orders different from others, we call it multi-stable. In this case, the photo-absorption, being initially a single-electron excitation from one of the true ground states, can triggers local instabilities. The photo-absorption can induce low-lying collective excitations during the lattice relaxation, and can finally produce another ground state at the expense to create boundaries between the two states.

Springer Series in Solid-State Sciences, Volume 124
Relaxations of Excited States and Photo-Induced Structural Phase Transitions
Ed.: K. Nasu

Thus a local excited domain appears.

The origin of the perfect degeneracy is already well known. The macroscopic ground state of a quantum many-particle system always appears through some broken symmetry, and it is the origin of the perfect degeneracy. In this case, the the ground state is composed of several true ones, all of which are not the eigenstates of the total Hamiltonian.

While, the origin of the pseudo-degeneracy can be understood from the conceptional point of view to calculate the total cohesive energy of a solid. The solid is composed of macroscopic numbers of few kinds of atoms and molecules with a definite composition ratio. However, even if the constituent atoms(molecules) are defined, the structural and electronic orders which will be realized in the macroscopic ground state, are not always determined straightforwardly. For example, in the case of alkali halide, the ionic state and the covalent one are two well-known candidates for the ground state. Even in other cases, there are usually several candidates.

Moreover each low-lying candidate can not inherit all the natures of the total Hamiltonian of this system evenly, but does only a few particular elements of it. Hence, if there are two predominant but mutually conflicting elements in the original Hamiltonian, there appear two candidates that sensitively inherit this conflict. Thus, we can get the multi-stability or the pseudo-degeneracy with the true and false ground states.

As mentioned before, even the true ground state can inherit only a few particular elements of the original total Hamiltonian, and it is but one possibile state among various possibilities coming from this Hamiltonian. In other words, if we pay our attention only to the ground state, we will have a great prejudice against other possible states. Consequently, in order to fully clarify the natures of our system without prejudice, we have to study at least other candidates than the true ground state. As mentioned occasionally, we can find them through the photo-excitations and the lattice relaxations therefrom.

It is also very important to see the relation between photo-induced phase transition and the ordinary one by the thermal excitation. Let us consider the aforementioned pseudo-degenerate case. When the energy of the false ground state relative to the true one is low enough to be excited by the thermal energy of the order of room-temperature, we may get the ordinary phase transition. As far as this ordinary phase transition can occur, the resultant state will be almost same as that we can get through the photo-excitation. However, there will be various other cases that the energy of the false ground state is too high to be excited thermally. Even in this case, we can create it by the photo-excitation and the lattice relaxation therefrom.

2. One-Dimensional Case

Let us take halogen-bridged mixed-valence metal complexes (HMMC's) [1],

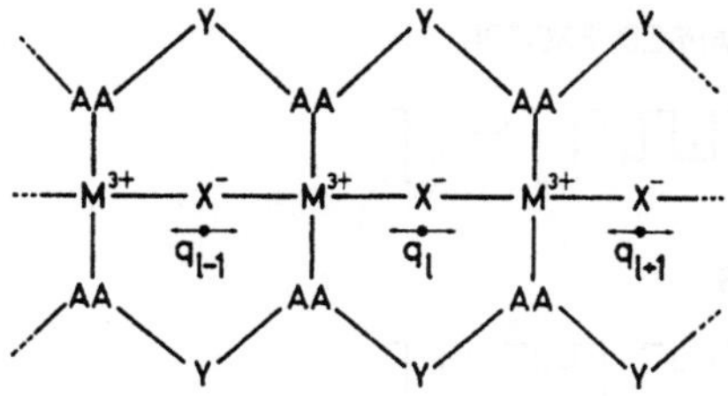

Fig. 1 Schematical structure of HMMC. M=Pt, Pd or Ni. X=Cl, Br or I. AA and Y denote a ligand and a counter-ion, respectively.

as one of typical examples for quasi one-dimensional cases, and discuss the lattice relaxation of an exciton in this system. The HMMC is composed of trivalent transition metal ion($\equiv M^{3+}$, M=Pt, Pd, Ni)'s, bridged by halogen ion($\equiv X^-$, X=Cl, Br, I)'s, as schematically shown in Fig. 1. M^{3+} is coordinated by a ligand($\equiv$AA), where AA is a planer organic molecule such as ethylenediamine, ethylamine or cyclohexanediamine. These AA's are bound together through the counter ion ($\equiv$Y)'s, where Y is ClO^{4-} or BF^{4-}. These Y's bring about a weak interchain interaction, and contribute to stabilize the chain structure of the HMMC. The essential factor that determines the electronic properties of the HMMC is the strong coupling between an unpaired electron in the d_{z^2} orbital of M^{3+} and the phonon modes of the X^- (, the z axis is parallel to the chain). This strong coupling brings about various collective and nonlinear phenomena in the lattice relaxation process of exciton.

Since we have an unpaired electron in the d_{z^2} orbital, we can expect that this system will be a metallic. As is well known, however, the one-dimensional metallic state of electrons is unstable against the electron-phonon(e-ph) interaction, and it results in the Peierls transition[2]. In the case of HMMC, it has been confirmed by various experiments that the charge transfer occurs between two neighboring M's so as to give the following mixed-valence state:

$$\text{▬}X^- \text{----} M^{+(3-\delta)} \text{----} X^- \text{▬} M^{+(3+\delta)} \text{▬} X^- \text{----} M^{+(3-\delta)} \text{----} X^- \text{▬}$$

$$\text{----} X^- \text{▬} M^{+(3+\delta)} \text{▬} X^- \text{----} M^{+(3-\delta)} \text{----} X^- \text{▬} M^{+(3+\delta)} \text{▬} X^- \text{----}$$

where X^-'s have been distorted with a new periodicity which is twice the period of the original lattice. δ denotes the degree of charge transfer. This is nothing but the charge density wave(CDW) state which is doubly degenerate, according to the two possible phases of the Peierls distortion as seen above. In this CDW state, the d_{z^2} electrons are influenced by the potential energy as schematically shown in Fig. 2a). Because of the electrostatic repulsion of X^-, the energy levels of the d_{z^2} orbitals become alternately higher and lower as X^- approaches or leaves the neighboring M.

In Fig. 3, we have shown the relevant adiabatic energies quite schematically as a function of an order parameter($\equiv$Q), which denotes the displacement of X^- from the middle point of the adjacent two M's. It is nothing but the Peierls type lattice distortion. The solid line denotes the ground state, and the dashed line denotes the excited one. The meaning of the abscissa Q in the case of the excited state is rather conceptional, and it expresses the local lattice distortion around the exciton. As shown in this figure, there are three energy minimum points in the ground state.

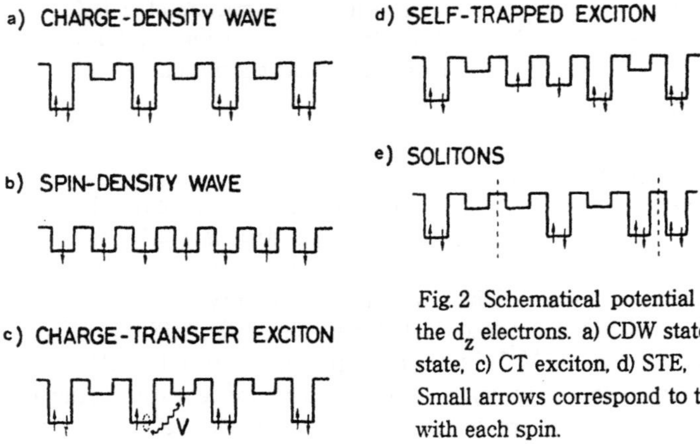

Fig. 2 Schematical potential energy for the d_z electrons. a) CDW state, b) SDW state, c) CT exciton, d) STE, e) solitons. Small arrows correspond to the electron with each spin.

Two minima where Q is equal to Q_c or $-Q_c$ correspond to the CDW states. They have different phases of the Peierls distortion of X^-, as described before. When the electron-electron Coulombic repulsion is taken into account, we get another energy minimum at Q=0, in addition to these CDW states. In this configuration, the up- and the down-spin are arranged alternately along the metallic sites M's to make this repulsion weak, as shown in Fig. 2b). This state is usually called a spin-density wave (SDW) state. The relative stabiltity between this SDW state and the aforementioned CDW one is determined by the competition between the e-ph interaction and the Coulombic repulsion[3]. In case of the HMMC's, it has been confirmed by various experiments that the ground state is the CDW when M is Pt or Pd, while, it is the SDW when M is Ni. Thus, we can see that the HMMC has a multi-stability, and if the e-ph interaction exceeds the repulsion only a little, this system has one false ground state and two true ones.

Referring Fig. 2 and Fig. 3, let us briefly foresee the lattice lattice relaxation processes of exciton together

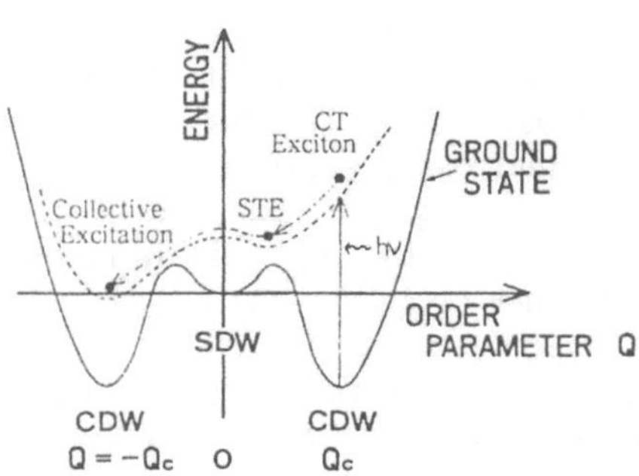

Fig. 3 Shematical potential energy surfaces of ground and excited states as a function of the CDW- type order parameter.

with its nonlinearities and collectivities. The HMMC has a strong light absorption band in the visible region[4], which corresponds to the charge-transfer (CT) excitation of an electron from the occupied d_{z^2} orbital to the vacant ones, as shown in Fig. 2c). Thus, an electron and a hole are created, and they attract with each other through the inter-orbital Coulombic force. Such a state is usually called a CT exciton. In Fig. 3, this type photo-excitation from one of the CDW ground states is represented by a vertical arrow. Once the exciton has been created, it will relax down through the potential curve of the excited state, and induces a local lattice distortion around it.

This photo-excitation is nothing but a backward charge transfer, and the electron number per orbital becomes almost equal in this local region (Fig. 2c), and therefore, the Peierls distortion has lost the reason for its presence. Hence, it tends to disappear, as seen from Fig. 2d). Thus, the exciton self-induces a local lattice distortion, and it is trapped in it. This is usually called a self-trapped exciton (STE). However, since this STE appears near the point with Q=0 as shown in Fig. 3, the SDW type spin order is also expected to appear in this localized state. In that sense, the STE appeared here has a strong collective nature, being quite different from the one in the ordinary insulators.

As seen from Fig. 3, the STE is expected to undergo further relaxations. In the CDW, as is well known, we have a low-lying excited state with a collective nature. That is, one true ground state can appear locally in the other true ground state, at the expense to create boundaries between two states. This boundary is usually called a soliton, and the exciton is expected to finally relax down to the state with a pair of solitons, as shown in Fig. 2e). In contrast to the initial single charge transfer occurred from the true ground state, a large number of charges have now been transferred to make this pair. That is, the exciton in the CDW proliferates through the relaxation, and it finally relaxes down to the collective excited state, in which a great deal of excitons or electron-hole pairs have condensed. Thus, we can get an excited domain

2.1 One-Dimensional Extended Peierls-Hubbard Model

In order to clarify the aforementioned nonlinearity much more in detail, let us study the following quasi one-dimensional extended Peierls-Hubbard model ($\equiv$ H), ($\hbar$ =1),

$$H = -T\sum_{\ell,\ell',\sigma} (a^+_{\ell\sigma}a_{\ell'\sigma} + h.c.) + \omega\sum_{\ell} q_\ell^2/2 + \sqrt{S\omega}\sum_{\ell\ell',\sigma} (q_\ell - q_{\ell'})n_{\ell\sigma}$$

$$+ U\sum_{\ell} n_{\ell\alpha}n_{\ell\beta} + V\sum_{\ell,\ell',\sigma} n_{\ell,\sigma}n_{\ell',\sigma'}, \qquad n_{\ell\sigma} \equiv a^+_{\ell\sigma}a_{\ell\sigma}, \tag{1}$$

where ℓ' denotes a nearest neighboring site of ℓ. $a^+_{\ell\sigma}$ is the creation operator of an electron localized at a metallic site ℓ with spin σ (=α, β). T is the transfer energy between nearest neighboring orbitals. q_ℓ is the dimensionless coordinate of the displacement of the lattice localized at the metallic site ℓ, and ω is the energy of this phonon mode. Its kinetic energy is neglected under the adiabatic approximation. S is the site-diagonal electron-phonon coupling energy. U and V denote the intra-site and the inter-site Coulombic repulsive energies, respectively. We also use dimensionless coordinate $Q_\ell [\equiv (S\omega)^{1/2} q_\ell]$ to obtain a convenient form of the Hamiltonian.

Let us now explain our method to calculate the ground and excited states. In the case of the ground state, we use mean-field theory for interelectron interactions, and the adiabatic approximation for phonons. Within mean-field theory, we can approximate H by the mean-field Hamiltonian($\equiv H_{HF}$), in which $n_{\ell\sigma}$ and $(a^+_{\ell+1\sigma} a_{\ell\sigma})$ are replaced by their averages so that

$$n_{\ell\sigma} \Rightarrow \langle n_{\ell\sigma} \rangle , \qquad (a^+_{\ell+1\sigma} a_{\ell\sigma}) \Rightarrow \langle a^+_{\ell+1\sigma} a_{\ell\sigma} \rangle .$$

These $\langle n_{\ell\sigma} \rangle$ and $\langle a^+_{\ell+1\sigma} a_{\ell\sigma} \rangle$ are unknown parameters to be determined later self-consistently. By diagonalizing this H_{HF}, we can obtain energies of the ground and excited states. In order to take the electron-hole correlation in the excited state into account, we define the difference($\equiv \Delta H$) between the mean-field Hamiltonian H_{HF} and the true one H as

$$\Delta H = H - H_{HF} ,$$

and diagonalize it within the basis of the one-electron excited states obtained from H_{HF}. This is first-order perturbation theory for the electron-hole correlation. Then, we can determine the new energies of excited states and their wave functions, including the exciton effects. By working this process out for various lattice configurations, we can obtain adiabatic potential energy surfaces.

As mentioned occasionally, our system is highly nonlinear in the sense that the electrons and phonons are strongly correlated with each other. In order to study this nonlinearity from quantitative point of view, let us now define an index which expresses the degree of nonlinearity of relaxation. We enumerate the number of electrons($\equiv N_e$), which is force to go out from one of the true ground states, by the optical excitation and the subsequent lattice distortions. It is written as,

$$N_e = \sum_{j=1}^{N/2} \langle \varphi_j | \overline{N}_e | \varphi_j \rangle , \qquad (2)$$

where the operator $\overline{N}_e$ is defined as

$$\overline{N}_e \equiv \sum_{i=N/2+1}^{N} | \phi_i \rangle \langle \phi_i | . \qquad (3)$$

Here, N is the total number of electrons in the system. $|\phi_i\rangle$ is the i th one-electron wavefunction of the true ground state, while, $|\varphi_j\rangle$ is the j th wavefunction of the relaxed excited state. The summation for j is taken over the occupied states, while, i is taken over the unoccupied states. Thus, N_e gives the total number of excited electrons.

Although N_e is almost unity in the Franck-Condon state, it will increase as the lattice relaxation proceeds. This increase is nothing but the nonlinearity of the relaxation. We can also schematically examine the variation of N_e from Fig. 2, as follows. N_e is equal to unity in the case of the CT exciton as shown in Fig. 2c). In the case of soliton pair described in Fig. 2e), however, there are four electrons transferred from the true ground state. That is, N_e increases from unity to four as a result of the lattice relaxation. Thus, N_e is the pertinent index that can properly express the degree of the nonlinearity of relaxation.

2.2 Relaxation in One-Dimensional CDW

Using the theoretical framework mentioned before, let us now consider the nonlinear lattice relaxation of exciton in one-dimensional CDW. To describe the relaxation path from the exciton to the soliton pair through the STE, we use the following variational function for Q_ℓ [5, 6]:

$$Q_\ell = (-1)^\ell Q \left\{ 1 + \Delta Q[\tanh\theta(|\ell| - \ell_0/2) - 1] \right\}, \qquad (4)$$

where $(-1)^\ell Q$ denotes the Peierls distortion in the CDW ground state, and this Q should be determined beforehand within mean-field theory. The curly bracket { ... } denotes the local lattice displacement from this ground state. ΔQ is its amplitude and θ corresponds to the reciprocal width of a soliton. ℓ_0 denotes the inter-soliton distance. When $\ell_0 = 0$, this pattern just corresponds to the STE-type local lattice distortion. On the other hand, when $\ell_0 \gg 1$ and $\Delta Q = 1$, the phase of the Peierls distortion is completely inverted in the region $-\ell_0/2 < \ell < \ell_0/2$. This situation corresponds to the soliton-antisoliton pair.

As for the values of parameters, we will investigate the intermediate coupling case, where the inter-site repulsion V and T are relatively large as compared with S.

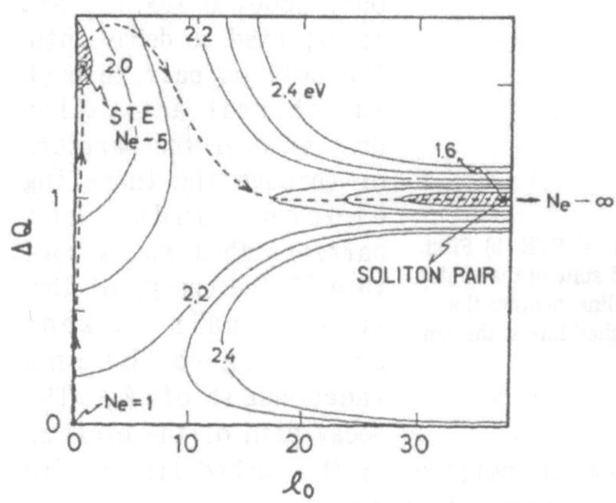

Fig.4 The contour map of the adiabatic potential energy surface of exciton for one-dimensional CDW. U=1.43eV, V=0.79eV, S=0.29eV, and T=1eV.

We have thus calculated the contour map of the adiabatic potential-energy surface of the exciton, and it is shown in Fig. 4, by using the $\ell_0 - \Delta Q$ plane. The point $\Delta Q = 0$ and $\ell_0 = 0$ corresponds to the Franck-Condon state, which is just made by photon from the true ground state. In this figure, we can find two energy minima. The minimum at $\ell_0=0$ and $\Delta Q=1.6$ corresponds to the STE, and another minimum at $\Delta Q=1$ corresponds to the pair of solitons, which is the lowest relaxed excited state of this system. The charge and the spin density profiles around the STE and the pair of solitons are shown in Figs. 5a) and b), respectively. Fig. 5a) shows that the SDW type spin order has appeared in the STE over about five metallic sites. In other words, the STE is equal to the local SDW domain. The appearance of this false ground state is due to the pseudo-degeneracy, or the severe conflict between the e-ph coupling and the Coulombic repulsion.

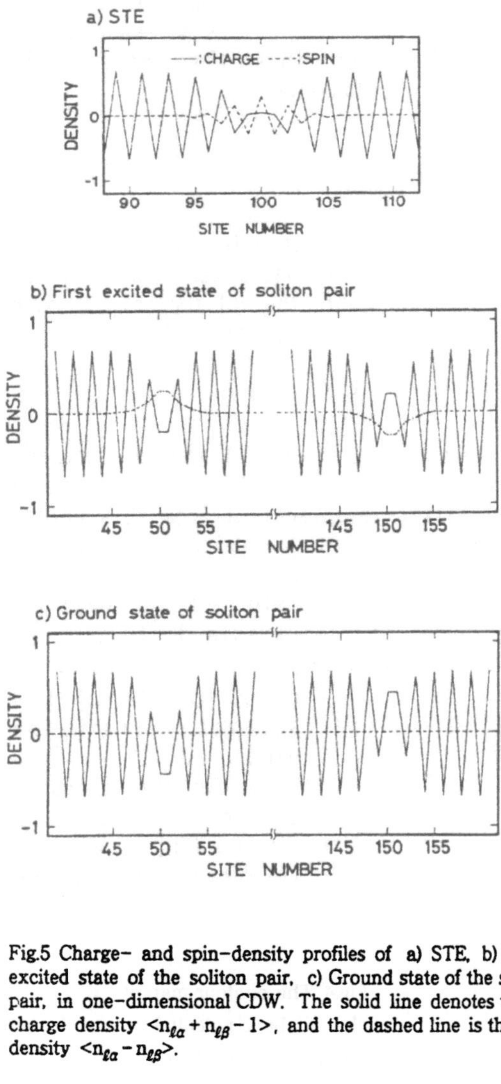

Fig.5 Charge- and spin-density profiles of a) STE, b) First excited state of the soliton pair, c) Ground state of the soliton pair, in one-dimensional CDW. The solid line denotes the charge density $\langle n_{\ell\alpha}+n_{\ell\beta}-1\rangle$, and the dashed line is the spin density $\langle n_{\ell\alpha}-n_{\ell\beta}\rangle$.

Since the energy barrier between the STE and the soliton pair is only about 0.1eV, the STE is expected to decay into the soliton pair, through the thermal activation process over the barrier, or through the tunneling process under the barrier. When ℓ_0 is more than 30, the energy of the soliton pair is kept unchanged, being independent of ℓ_0. The decay path of the exciton on the adiabatic potential surface is indicated by the dashed line in Fig. 4. This figure also shows that N_e increases from unity to about five, as the lattice relaxation proceeds from the Franck-Condon state to the STE. From this STE, the relaxation proceeds further with the increase of N_e, and finally, the soliton pair is created as the relaxed excited state. In order to visualize the decay path, we have shown in Fig. 6 the adiabatic potential energy along the decay path which is indicated by the dashed line in Fig.

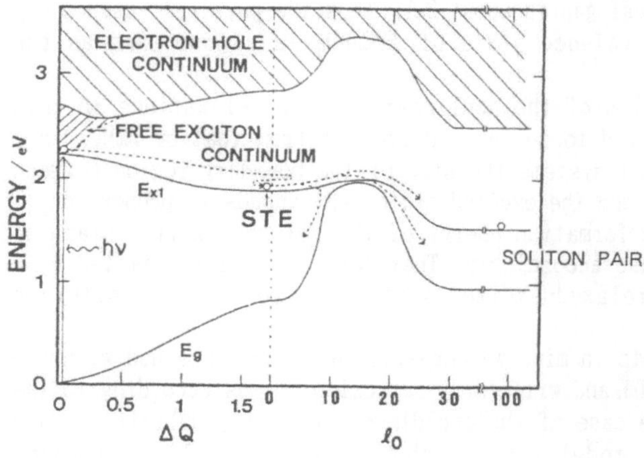

Fig. 6 Adiabatic potential energy surface along the decay path indicated in Fig. 4 as a function of ΔQ and ℓ_0.

4. In this figure, E_{X1} denotes the energy of the first excited state, namely, the energy of the exciton, and E_g denotes that of the ground state. From Fig. 6, we can see there are three decay channels of the STE. It can relax down to the excited state or the ground state of the soliton pair shown in Figs. 5b) and c), and it can also decay into the starting true ground state, nonradiatively.

The formation energy of the soliton pair depends on only the strain energy to create boundaries between the different phases. In the case of one-dimensional CDW, the boundary region is always no more than microscopic, namely, "points". Hence, as seen from Figs. 4 and 6, once the soliton has been created, its energy is kept unchanged even if ℓ_0 increases. Therefore, the two solitons can depart from each other entirely, and N_e will become infinite without any resistance. It means that, in the case of one-dimensional CDW, another true ground state can be created from the starting true ground state only by one photon. This is one of the distinctive features of the nonlinear lattice relaxation in one-dimension.

3. Relaxation in Two-Dimensional CDW

Let us now proceed to lattice relaxations in higher dimensional CDW's than the previous one dimension. One of the typical examples for higher dimensional CDW states is $BaBiO_3$, which has a three-dimensional perovskite structure[7, 8, 9, 10]. In this compound, the electrons in the 6s orbitals of Bi interact with the phonon modes of O's, which are bridging neibouring two Bi's. This e-ph interaction causes the mixed valence state composed of Bi^{3+} and Bi^{5+}. In this mixed valence state, O's approach to Bi^{5+} sites and leave Bi^{3+} sites. This new structure appears in all Cartesian coordinate axes (x, y and z) with twice the period of the original lattice. Consequently, we get an insulator whose optical gap is about 2eV. Thus, the physical essence why we can get this mixed valence state of $BaBiO_3$ is almost same as the aforementioned HMMC.

However, the natures of the nonlinear lattice relaxations in this type material are expected to be quite different from that of HMMC. In a two- or three-dimensional systems, the size of the boundary region between the ground state domain and the excited state one expands in proportion to ℓ_0 or ℓ_0^2, and hence, the formation energy of the relaxed excited state is also expected to increase accordingly. Therefore, N_e will be limited in a finite value, and the relaxed excited state is also confined within a limited region.

Keeping these points in mind, we hereafter will be concerned with the quasi two-dimensional CDW, and will carry out calculations according to the same priciple as in the case of the one-dimension. For simplicity, we use the molecular crystal model for the phonons, with a following model Hamiltonian($\equiv H'$),

$$H' = -T\sum_{\ell,\ell',\sigma}(a^{+}_{\ell\sigma}a_{\ell'\sigma} + \text{h.c.}) + \omega\sum_{\ell} q_\ell^2/2$$
$$+\sqrt{S\omega}\sum_{\ell,\sigma} q_\ell(n_{\ell\sigma} - 1/2) + U\sum_{\ell} n_{\ell\alpha}n_{\ell\beta} + V\sum_{\ell,\ell',\sigma} n_{\ell,\sigma}n_{\ell',\sigma'}. \quad (5)$$

As for the values of parameters S, U and V appeared in this model Hamiltonian, we will also study the case of the intermediate interaction mentioned before.

If we solve the mean-field theory equation to determine the ground state of this model Hamiltonian H' under the condition that the e-ph coupling exceeds the repulsion only a little, we get the SDW type false ground state and the two CDW type true ones. This situation is almost similar to Fig. 3. Bearing this new multi-stability in mind, let us proceed to the excited state and the relaxation of the exciton in this two-dimensional CDW.

As for the displacement Q_ℓ which describes the lattice relaxation path of the excited state, we use the following variational function:

$$Q_\ell = e^{i\boldsymbol{\ell}\cdot\mathbf{q}}Q\left\{1 + \Delta Q[\tanh\theta(|\boldsymbol{\ell}| - \ell_0/2) - 1]\right\}, \quad (6)$$

Here, Q_ℓ is the dimensionless coordinate. This function gives a lattice distortion with a circular pattern centered at a metallic site. $\boldsymbol{\ell}$ is now a vector which expresses the lattice points in two-dimensional square lattice, and $\mathbf{q}$ is the wave vector of the Peierls distortion:

$$\boldsymbol{\ell} = (\ell_x, \ell_y), \qquad \mathbf{q} = (\pi, \pi),$$

where the unit of length is the lattice constant. The first factor $e^{i\boldsymbol{\ell}\cdot\mathbf{q}}Q$ of eq. (6) denotes the Peierls distortion in the CDW state, with twice the period of the original lattice in both x and y axes. ℓ_0 is a diameter of the circular distortion.

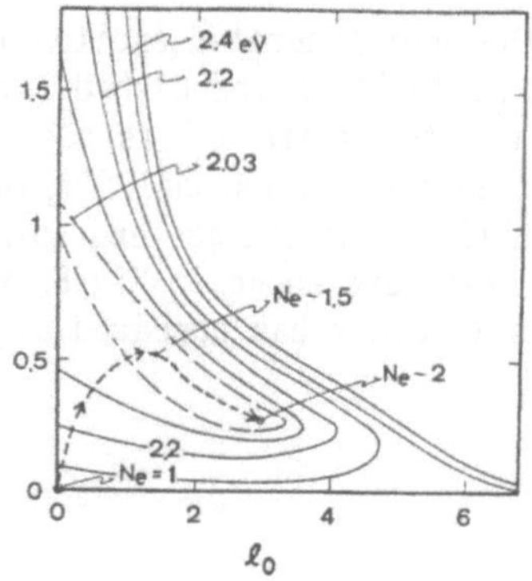

Fig. 7 The contour map of the adiabatic potential energy surface of exciton in two-dimensional CDW. U=0.5 eV, V=0.27 eV, S=0.65 eV, and T=0.5 eV.

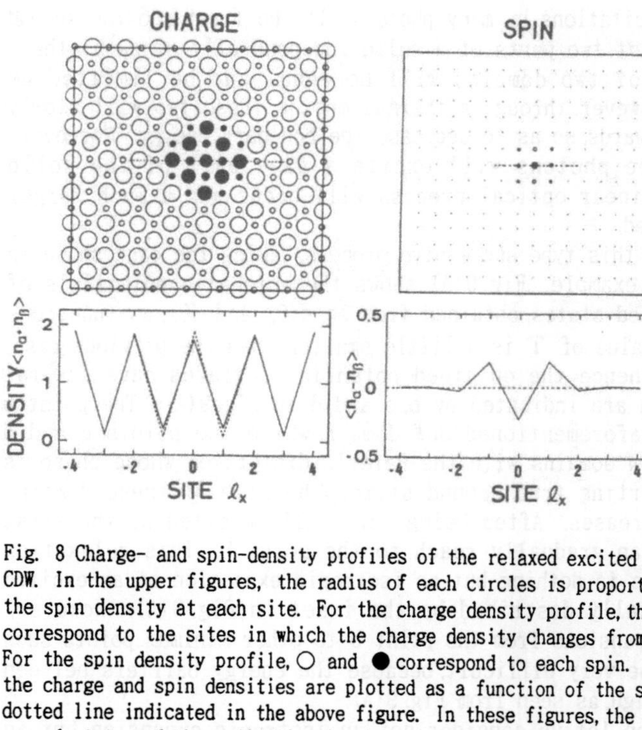

Fig. 8 Charge- and spin-density profiles of the relaxed excited state in two-dimensional CDW. In the upper figures, the radius of each circles is proportional to the charge or the spin density at each site. For the charge density profile, the black-lacquered circles correspond to the sites in which the charge density changes from the CDW ground state. For the spin density profile, ○ and ● correspond to each spin. In the lower figures, the charge and spin densities are plotted as a function of the site number ℓ_x along the dotted line indicated in the above figure. In these figures, the dashed line denotes the ground state and the solid line is the relaxed excited state.

Thus, we can calculate the contour map of the adiabatic potential energy surface of the exciton, and it is shown in Fig. 7, by using the $\ell_0 - \Delta Q$ plane, where θ is determined to minimize the energy at each point of the surface. In this figure, we can find the minimum point in the vicinity of ΔQ=0.3 and ℓ_0=3. Since there is no barrier between the

Franck-Condon state and this minimum point, the photogenerated exciton is expected to go down to this relaxed excited state along the dashed line indicated in this figure. In this relaxed excited state, N_e is nearly equal to two. Thus, we can say that, even in the two-dimensional CDW, the exciton proliferates during the lattice relaxation. The charge- and spin-density profiles around the relaxed excited state are shown in Fig. 8. We can see from this figure that the SDW type spin order has appeared over about nine metallic sites.

4. Multiphoton Domain Growth

As mentioned in the previous section, in the case of two and three dimensions, the size of the excited domain which can be created by a single photon is limitted by the energy increase due to the boundary formation. For this reason, let us now consider a possible increase of the domain size due to successive excitations by many photons. If the first and the second photons excite distant two parts of a solid independently of each other, the initial sizes of two domains will be equal and be limitted as mentioned before. However, through a thermal motion, these two will slowly merge into one afterwards, so as to decrease the boundary energy. Moreover, if the two or three photons will excite a same part of the solid successively, a nonlinear optical process will occur, and a much larger domain will be created.

Let us examine this type successive process, taking the case shown in Fig. 9, as an typical example. Fig. 9(A) shows the adiabatic potentials of the ground and excited states, obtained from eq. s(5) and (6), as functions of ℓ_0. However, the value of T is a little smaller than the previous case shown in Fig. 7, and hence the obtained potential surfaces have several minima, some of which are indicated by b, c and d in Fig. 9(A). The point b corresponds to the aforementioned SDW domain, while the points c and d correspond to the CDW domains with the Peierls distortion whose phase is opposite to the starting true ground state. The size of these domains increases as ℓ_0 increases. After being vertically excited by the first photon, our systen can gradually reach to the point b. This relaxation motion of the lattice is nothing but an isotropic expansion of an excited domain as schematically described by the inset in Fig. 9(A). However, further isotropic expansions from the point b to other minimum points such as c and d seem to be very difficult, because the energy barriers between b, c and d are very high, as seen from Fig. 9(A).

For this reason, let us consider not an isotropic expansion but an anisotropic one. The excited domain can also increase not radially but one dimensionally along the domain boundary of the state b, as schematically described by the two insets in Fig. 9(B). Here $\mu(,0 \leqq \mu \leqq 4\pi)$ denotes the angle of the newly expanded region along the boundary. In this figure, we have also shown the adiabatic potentials as functions of μ. In contrast to the Fig. 9(A), we can see that the potential barriers along this path are low enough to be overcomed by the second and third photons. Thus our system can reach to the minima c and d after one and two rotations along the boundary. This is nothing but the photo-induced domain growth.

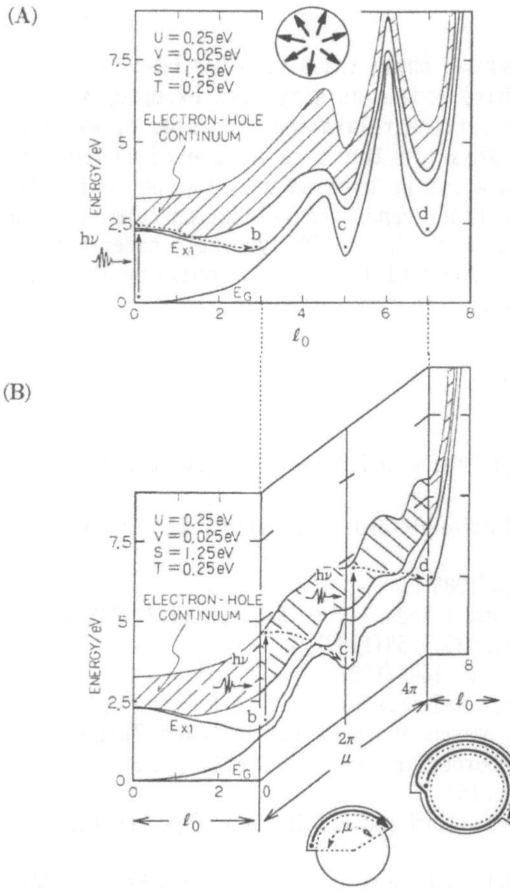

Fig.9 The adiabatic potentials of the ground and the excited states of the two-dimensional CDW. The minimum b correspond to the SDW domain, while c and d correspond to the CDW domains with an opposite phase. (A) The isotropic expansion. (B) The one-dimensional expansion along the boundary line of the state b. μ is the angle of the new domain.

Using the photo-induced absorption technique, we can easily confirm the increase of the domain size, in contrast to the simple increase of the number of domains with a definite size. As the size increases, both the peak and the intensity of the photo-induced absorption band will change, while, in the case of the simple increase of the number of domains, only the intensity will increase with no change in its spectral shape. This type phenomenon has been observed very recently by Nakamura et al[11].

5. Conclusion

The concept of the photo-induced phase transition has been developed. Using the extended Peierls-Hubbard models as a typical example, we have also investigated the nonlinear lattice relaxation processes of excitons in one- and two-dimensional CDW's. In the case of one-dimension, the macroscopic relaxed excited state can be created only by one photon. In the case of two-dimension, on the other hand, the relaxation terminates up to the domain like excited state, in which the SDW type spin order appears over several metallic sites. The growth of the excited domain by the multi photon excitation is also clarified.

References

[1] M. Yamashita, Y. Nonaka, S. Kida, Y. Hamaue and R. Aoki, Inorg. Chem. Acta, **52**, 43 (1981).

[2] R. Peierls, *Quantum Theory of Solid* (Oxford University Press, Oxford, 1933), p. 108.

[3] K. Nasu, J. Phys. Soc. Jpn, **52**, 3865 (1983).

[4] Y. Wada, T. Mitani, M. Yamashita, and T. Koda, J. Phys. Soc. Jpn. **54**, 3143 (1985).

[5] A. Mishima and K. Nasu, Phys. Rev. B**39**, 5758 (1989).

[6] M. Suzuki and K. Nasu, Phys. Rev. B**45**, 1605 (1992).

[7] D. Cox and A. Sleight, Acta Cryst. B**35**, 1 (1979).

[8] G. Ruani, A. J. Pal, C. Taliani, R. Zamboni and R. Masini, in "*High Temperature Superconductivity*", eds by C. Ferdeghini and A. S. Siri (World Scientific Publishing, Singapore, 1990), p. 164.

[9] J. F. Federici, B. I. Greene, E. H. Hartford and E. S. Hellman, Phys. Rev. B**42**, 923 (1990).

[10] J. F. Federici, B. I. Greene, D. H. Rapkine and J. Orenstein, Phys. Rev. B**43**, 8617 (1991).

[11] M. Sakai, M. Ichida, T. Yajima, A. Nakamura and H. Shinohara, Proceedings of the Annual Conference of the Physical Society of Japan (Kanazawa, Japan, March 31~April 3, 1996, in Japanese), Vo. **2**, p. 354.

The Instabilities of Excitons in Deformable Lattice and the Photo-Induced Phase Transition

Y. Toyozawa

Department of Physics, Chuo University,
1-13-27 Kasuga, Bunkyo-ku, Tokyo 112, Japan

Abstract. Possible instabilities of excitons in deformable lattice are explored in the increasing order of lattice relaxation. Particular attention is paid to the effect of dimensionality on self-trapping, the modes of symmetry breaking and the radiative versus nonradiative decay channels of an exciton. As the ultimate instability, we consider a catastrophic situation in which the self-trapped excitons are destabilized against spontaneous generation and condensation through the attractive interactions, resulting in the electronic-structural phase transition of the entire lattice. Such a view-point is shown to be useful for some quasi-one dimensional materials which are subject to instability against charge density wave (Wolffram's red) or to the neutral-ionic phase transition (organic charge transfer compounds). We apply this view-point to the latter system to interrelate the photo-induced phase transition and the phase transition under thermal equilibrium, with the use of a microscopic model with dipole-dipole interaction between the parity-broken self-trapped excitons. It is emphasized that the study of photo-induced phase transition will reveal the dynamical aspect of phase transition, thus shedding a new light on it.

1 Introduction

The aggregation of atoms to form molecules and condensed matter originates in the interatomic forces mediated by the valence electrons, namely the outermost electrons of constituent atoms extended to, or shared by, the nearby atoms. As a matter of course, the atomic configuration on the one hand and the electronic state on the other hand are intimately related with each other. However, these interatomic forces, quite large themselves, are carefully concealed behind their balance in the ground state (GS) and there is no way of observing them directly as far as the system stays in the ground or low-lying excited states.

Once a valence electron is brought into an excited state by light, however, the interatomic forces may reveal themselves through the sudden change of charge distribution or chemical bonds which naturally destroy their balance, especially significantly in nonmetallic solids with no free electrons to screen this change. Thus the atoms will seek for their new equilibrium configurations (not necessarily unique since there can be more than one minima of the adiabatic potential for the excited state), resulting in their rearrangement, or "lattice relaxation" in the case of solids.

This means that the nature of the interatomic forces which governed the ground state is revealed only in the excited state where their balance is lost, namely, through the new equilibrium configuration of atoms in the excited electronic state which we call the "relaxed excited state" (RES). The dynamics of this lattice relaxation is reflected in the correlated absorption and emission spectra, or more exactly speaking, in the second order optical spectra.

The two kinds of electronic-atomic states, GS and RES's, both of which are to be determined through the self-consistent optimization — minimization of the adiabatic energy — of the electronic-atomic system, should be considered *on an equal footing* from

Springer Series in Solid-State Sciences, Volume 124
Relaxations of Excited States
and Photo-Induced Structural Phase Transitions
Ed.: K. Nasu

the fundamental view-point: only that state which happened to have the lowest energy among others has been chosen to be realized as GS. In other words, the nature might have chosen one of RES's, instead of "the GS we know", as the true ground state if the values of relevant material constants were different.

In this way, one can learn, through the optical study of the electronic excitation and the consequent lattice relaxation, why this matter, the given set of constituent atoms, has chosen the electronic state and the lattice structure we know, from among the possible candidates. That is why the study of the RES's provides us with more profound understanding of the GS.

In the case of perfect crystal, it is the self-trapped exciton (STE) which are the candidate RES's to struggle for stability with the GS (although not all insulators have STE as RES). The STE is essentially a localized excited state although it can be located at any lattice site due to the translational symmetry. In order that such a localized entity be destabilized to form a new GS which should have its own lattice structure with *translational symmetry*, the system must be subject to *spontaneous generation* of STE's *at all lattice sites* [1]. Naturally, the STE's on nearby sites have significant interaction consisting of electronic and lattice contributions. If the interaction is attractive and so large that the cohesive energy W_1 per one STE exceeds the energy E_R of one STE (the energy of this RES measured from that of the original GS), the STE's will be spontaneously generated everywhere without any optical excitation, and be condensed to form a new electronic state on a new lattice structure [2]. Namely, E_R needs not be so small as to vanish in order that STE is destabilized against spontaneous generation and condensation.

Such an electronic-structural phase transition, which is the first order if any, can be realized in two different ways. One is to vary the ambient temperature or pressure which may in turn changes the relevant material constants such as E_R and W_1, e.g. through expansion or contraction of the lattice. In contrast to this spatially uniform phase transition under thermal equilibrium, the second is the photo-induced *transient* phase transition of a *limited region* of the material which has been studied by Koshihara et al. in some quasi-one dimensional systems [3]. This occurs most favorably when the system is on *slightly lower* side of the transition temperature T_C where the free energy difference between the two phases per one unit cell is much smaller than the photon energy: one photon absorbed for one electron excitation is enough to generate many STE's converting the lattice region of as many unit cells into the metastable phase if the energy is efficiently transferred from the electronic to the lattice system. This "metastable" phase, realized within this locally heated region of the lattice, will stay to be transiently stable (not metastable!) as long as the local temperature stays above T_C [4].

The first purpose of the present paper is to describe various stages of relaxation of an exciton in deformable lattice including the ultimate, catastrophic stage of spontaneous generation and condensation of STE's. This will bring us closer to the second purpose: to elucidate the microscopic mechanism of photo-induced phase transition and to shed a new light on the phase transition in thermal equilibrium. An emphasis will be put on one dimensional systems which are considered to be most favorable for self-trapping as well as condensation of excitons.

2 Self-trapping of an exciton and dimensionality of the lattice

An exciton will prefer self-trapping when E_{LR}, the energy lowering due to the lattice relaxation, overcomes B, the increase of kinetic energy due to the localization. To be more exact, one has to introduce a, the orbital radius (of the center-of-mass motion) of the exciton as a variable parameter. a should not be smaller than a_0, the lattice constant. The energy change of an exciton in d-dimensional deformable lattice due to an intermediate localization ($0 < \lambda < 1$, where $\lambda \equiv a_0/a$ is the "degree of localization") is then given by [5]

$$E = B\lambda^2 - E_{\rm LR}\lambda^d, \tag{1}$$

which reduces to the above statement in the limit of complete localization ($\lambda = 1$). The power-dependence of the first term comes from the uncertainty principle, while that of the second from the following argument. Suppose the region of volume a^d of the d-dimensional medium with elastic constant C is subject to uniform dilation: $\Delta = \mathbf{div}\, u(r)$ where $u(r)$ is the displacement of the lattice at r. The elastic energy, $(C\Delta^2/2)\cdot a^d$ must be supplied for this. The exciton assumed to be mostly confined within this region is subject to potential energy $\Xi\Delta$ where Ξ is the deformation potential constant. Minimization of the sum of these two energies with respect to Δ leads to the second term of eq.(1), with $E_{\rm LR} \equiv \Xi^2/(2Ca_0^d)$. A more sound argument is found in ref.[5].

As a function of λ, eq.(1) behaves differently for $d = 3$ and 1. In 3-dimensional system, the stable state changes from the free state: F($\lambda = 0$) to the self-trapped state S($\lambda = 1$) *abruptly* when the electron-phonon coupling constant g, defined by $g \equiv E_{\rm LR}/B$, exceeds unity [5]. S is metastable when $2/3 < g < 1$ while F is metastable when $g > 1$. F and S are different state, being separated by adiabatic potential barrier. In contrast, there is the only minimum at $\lambda_m = $Min.$\{(g/2), 1\}$ in 1-dimensional system, namely the stable state changes its nature gradually from F-like to S-like as g increases, and there is no clear cut distinction between the free state and the self-trapped state [6]. The RES may better be called "polaron state" since the situation resembles an electron interacting with polar mode in three dimensional lattice. The optically created free exciton ($\lambda = 0$) will straightforwardly relax to the RES ($\lambda = \lambda_m$) with stabilization energy $E_{\rm ST} = B\cdot$Min.$\{g^2/4, (g-1)\}$ in the case of $d = 1$, while it must thermally overcome (or quantum mechanically tunnel through) the potential barrier in order to be finally relaxed to STE in the case of $d = 3$ even when the binding energy of STE, $E_{\rm ST} = E_{\rm LR} - B = (g-1)B$, is positive. For this reason, one may state that $d = 1$ is more favorable for self-trapping to occur than $d = 3$ as far as the strong coupling regime is concerned.

As a matter of fact, the relaxed excitons in 3-dimensional crystals so far studied can be distinctly classified into the two types through the emission spectra: a resonant sharp line from the F-type exciton and a Stokes shifted broad band from the S-type exciton [5][7], as will be more elaborated in Section 3.

So far we have considered the effect of lattice relaxation on the center-of-mass motion only. How about the relative motion? The continuum model applied to an electron-hole pair in deformable lattice predicted [8] that a *self-decomposed* state with electron and hole being self-trapped at a certain distance can be most stable when Ξ_e and Ξ_h, the deformation potential constants for the electron and the hole, have *opposite signs.* The mean distance between the electron and the hole is determined by the balance of long range Coulomb attraction and the phonon-mediated short range repulsion between them. Self-trapped excitons in majority of alkali halides were found later to be in such states [9][10]. We have thus seen that self-trapping of an exciton not only breaks its *translational symmetry*, but sometimes breaks its *parity* so as to form nearly separated electron and hole. This fact will play an important role in later Sections.

3 Radiative and nonradiative decay of an exciton

Let us now consider the situation in which the crossing of the adiabatic potentials of the relaxing exciton and the ground state comes into play [11]. We use the configuration coordinate model to plot the energies of free and localized states of an exciton against an appropriate localized mode Q, as shown schematically in Figure 1 by the hatched continuum and the solid line split off therefrom. We are considering here the situation in which the STE at Q_m is lower than the bottom F of the free exciton band ($E_{\rm LR} - B = E_{\rm ST} > 0$). The adiabatic potential for the GS is situated lower than the bottom of the exciton band by E_0 which correspond to the optical absorption peak in the case of direct band gap. In Figure 1 we plot the adiabatic potentials I, II, III and U corresponding

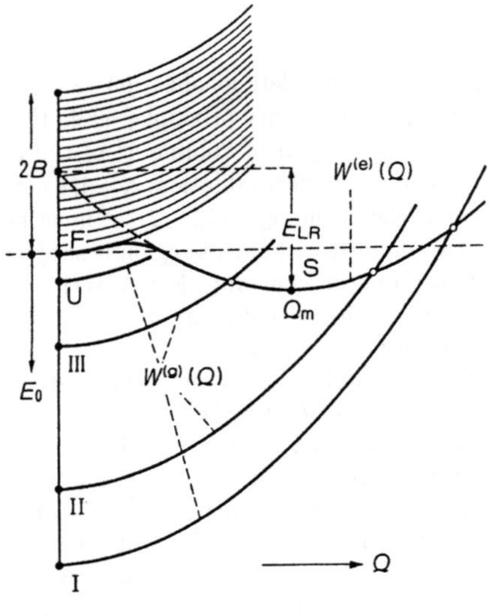

Figure 1: Adiabatic potentials for the groundstate and the free and self-trapped excitation for various values of E_0, the energy of the optical exciton.

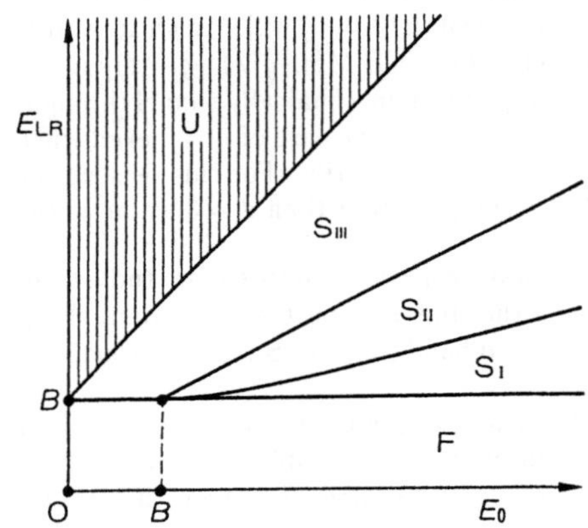

Figure 2: Classification of the relaxed excited states. F: the free exciton ; $S_{I,II,III}$: the self-trapped exciton of three types ; U: unstable against self-proliferation.

to four typical values of E_0 (in decreasing order) with a fixed set of values of B and E_{LR}. Figure 2 shows the regions on the (E_0, E_{LR})-plane corresponding to these four cases, in addition to the free exciton region F.

We resort to the Dexter-Klick-Russel criterion [12] on the competition of radiative and nonradiative processes. In case I in which the adiabatic crossing point of the ground state and the localized exciton are situated higher than F, the energy of the optically created exciton, the nonradiative transition is improbable except in very high temperature. The quantum yield for emission is nearly unity. Because of long radiative lifetime, a quasi-thermal equilibrium of population may be established between F and S excitons which allows the estimation of E_{ST}, the F-S energy difference, from the temperature dependent intensity ratio of F- and S-type emission bands, as was done for some aromatic molecular crystals [13].

In case II in which the crossing point is situated lower than F but its abscissa is still beyond the minimum point Q_m, it is energetically possible for the relaxing exciton to reach this point (after overshooting Q_m) where it will make nonradiative transition to the ground state with significant probability. The opening of this nonradiative channel affects most directly the steady state population of STE's, making the quantum yield of emission much lower than in case I. The temperature dependence of the intensity of the F-type emission which is less affected allows one to estimate the height of the potential barrier for self-trapping [14]. The STE's in alkali halides which have been most extensively studied belong to this case.

In case III of Figure 1, the RES cannot emit a photon since there is no lower electronic state at Q_m. It may not be appropriate to call such RES as STE if all that is left there is the lattice distortion or the Frenkel defect, with no electron hole pair. It is possible, however, that the exciton has been decomposed into an electron and a hole situated on nearby defects by the parity breaking mechanism mentioned in Section 2, as is exemplified by the formation of a pair of an F-center and an H-center in ultraviolet irradiated alkali halides [15][16]. The parity-broken STE in alkali halides which was mentioned in that Section is considered nowadays to be a precursor of this F-H pair. In general, the RES of case III can return to the original configuration ($Q = 0$) by thermal activation over the crossing point, or by optical excitation to that electronic state which was the ground state at our starting point. Some of the photo-structural changes observed by Tanaka et al. [17] and Kimura and Ninomiya [18] in a variety of materials seem to belong to this case III.

The case U, as hatched in Figure 2, is a fictitious situation since $E_{\mathrm{R}} = E_0 - E_{\mathrm{ST}}$, the energy of RES or STE whichever it may be called, is negative, namely the ground state from which we started is unstable. However, we should mention an illuminating example of material of which we can start with a *conceptually simple* but *actually false* ground state and get to a *real* but *complicated* ground state [19]. It is the Wolffram's red, a quasi-one dimensional material of which the skeleton chain consists of alternate array: $(\mathrm{Cl}^- — \mathrm{Pt}^{3+} —)^{2n}$, which is simple (and metallic) but a false ground state. The real ground state is an insulator with a complicated structure of doubled period: $(\mathrm{Cl}^- — \mathrm{Pt}^{4+} — \mathrm{Cl}^- —— \mathrm{Pt}^{2+} ——)^n$, which can be reached from the former through the Peierls transition with charge density wave of large amplitude. In the context of our argument, we can consider this transition as the condensation of self-decomposed STE's spontaneously generated on all unit cells.

4 Condensation of self-trapped excitons and the electronic-structural phase transition

In the argument of the case U of the preceding Section, we have ignored the interaction among the STE's, which, however, will turn out to play an important role in the phase transition through the condensation of STE's [2][4]. Assuming for simplicity that the interaction energy among them consists of the sum of the pair interactions v_{ij}, one can write down the energy of the cluster as

$$E_{\mathrm{R}[n]} = nE_{\mathrm{R}} + V_{[n]}, \quad V_{[n]} \equiv \sum_{i<j} v_{ij}, \tag{2}$$

where the summation over i and j is to be extended over the sites occupied by STE's. The shape of an energetically favorable cluster for a given size n should be that which makes best use of the attractive interactions ($v_{ij} < 0$) to maximize the binding energy $-V_{[n]}$. If the inequality

$$E_0 > E_{\mathrm{R}[n]} \tag{3}$$

holds, it is energetically possible for an optical exciton to relax in the deformable lattice and to proliferate so as to form such a cluster. If this inequality holds even in the limit of $n \to \infty$ with the cluster assumed to cover the entire lattice, namely if the inequality

$$\lim_{n\to\infty} E_0/n = 0 > E_{\mathrm{R}} - W_1, \tag{4}$$

$$W_1 = \lim_{n\to\infty}(-V_n/n) = \sum_{j\neq 0} v_{0j}/2, \tag{5}$$

holds, it must occur that the STE is spontaneously generated on every site, even without the optical excitation as is also obvious from the fact that E_0 plays no role in the inequality (4). In eq.(5), we assumed the existence of the limit — a finite value of the cohesive energy per one STE, and made use of the translational symmetry of the lattice on the right hand side. Eq.(4) means that the phase transition takes place when the STE energy becomes smaller than the cohesive energy as mentioned in Section 1. Therefore, the borderline between the regions III and U must be shifted towards right by W_1 (note that eq.(4) reads as $E_{\rm LR} - B > E_0 - W_1$).

The guiding principle to seek for those materials which are close to instability is therefore "smaller $E_{\rm R}$ and larger W_1". As for the second criterion, the conceivably largest attractive interaction is the electrostatic dipole-dipole interaction between parity-broken STE's with their dipole moments aligned along their distance. A conceivable candidate in 3-dimensional material is the 1-dimensional array, along a [110] halide line of alkali halide, of the parity-broken STE's, although this possibility does not seem to have been sought seriously.

A more realistic example is an array of parity-broken STE's in quasi-one dimensional charge transfer compounds with mix-stacked donor (D) and acceptor (A) molecules. As such STE, a D^+A^- pair in the neutral (N) phase and a D^0A^0 pair in the ionic (I) phase have effective dipole moments μ of the order of $\mp ea$, respectively, where a is the DA distance. The dipole-dipole interaction between the STE's on the i-th and j-th unit cell is then given by $v_{ij} = v_{01}|i-j|^{-3}$ where $v_{01} = -2\mu^2/(2a)^3$. Since the binding energy of the cluster of n connected STE's is given by $V_n = \sum_{j=1}^{n-1}(n-j)v_{0j}$, the cohesive energy per one STE is given, according to eq.(5), by

$$W_1 = -\sum_{j=1}^{\infty} v_{0j} = \alpha_3|v_{01}|, \quad \alpha_3 = \sum_{j=1}^{\infty} j^{-3} = 1.202. \tag{6}$$

With the use of the stability parameter γ defined by

$$\gamma = E_{\rm R}/W_1 \quad (\gamma \gtrless 1 \text{ according as the ground state is stable or not}), \tag{7}$$

one can write the cluster energy of eq.(3) as

$$E_{{\rm R}n} = n(\gamma-1)W_1 + W_n, \quad W_n = nW_1 + V_n. \tag{8}$$

With increasing n, W_n increases monotonically from W_1 to

$$W_\infty = -\sum_{j=1}^{\infty} j v_{0j} = \alpha_2|v_{01}|, \quad \alpha_2 = \sum_{j=1}^{\infty} j^{-2} = 1.645. \tag{9}$$

In Figure 3, we plot $E_{{\rm R}n}$ against n for typical values of the stability parameter, $\gamma = 1.1$ (stable phase), $\gamma = 1.0$ (transition point) and $\gamma = 0.9$ (unstable phase). If a photon with energy E ($\geq E_0 > E_{\rm R}$) is absorbed by the electronic system, it is energetically possible that the STE cluster (the domain of metastable phase) of size n is generated provided $E > E_{{\rm R}n}$, in the case of most efficient conversion of the electronic excitation energy to the growth of the cluster. One can see from Figure 3 that the increasing number of unit cells ($n \to (E - W_\infty)/\{(\gamma-1)W_1\}$ asymptotically) can be converted to the metastable phase as one approaches the transition point ($\gamma \to 1+0$), and as one uses higher energy photon. These tendencies, though predicted in ideal situation, are in fact consistent with what have been observed in the I-phase (the low temperature phase) of tetrathiafulvalene-chloranil (TTF-CA). More than a hundred of unit cells per one ultraviolet photon have been converted to the N-phase at temperature slightly below T_C, but less for lower temperature or lower photon energy [3][20].

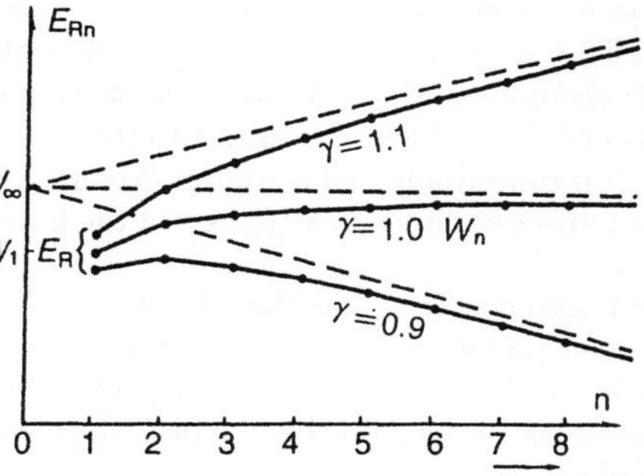

Figure 3: The energy of the cluster of self-trapped excitons as a function of its size n. γ is the stability parameter.

The fact that the metastable I-phase formed by photo-excitation of N-phase is undetectably short-lived, in contrast to the above mentioned case in which the metastable N-phase lives as long as milliseconds, can be explained by the thermodynamic argument as follows. Let us imagine the free energy versus temperature curves for the N- and I-phases which cross each other at T_C as $F_N - F_I = C(T_C - T)$. Photo-excitation of the I-phase below T_C will generate a region of N-phase, which, however, may have been locally heated above T_C during the conversion of the electronic excitation energy to the lattice energy and hence be in the stable state until it is cooled again below T_C by slow outflow of heat. In contrast the transient I-phase region generated by photo-excitation of the N-phase is even more above T_C, being in the metastable electronic state which is very short-lived.

5 Speculations on dynamics of photo-induced phase transition

We will now challenge the core problem of our subject, namely the study of microscopic dynamics of the photo-induced phase transition of one-dimensional charge transfer compounds which have been subject to intensive experimental studies for these years. It is a difficult but worthwhile task since it is expected to shed a new light on the electronic-structural phase transitions which have so far been studied in equilibrium situation. At the present stage, however, I shall have to be highly speculative, far from convincing.

The movement of the phase boundary which was discussed Nagaosa and Takimoto [21] seems to be a useful concept to describe phenomenologically the photo-induced phase transition in one dimensional system. From the microscopic point of view, the problem is how the photon energy, absorbed primarily to excite the electronic system, is transferred to the lattice system so efficiently as to generate a domain of metastable phase which can be of tremendous size. A microscopic description of the proliferation of relaxing excitons in some organic materials was made by Nasu and collaborators [22]. Hanamura and Nagaosa described the photo-induced phase transition in terms of the crossing of the adiabatic potentials of successive molecules to be rearranged [23].

Out of these studies, I have extracted the electrostatic dipole field from the parity-broken STE's as the most important incentive to their proliferation and the spatial growth of the domain of the new phase. This is also in conformity with the view-point that the electrostatic interaction plays a dominant role not only for the neutral-ionic transition but also for the dimerization [24]. The dipole field is strongly dependent upon the positional direction, and takes the greatest value along the line coaxial with the source dipole moment μ. This means that the one dimensional system with STE with its parity being broken along the axis is especially favorable for the proliferation of the STE's on successive neighbors. This is the same mechanism as the domain growth of spontaneous polarization in ferroelectric materials. According to Figure 3, the energy needed to induce another STE is a decreasing function of size n, tending to a very small value near the transition temperature, which will be readily supplied from the lattice which is under relaxation.

Yet there may be an initial critical size beyond which the growth can proceed quasi-automatically. The recent observation on TTF-CA by Tanimura et al. [20] show the existence of a threshold of laser fluence for the photo-induced I-to-N transition in the case of direct excitation of charge transfer excitons by infrared light, in contrast to the case of intramolecular excitation by ultraviolet light. It is conceivable that the latter excitation is immediately converted to two or more charge transfer excitons so as to exceed the initial critical size before the lattice relaxation starts.

A completely different mechanism which is also conceivable is that the electric field due to a moderately separated electron hole pair, possibly generated under ultraviolet excitation, induces polarization or depolarization of the region between them, converting it into another phase. The result is a pair of polarons on the phase boundaries, which will finally annihilate as the transient metastable phase disappears.

To conclude, I would like to emphasize that everything is ready to occur and proceed with small supply of energy in this close neighborhood of phase transition point. Likewise, any new observation, idea and proposal will be useful and hence should be encouraged for the development of this new field which is almost on its starting point.

References

[1] Y. Toyozawa: Physica **117&118** B+C, 23 (1983).

[2] Y. Toyozawa: Solid State Commun. **84**, 255 (1992).

[3] S. Koshihara, Y. Tokura, T. Mitani, G. Saito and T. Koda: Phys. Rev. B **42**, 6853 (1990).

[4] Y. Toyozawa: Acta Physica Polonica **87**, 47 (1995).

[5] Y. Toyozawa: Prog. Theor. Phys. **26**, 29 (1961), see also, Y. Toyozawa: in "*Relaxation of Elementary Excitations*", ed. R. Kubo and E. Hanamura (Proc. Taniguchi Int. Symp. 1979, Springer 1980), p.3.

[6] D. Emin and T. Holstein: Phys. Rev. Lett. **36**, 323 (1976).

[7] M. Ueta, H. Kanzaki, K. Kobayashi, Y. Toyozawa and E. Hanamura: "*Excitonic Processes in Solids*" (Springer, 1984).

[8] A. Sumi: J. Phys. Soc. Jpn. **43**, 1286 (1977).

[9] K. S. Song and R. T. Williams: "*Self-Trapped Excitons*" (Springer, 1993).

[10] T. Matsumoto, T. Kawata, A. Miyamoto and K. Kan'no: J. Phys. Soc. Jpn. **61**, 4229 (1992).

[11] Y. Toyozawa: J. Phys. Soc. Jpn. **58**, 2629 (1989).

[12] D. L. Dexter, C. C. Klick and G. A. Russel: Phys. Rev. **100**, 603 (1955).

[13] A. Matsui, K. Mizuno, N. Tamai and I. Yamazaki: Chem. Phys. **113**, 111 (1987).

[14] Ch. B. Lushchik; in "*Excitons*", ed. E. I. Rashba and M. Sturge (North-Holland,1982), p.505; T. Hayashi, T. Ohata and S. Koshino: J. Phys. Soc. Jpn. **42**, 1647 (1977); H, Nishimura, C. Ohhigashi, Y. Tanaka and M. Tomura: J. Phys. Soc. Jpn. **43**, 157 (1977).

[15] M. Hirai, Y. Kondo, T. Yoshinari and M. Ueta: J. Phys. Soc. Jpn. **30**, 440 (1971).

[16] N. Itoh: Adv. Phys. **31**, 491 (1982).

[17] K. Tanaka: in "*Fundamental Physics of Amorphous Semiconductors*", ed. F. Yonezawa (Springer, 1981), p.104.

[18] K. Kimura, H. Nakata, K. Murayama and T. Ninomiya: Solid State Commun. **40**, 551 (1981).

[19] H. Tanino and K. Kobayashi: J. Phys. Soc. Jpn. **52**, 1446 (1983).

[20] K. Tanimura: private communication.

[21] N. Nagaosa and J. Takimoto: J. Phys. Soc. Jpn. **55**, 2735 (1986).

[22] A. Mishima and K. Nasu: Phys. Rev. B **39**, 5758; 5763 (1989); M. Suzuki and K. Nasu, Phys. Rev. B **45**, 1605 (1992).

[23] E. Hanamura and N. Nagaosa: Solid State Commun. **62**, 5 (1987).

[24] T. Iizuka-Sakano and Y. Toyozawa: J. Phys. Soc. Jpn. **65**, 671 (1996).

Hierarchy of Formation of Nanometer Clusters and Periodic Defect-Deformational Structures in Solids Under the Action of Laser Beam

V.I. Emel'yanov

International Laser Center, Moscow State University,
119899 Moscow, Russia

Abstract. The defect-deformational theory of formation of nanometer clusters and periodic structures of point defects generated by the action of energy beams in solids is developed. A good correspondence of theoretical and experimental results is obtained.

1. Introduction

The selforganization of laser-induced defects (interstitials, vacancies and electron-pairs), interacting with each other through the long-range field of elastic deformation was investigated previously on surfaces of solids (see review [1]). The characteristic scale of surface periodic defect-deformational (DD)-structures, formed as a result of laser action on strongly absorbing solids (semiconductors and metals), is proportional to the thickness of defect enriched subsurface layer h [1-3], which is of order of $10^{-4} - 10^{-5}$ cm (large scale DD-structures).

The consideration of formation of small scale (nanometer) periodic and localized defect-deformational structures in solids subjected to action of energy (laser) beams is interesting for a number of practically important effects such as laser-induced amorphization of semiconductors [4-8] and formation of extended defects [8], formation of nanometer period lattices of voids in dielectric optical materials [9] and metals [10], subjected to irradiation by particle beams

In this work, dedicated to this problem, we on the base of the mechanism of DD-instability [1] derive the nonlinear DD-equation for the selfconsistent deformation (strain) in solid, caused by stationary distribution of laser-generated point defects. The exact solution of this equation is found and it is shown that depending on the relative value of the characteristic lengths l_0 (the length of interatomic interaction) and l_d (the length of defect-atom interaction), and also on the exceeding over the threshold value of control parameter (the concentration of laser-generated defects n_{d0}) either localized DD-structures (clusters) or periodic DD-structures are formed. The conditions of their formation and their characteristics are determined: cluster size, period of DD-structure, the spatial distribution of deformation and defect concentration in DD-structures.

The correspondence of the predicted hierarchy of defect selforganization processes with observation of the sequence of laser-induced structural modification of subsurface layers of semiconductors is shown. The comparison of predicted temperature dependency of the period of DD-structure with the corresponding experimental dependence of void lattice formed in metals under irradiation by particle beams is carried out and good correspondence is found.

Springer Series in Solid-State Sciences, Volume 124
Relaxations of Excited States
and Photo-Induced Structural Phase Transitions
Ed.: K. Nasu

2. Coupled equations for the medium deformation and defect concentration

The free energy density of unidimensional anharmonic crystal with defects, with allowance for nonlocal interaction of atoms in crystal lattice with each other and with defects, has the form [11,12]:

$$F = \frac{c_l^2 \xi^2}{2} + \rho c_l^2 l_0^2 \frac{\partial^2 \xi}{\partial x^2} \xi + \frac{\beta \rho c_l^2 \xi^4}{4} - \theta_d n_d \xi - \theta_d n_d l_d^2 \frac{\partial^2 \xi}{\partial x^2}. \quad (1)$$

Here the strain of the medium $\xi(x) = \partial u_x(x)/\partial x$ (u_x is the component of the vector of displacement of the medium) and point defect concentration $n_d = n_d(x)$ are the variables, ρ is the medium density, c_l^2 is longitudinal sound velocity, β is the constant of anharmonicity, l_0 and l_d are correspondingly the characteristic lengths of interaction of atoms with each other and with defects, defect deformation potential $\theta_d = \Omega_d K_d$ (Ω_d is crystal volume change under production of a single defect, K is the bulk elastic modules).

With the help of (1) we derive the equation for the strain

$$\frac{\partial^2 \xi}{\partial t^2} = c_l^2 \frac{\partial^2 \xi}{\partial x^2} + c_l^2 l_0^2 \frac{\partial^4 \xi}{\partial x^4} + \beta c_l^2 \frac{\partial^2 \xi^3}{\partial x^2} - \frac{\theta_d}{\rho} \frac{\partial^2 n_d}{\partial x^2}. \quad (2)$$

Using the expression for the defect-strain interaction energy, following from (1), we obtain the nonlocal force, that gives rise to the defect flux

$$j_d = \frac{D_d}{k_B T} n_d(x) \theta_d \left[\frac{\partial \xi(x)}{\partial x} + l_d^2 \frac{\partial^3 \xi(x)}{\partial x^3} \right], \quad (3)$$

where D_d is the coefficient of diffusion, T is the temperature.

With the help of (3) we obtain the equation for the defect concentration

$$\frac{\partial n_d}{\partial t} = D_d \frac{\partial^2 n_d}{\partial x^2} - \frac{D_d \theta_d}{k_B T} \frac{\partial}{\partial x} \left[n_d(x) \left(\frac{\partial \xi(x)}{\partial x} + l_d^2 \frac{\partial^3 \xi(x)}{\partial x^3} \right) \right] + G_d - \frac{n_d}{\tau_d} \quad (4)$$

where G_d is the rate of defect generation by external energy source, τ_d is defect lifetime.

The system of equations (2), (4) has the spatially uniform stationary solution $n_{d0} = G_d/\tau_d$, $\xi = \xi_0$. As we show below this spatially uniform solution is unstable under exceeding of certain critical defect concentration and a new spatially inhomogeneous stationary state arises. In this state

$$n_{d0} = n_{d0} + n_{d1}(x) \quad (5)$$

where $n_{d1}(x)$ is the spatially nonuniform part of the defect concentration. From (4) and (5), neglecting defect recombination ($\tau_d^{-1} = 0$), we have the equation for n_{d1}:

$$\frac{\partial n_{d1}}{\partial t} = D_d \frac{\partial^2 n_{d1}}{\partial x^2} - \frac{D_d \theta_d n_{d0}}{k_B T}\left(\frac{\partial^2 \xi(x)}{\partial x^2} + l_d^2 \frac{\partial^4 \xi(x)}{\partial x^4}\right) \qquad (6)$$

Here and below we designate: $\xi = \xi_1$, where ξ_1 is the selfconsistent spatially nonuniform deformation of the medium, described by the equation following from (2)

$$\frac{\partial^2 \xi}{\partial t^2} = c_l^2 \frac{\partial^2 \xi}{\partial x^2} + c_l^2 l_0^2 \frac{\partial^4 \xi}{\partial x^4} + \beta c_l^2 \frac{\partial^2 \xi^3}{\partial x^2} - \frac{\theta_d}{\rho} \frac{\partial^2 n_{d1}}{\partial x^2} \qquad (7)$$

From (7) it follows that $\mathbf{sign}\, \xi = \mathbf{sign}\theta_d$.

3. Formation of cluster and periodic defect-deformational structures

We consider the system of Eqs. (6), (7) in the stationary case, when $\partial n_d / \partial t = 0$, $\partial^2 \xi / \partial t^2 = 0$. Integrating (6), we have for the spatially nonuniform defect concentration

$$n_{d1} = n_{d0} \frac{\theta_d}{k_B T}\left(\xi + l_d^2 \frac{\partial^2 \xi}{\partial x^2}\right), \qquad (8)$$

where we put two integration constants equal to zero. Substituting (8) in (7) and integrating the resultant equation, we obtain

$$\frac{\partial^2 \xi}{\partial x^2} - a\xi - c\xi^3 = 0 \qquad (9)$$

where

$$a = \left(1 - n_{d0}/n_{dc}\right)\left(l_d^2 n_{d0}/n_{dc} - l_0^2\right)^{-1}, \quad c = \beta\left(l_d^2 n_{d0}/n_{dc} - l_0^2\right)^{-1},$$

and the critical defect concentration is introduced:

$$n_{dc} = \rho c_l^2 k_B T / \theta_d^2 . \qquad (10)$$

In derivation of Eq. (9) we put two integration constants equal to zero, that met the condition $\xi(x) = 0$ for $x \to \pm\infty$ (cluster solution) or the condition of translation invariance: $\xi(x) = \xi(x + d)$, where d is the translation period (periodic solution).

The inexplicit solution of Eq. (9) is

$$x = \int d\xi \Big/ \left(a\xi^2 + \frac{c}{2}\xi^4 + d \right)^{-1/2} , \qquad (11)$$

where d is an integration constant. At $d \neq 0$ the solution of Eq.(11) is expressed through elliptic integrals.

Let us consider the class of solutions corresponding to the case $d = 0$. Then from (11) we have:

$$x = \int d\xi \Big/ \left(a\xi^2 + \frac{c}{2}\xi^4 \right)^{-1/2} , \qquad (11a)$$

The integral (11a) can be expressed through analytic functions, the particular type of which is determined by the particular combination of signs of coefficients a and c. We focus on two types of solutions, describing two qualitatively different types of DD-structures: clusters and periodic structures.

3.1. Defect cluster in selfconsistent deformation well

Depending on the relative values of characteristic lengths l_d and l_0 two types of clusters can be formed.

a) antisymmetric DD-cluster ($l_d > l_0$). We divide the region of variation of n_{d0} into three intervals, in which combinations of signs of coefficients a and c are different: $n_{d0} / n_{dc} < (l_0/l_d)^2$, $(l_0/l_d)^2 < n_{d0}/n_{dc} < 1$ and $n_{d0}/n_{dc} > 1$.

In the region of low defect concentration, where

$$n_{d0} / n_{dc} < (l_0/l_d)^2 \qquad (a < 0, c < 0)$$

the integral (11a) can not be expressed in analytic functions , so $\xi = 0$ and $\partial^2\xi / \partial x^2 = 0$. From (8) then we have $n_{d1} = 0$. Thus at low defect concentration there is no defect selforganization.

In the region of intermediate defect concentration, where

$$(l_0/l_d)^2 < n_{d0}/n_{dc} < 1 \qquad (a < 0, c > 0)$$

from (11a) one has for the selfconsistent strain:

$$\xi(x) = \mathrm{sign}\,\theta_d \left(\frac{2}{\beta} \left(\left| \frac{n_{d0}}{n_{dc}} - 1 \right| \right) \right)^{1/2} \mathrm{sh}^{-1} \left\{ \left(\frac{1 - n_{d0}/n_{dc}}{l_d^2\, n_{d0}/n_{dc} - l_0^2} \right)^{1/2} x \right\} \qquad (12)$$

In accordance with (8), the antisymmetric defect cluster is selfconsistently coupled with antissymetric deformation well (12) with defect localized either in the region with $\xi > 0$ in the case $\theta_d > 0$ (interstitials), or in the region with $\xi < 0$ in the case $\theta_d < 0$ (vacancies). The characteristic size of the cluster d_{clust} as it is seen from (12) is

$$d_{\text{clust}} = \frac{l_d}{2}\left(\frac{n_{d0}/n_{dc} - l_0^2/l_d^2}{1 - n_{d0}/n_{dc}}\right)^{1/2} \tag{13}$$

In the region of high defect concentration

$$n_{d0}/n_{dc} > 1 \qquad (a > 0, c > 0)$$

the periodic DD-structure is formed (see Sec. 3.2).

b) symmetric DD-cluster ($l_d < l_0$). At low defect concentration, when $n_{d0}/n_{dc} < 1$ and $a < 0$, $c < 0$ there is no defect selforganization. At intermediate defect concentrations, after exceeding the threshold value n_{dc}, in the region $1 < n_{d0}/n_{dc} < (l_0/l_d)^2$ ($a > 0$, $c < 0$) the solution of Eq.(11a) is

$$\xi(x) = \operatorname{sign}\theta_d\left(\frac{2}{\beta}\left(\frac{n_{d0}}{n_{dc}} - 1\right)\right)^{1/2} ch^{-1}\left\{\left(\frac{1 - n_{d0}/n_{dc}}{l_d^2\, n_{d0}/n_{dc} - l_0^2}\right)^{1/2} x\right\} \tag{14}$$

In accordance with (8), the symmetric defect cluster is selfconsistently coupled with this symmetric deformation well (14) with the characteristic size

$$d_{\text{clust}} = l_d\left(\frac{n_{d0}/n_{dc} - l_0^2/l_d^2}{1 - n_{d0}/n_{dc}}\right)^{1/2} \tag{14a}$$

At high defect concentration $n_{d0}/n_{dc} > (l_0/l_d)^2$ ($a > 0$, $c > 0$) the periodic DD-structure is formed (see Sec. 3.2).

3.2. Periodic defect-deformational structure

Consider again two possible cases of inequalities between the values of characteristic lengths l_d and l_0.

a) $l_d > l_0$.

The case of low and intermediate values of n_{d0} was investigated in Sec.3.1a. At high defect concentration in the region $n_{d0}/n_{dc} > 1$ ($a > 0, c > 0$) one has the following solution of Eq.(11a) for the selfconsistent deformation

$$\xi(x) = \operatorname{sign}\theta_d\left(\frac{2}{\beta}\left(\frac{n_{d0}}{n_{dc}} - 1\right)\right)^{1/2} \cos^{-1}\left\{\left(\frac{n_{d0}/n_{dc} - 1}{l_d^2\, n_{d0}/n_{dc} - l_0^2}\right)^{1/2} x\right\} \tag{15}$$

The graph of the highly anharmonic periodic function (15) is shown in Fig.1. The

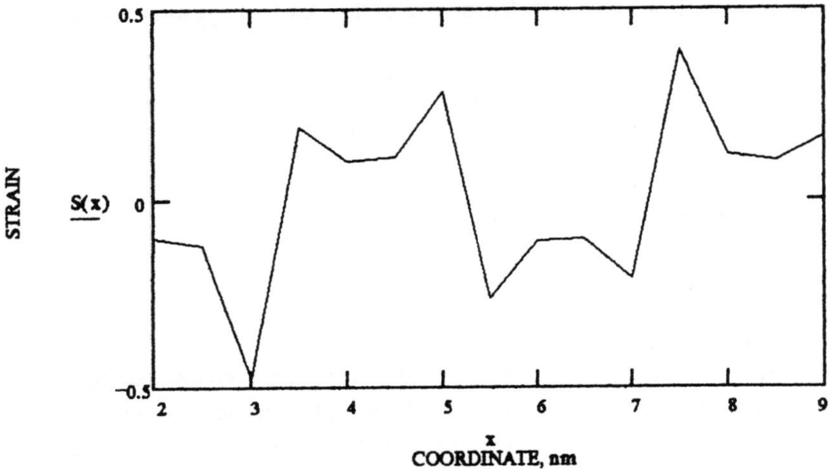

Fig. 1. The spatial distribution of selfconsistent strain in periodic DD - structure. Mathcad drawing of the function $S(x) = \xi(x) = 10^{-1}\cos^{-1}(2\pi\, x/d)$, Eq.(15) with $d = 4.2{*}10^{-7}\,\mathbf{cm}$. Computer smoothes out the divergent function, effectively taking into account higher order derivatives.

divergence of the solution (12) and (15) is due to neglect of higher order derivatives in Eq.(1) (i.e. in the expansion of operators of interatomic interactions (see [11,12]).

In accordance with (8), interstitials ($\theta_d > 0$) are localized in regions with $\xi > 0$, while vacancies ($\theta_d < 0$) are piled up in the regions with $\xi < 0$. The period of this DD-structure is

$$d = 2\pi\, l_d \left(\frac{n_{d0}/n_{dc} - l_0^{\,2}/l_d^{\,2}}{n_{d0}/n_{dc} - 1} \right)^{1/2} \tag{16}$$

b) $l_d < l_0$

The cases of low and intermediate values of n_{d0} were investigated in Sec. 3.1b. At exceeding the threshold $n_{dc1} = n_{dc}(l_0/l_d)^2$ in the region $n_{d0}/n_{dc} > (l_0/l_d)^2$ where $a > 0$, $c > 0$, again the periodic DD-structure is formed, described by Eqs.(15) and (16).

4. Discussion and comparison with experimental results

The above obtained results predict the following hierarchy of processes of nanometer selforganization of point defects pumped into a solid by external energy (laser) beam. With increase of irradiation dose (laser fluency) the spatially uniform defect concentration n_{d0} (control parameter) is increased. There are two threshold

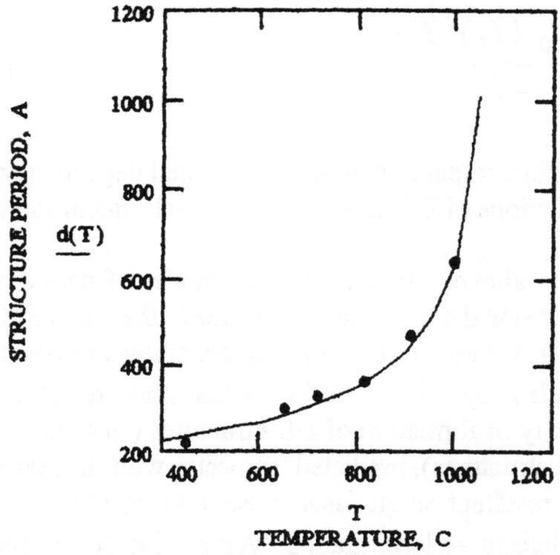

Fig 2. The dependence of void lattice period on temperature.
Points: the experimental data for Mo, irradiated by neutron beam with energy > 0,1 MeV at irradiation dose 10^{22} cm^{-2}. The curve is calculated with the help of Eq.(17) at $T_c = 1084\ ^0\mathrm{C}, l_d = 29\,\mathrm{A},\ l_0 = 5\,\mathrm{A}$.

values of control parameter: n_{dc} - (10) and $n_{dc1} = (l_0/l_c)^2\, n_{dc}$, after exceeding of which two qualitatively different selforganization processes occur. In the case $l_d > l_0$ (see Fig. 2) in the region $n_{d0} < n_{dc1}$ spatial distribution of defects remains uniform. In the region $n_{dc1} < n_{d0} < n_{dc}$ the antisymmetric selfconsistent defect clusters are formed., Eq.(12), with characteristic size $d_{clust} \sim l_d \leq 10^{-7}$ cm. In the region $n_{d0} > n_{dc}$ the highly anharmonic periodic DD-structure, Eq.(15), is formed, shown in Fig. 1. The period of this structure tends to $d = 2\pi\, l_d$ at high n_{d0}. The dependence of the amplitude of the periodic structure on n_{d0}, given by Eq.(15), shows that the formation of DD-structure occurs as a phase transition of the second kind.

We have considered the selforganization of defects of one type "d" while in crystals, irradiated by laser [13] or particle beams, point defects are created by pairs of interstitials ($d = i$) and vacancies ($d = v$). Taking into account that $\theta_i >> |\theta_v|$, we retain only one contribution in the free energy (1) with $\theta_d = \theta_i$.

When concentration of vacancies captured in potential wells of periodic interstitial - deformational structure (Fig. 1) becomes large, periodic lattice of voids is formed with the period, given by Eq.(16).

The formation of periodic void lattice was observed under particle beam irradiation of metals [10] and dielectrics [9]. From Eq.(16) one obtains the following temperature dependency of structure period

$$d = 2\pi\, l_d \left(\frac{T_c - (I_0 / I_d)^2 T}{T_c - T} \right)^{1/2} \tag{17}$$

The dependence (17) is shown in Fig. 2 in comparison with experimental dependency [10]. Good correspondence of the predictions of DD-theory of defect selforganization and experimental results is seen.

If one qualitatively generalizes above one-dimensional model of periodic DD-structure formation on three-dimensional case, then one comes to the picture of threshold crystal fragmentation. At high values of n_{d0} the fragmentation of crystal into nanometer crystal blocks divided by binary defect walls may result in amorphization [11,7]. Then the hierarchy of formation of DD-structures (clusters $\rightarrow$ periodic large period $\rightarrow$ small period structures), predicted to occur with increase of n_{d0} corresponds to hierarchy of resultant single laser pulse-induced structural modifications of Si surface (extended defects $\rightarrow$ large grain polycrystalline Si $\rightarrow$ fine grain polycrystalline Si $\rightarrow$ amorphous Si) observed with decrease of laser fluency W (J/cm^2), exceeding melting threshold [8]. In this case defects are captured into solids from the liquid phase during the process of solidification front movement. In doing so the smaller is W the higher is solidification front velocity and, correspondingly, higher are the values of n_{d0}.

In the case of multipulse irradiation of semiconductors at relatively low intensity (below melting threshold) high concentrations of interstitial-vacancy pairs are created in the process of recombination-enhanced defect reactions [14]. We showed in [7], that the experimental results of [6] on multipulse amorphization of Si surface can be quantitatively described in the frame of the model of point defect formation wave (DFW) and amorphization wave (AW), ignited and propagated jointly when the concentration of laser generated electron-hole (e-h) pairs exceeds certain critical value. The duration of e-h pair recombination-induced amorphization process in thin (10^{-6} cm) layer at the front of DFW and AW, steady propagated with velocity $v \geq 10^5$ cm$*$s^{-1}, may as short as 10^{-11} s [7]. Note that the model of DFW and AW in semiconductors is similar to the model of DFW and cluster nucleation wave, ignited and propagated jointly in transparent dielectrics under multiphoton excitation of e-h pairs [15,16].

The author is grateful to Russian Foundation for Fundamental Studies for support (grant 95-02-05303à).

References

1. V.I.Emel'yanov: Laser Physics, 2, 389 (1992)
2. V.I.Emel'yanov: in "Laser-Induced Damage in Optical Materials", SPIE vol. 2114, 580(1993).Eds. H.E.Bennet, L.L.Chase, A.H.Guenther, B.E.Newnam and M.J.Soileau.
3. V.I.Emel'yanov, M.Ledgerwood and H.M. van Driel: in "Nonlinear optics of low-dimensional structures and new materials", SPIE vol. 2801(1996). Eds. V.I.Emel'yanov and V. Ya. Panchenko.

4. Laser and Electron Beam Interactions with Solids, Eds. B.R.Appleton. and G.K. Celler, 1982 (Amsterdam: North Holland).
5. "Crystallization and related phenomenon in amorphous materials", Symposium - Boston, MA, USA, 29.11-02.12.1993, (Pittsburg PA, USA, MRS, 1994).
6. P.M Fauchet: Phys.Lett. 93A, 155 (1983)
7.V.I.Emel'yanov: "Ignition and propagation of amorphization wave in semiconductors under intense generation of e-h pairs by short laser pulses", Laser Physics (in press).
8.Komarov, A.P.Novikov, V.S.Solov'yov, S.Yu.Shiryaev: Structural defects in ion-implanted silicon, Minsk, Universitetskoye Publ.(1990)
9. L.T.Chadderton, E.Johnson, T.Wohlenberg: Radiation Effects, 28, 111(1976).
10. V.K.Sikka, J.Moteff: J.Nucl.Mat. 54, 325 (1974)
11. V.I.Emel'yanov: Laser Physics, 6, N2, 423 (1996)
12. V.I.Emel'yanov and I.M.Panin: Laser Physics (in press).
13. K.S.Song , R.T. Wiliams: Self-Trapped Excitons. Berlin: Springer (1993).
14. H. Sumi: Phys.Rev.B, 29, .4616 (1984)
15. V.I.Emel'yanov: Abstr. of XXY Simp. on Optical Materials for High Power Lasers, Boulder, USA, Sec. Fundamental Mechanisms (1993)
16. V.I.Emel'yanov: Laser Phys. 5, 1041 (1995)

Photo-Induced Structure Changes of Quasi-one-dimensional Organic Crystals

E. Hanamura

Department of Applied Physics, University of Tokyo,
7-3-1 Hongo, Bunkyo-ku, Tokyo 113, Japan

Abstract A new method is presented to describe structure-changes of a system far from thermal equilibrium, e.g., photo-induced structure changes of quasi-one-dimensional systems. In terms of this method, we try to theoretically understand photo-polymerization of diacetylene monomer crystals and photo-isomerization of polydiacetylene crystals. Excitonic n-strings in organic charge-transfer crystals are also discussed from present unified point of view.

1. Introduction

Photo-excited electron-hole pairs or excitons are stabilized into self-trapped excitons (STEs) accompanied with lattice deformation within several pico-seconds [1]. While some STEs decay into the ground state radiatively or nonradiatively, clusters of STEs beyond a critical size can persist as metastable state or stable state [2, 3]. A new lattice structure formed by clustering STEs is induced beyond some critical pumping. These phenomena have been observed in several polydiacethlene crystals: Blue-type structure is changed into red-type under irradiation of Ar^+ ion lasers [4]. Koshihara et al. succeeded in inducing the reversal structure change from the red- to the blue phase by properly choosing the pumping frequency and lattice temperature [5].

The linear chain system is treated in Section 2, corresponding to photo-polymerization of diacetylene [6, 7] and diolefin crystals [8] and photo-isomerization of polydiacetylene crystals [4]. Here it will be shown for the linear chain system how a cluster of photo-excited molecules is stabilized through the electron-lattice interaction in an electronic excited state. In Section 3, we discuss the metastability and absolute stability of the cluster of m neighboring excited molecules, and also the thermal or spontaneous expansion of the excited-molecule cluster. At the same time, the excitation energy of the molecule is shown to become lower when it is closer to the cluster. Therefore once the excitation transfer is finite, the optical excitation of a molecule is attracted to the nearest neighbor of the cluster so that this cluster expands effectively under the light irradiation. In Section 4, dynamics of photo-induced structure change is studied by numerical methods. We have found to be able to induce any degree of structure change depending upon the pump power for some value of exciton-lattice interaction.

Springer Series in Solid-State Sciences, Volume 124
Relaxations of Excited States
and Photo-Induced Structural Phase Transitions
Ed.: K. Nasu

Electrostatic interaction of charge-transfer excitons works so as to stabilize the excited clusters, which are called excitonic n-strings. We present theoretical results of excitonic n-strings referring to the charge-transfer excitons (CTEs) in Anthracene-PMDA crystals in Section 5 and how to observe a new family of these elementary excitations in Section 6. We will speculate on the excitonic n-strings associated with lattice distortions and on the exciton-mediated carrier transport of these systems in the last Section 7.

2. Hamiltonian

Hamiltonian of our system which is discussed in this paper is written as follows:

$$H=\frac{1}{2}\sum_i Q_i^2-\sum_{i\neq j}K_{ij}Q_iQ_j+\sum_l\left(\omega_0-\sqrt{S}Q_l\right)\left(\hat{s}_l^z+\frac{1}{2}\right)$$
$$-J\sum_l\left(\hat{s}_l^+\hat{s}_{l+1}^-+\hat{s}_l^-\hat{s}_{l+1}^+\right)-2J_z\sum_l\left(\hat{s}_l^z+\frac{1}{2}\right)\left(\hat{s}_{l+1}^z+\frac{1}{2}\right). \quad (1)$$

Here Q_i denotes the relevant displacement of the i-th molecule and K_{ij} the coupling constant between the displacements of the i-th and j-th molecules. Once the l-th molecule is optically excited with expense of Franck-Condon energy $\hbar\omega_0$, it is stabilized by the Stokes shift $\sqrt{S}Q_l$. Here and hereafter we put $\hbar=1$. The electronic structure of the l-th molecule is described by two-level model, and $\hat{s}_l^\pm\equiv\hat{s}_l^x\pm i\hat{s}_l^y$ the excitation and deexcitation operators and $\hat{s}_l^z+\frac{1}{2}$ represents the population at the excited or ground state. The fifth term proportional to J describes propagation of the excitation under the nearest-neighbor approximation and the last term (J_z) comes from the electrostatic interaction between two excitons when the excited state is accompanied with an induced static dipolemoment within a unit cell, e.g., within the donor-acceptor pair in the case of charge-transfer exciton.

For the photo-induced structure changes of polydiacetylene crystals, the excited state is not accompanied with the static dipolemoment so that J_z vanishes. Furthermore the self-trapping due to the third term has much stronger effect than propagation by the fourth term in Eq.(1). Therefore we discuss the photo-induced structure change first by using the following Hamiltonian:

$$H_0=\sum_{l\in E}\left(\omega_0-\sqrt{S}Q_l\right)+\frac{1}{2}\sum_i Q_i^2-\sum_{i\neq j}K_{ij}Q_iQ_j. \quad (2)$$

The last two terms in Eq.(2) work in both the electronic ground and excited states and the summation of the first term $l\in E$ is taken over the sites of the excited molecules. Let us consider that $l=1,2,\cdots m$-th molecules are optically excited and others are in the ground state. Then, in terms of the Fourier transform of the lattice

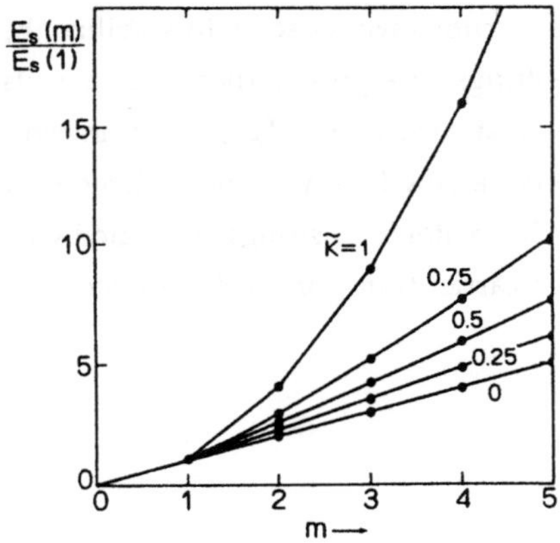

Figure 1. The stabilization energy $E_s(m)/E_s(1) \equiv J(m)$ of a cluster of m neighboring excited molecules is plotted as a function of m.

displacement $Q_l = N^{-1/2}\sum_k e^{ikl} Q_k$ and the lattice fluctuation $Q_k^{(m)} = Q_k - \Delta^{(m)}(k)$ around the stationary displacement $\Delta^{(m)}(k)$, H_0 is rewritten as

$$H^{(m)} = \frac{1}{2}\sum_k \omega_k^2 Q_k^{(m)} Q_{-k}^{(m)} + m\omega_0 - \frac{1}{2}\sum_k \omega_k^2 \Delta^{(m)}(k)\Delta^{(m)}(-k), \qquad (3)$$

where $\omega_k^2 = 1 - 2\sum_{j\neq 0} K_{j0} e^{ikj}$ and $\Delta^{(m)}(k) = \sqrt{\frac{S}{N}}\frac{1}{\omega_k^2}\left(\sum_{l=1}^{m} e^{ikl}\right)$. (3a,b)

Then the stabilization energy $-SJ(m)/2$, i.e., the third term of Eq.(3) denoted by $E_s(m)$ is analytically calculated and its absolute value is found to increases by stronger than the first power of m as shown in Fig.1. Here and hereafter we take the nearest neighbor appaoximation $K \equiv 4K_{i+1,j}$ with others $= 0$.
See ref.(2) for analytical derivation of the stabilization energy. Especially, for $K \equiv 4K_{j+1,j}$ close to one, it increases as almost square of m, so that the absolute value of the stabilization energy $-SJ(m)/2$ overcomes the Franck-Condon energy $m\omega_0$ for enough large m and the distorted state accompanied with the electronic excitations becomes more stabilized than the initial ground state of the system. More detailed discussion on these structure changes is given in the following Sections.

We have also stable clusters of excited molecules, e.g., in the donor-acceptor complexes. These come from mainly from the fourth and fifth terms in Eq.(1) and these complexes are called excitonic n-strings. When these complexes are accompanied with the lattice distortions, we may expect some phase in which the exciton-mediated carrier transport is expected. These will be discussed in Sections 6 and 7.

3. Stability of a Cluster of m Excited Molecules

Let us consider a cluster of m neighboring excited molecules ($l = 1, 2, \cdots, m$). We study the relative stability of the 0-th or $(m+1)$-th molecule in the electronic excited state to the electronic ground state. The former is described by

$$H_e^{(m+1)} = H^{(m)} + \omega_0 - \sqrt{S}Q_0 \qquad (4a)$$

$$= H^{(m)} + \omega_0 - \sqrt{\frac{S}{N}}\sum_k \Delta^{(m)}(k) - \sqrt{\frac{S}{N}}\sum_k \frac{1}{\omega_k} q_k . \qquad (4b)$$

Here the superscripts (m) and $(m+1)$ denote the numbers of excited molecules and the subscript g or e of H means whether the 0-th molecule is in the electronic ground or excited state. The lattice coordinate $q_k \equiv \omega_k Q_k^{(m)}$ is used and the interaction mode at the 0-th molecule q_0 is introduced by

$$q_0 = \varpi \frac{1}{\sqrt{N}} \sum_k \frac{1}{\omega_k} q_k \qquad \text{with} \qquad \frac{1}{\varpi^2} = \frac{1}{N}\sum_k \frac{1}{\omega_k^2} = \frac{1}{\sqrt{1-K^2}} . \qquad (5a, b)$$

Then Eqs.(3) and (4a) are rewritten as

$$H_g^{(m)} = \frac{1}{2} q_0^2 + E_g^{(m)} , \qquad (6)$$

$$H_e^{(m+1)} = \frac{1}{2} q_0^2 + E_g^{(m)} + \omega_0 - \frac{\sqrt{S}}{\varpi} q_0 - \delta_g(m) \qquad (7a)$$

$$= \frac{1}{2}\left(q_0 - \frac{\sqrt{S}}{\varpi} \right)^2 + E_g^{(m)} + \omega_0 + \frac{1}{2}\frac{S}{\varpi^2} - \delta_e(m+1), \qquad (7b)$$

where

$$\delta_g(m) = \sqrt{\frac{S}{N}} \sum_k \Delta^{(m)}(k) = S\frac{1}{N}\sum_k \sum_{l=1}^{m} e^{ikl} \frac{1}{\omega_k^2} = r\delta_e(m), \qquad (8)$$

$$\delta_e(m+1) = S\frac{1}{N}\sum_k \sum_{l=0}^{m} e^{ikl} \frac{1}{\omega_k^2} = \delta_g(m) + \frac{S}{\varpi^2} = \frac{S}{\sqrt{1-K^2}} \frac{1-r^{m+1}}{1-r}, \qquad (9)$$

with

$$r = \left\{ 1 - \sqrt{1-K^2} \right\} / K .$$

For the phonon dispersion relation $\omega_k^2 = 1 - K\cos k$, $\delta_g(m)$ and $\delta_e(m+1)$ are analytically integrated in terms of S and K. Here the reference energy $E_g^{(m)}$ consists of the electronic energy of m excited molecules and the lattice vibration energy besides the interaction mode q_0. Without explicit knowledge of $E_g^{(m)}$, we can discuss the stability of a cluster of m excited molecules as follows:

(1) $\omega_0 - \delta_e(m+1) > 0$:

The energy dispersion of this case is shown in Fig.2(1). Here the minimum energy $\omega_0 - \delta_e(m+1) + S/(2\varpi^2)$ of the excited state is still above the corresponding displaced ground state $E_g^{(m)} + S/(2\varpi^2)$. The excited state $|e\rangle$ at the 0-th and ($m+1$)-th molecules

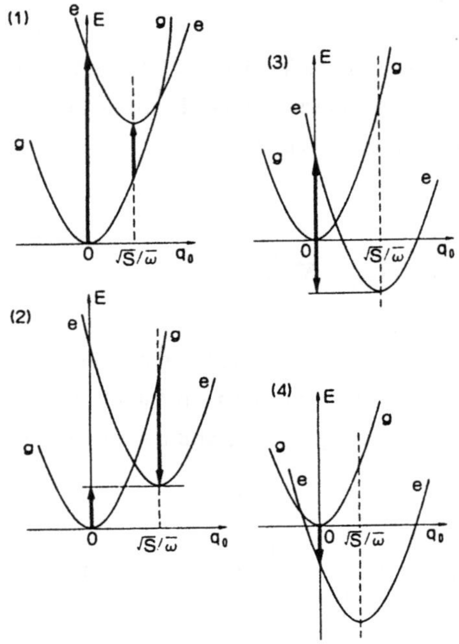

Figure 2. Energy dispersions of the 0-th molecule closest to the cluster of excited molecules: (1) The cluster of excited molecules is unstable, (2) metastable, (3) expands thermally, and (4) expands spontaneously without any additional optical pumping.

can radiatively decay into the ground state. We can show $\delta_e(m) < \delta_e(m+1)$ so that all the excited molecules can radiatively decay into the ground state.

(2) $\delta_e(m+1) \geq \omega_0 > [\delta_e(m+1) + \delta_g(m)]/2$:

The right-hand-side quantity is stabilization energy of electronic excited state due to the lattice distortion. The left-hand inequality means that the minimum energy of electronic excited state is below the ground state energy at the displaced coordinate $q_0 = \sqrt{S}/\varpi$, and the right-hand-side describes the absolute stability between the electronic ground and excited states at the 0-th molecule. This inequality means that $\omega_0 - \delta_e(m+1) + S/(2\varpi^2)$ is still higher than the minimum energy 0 of the electronic ground state at the 0-th molecule. This is expressed as $\omega_0 > [\delta_e(m+1) + \delta_g(m)]/2$ because $\delta_e(m+1) = \delta_g(m) + S/\varpi^2$. Under this condition, the electronic excitation can not radiatively decay but thermally into the ground state as shown in Fig.2(2). At such a low temperature as $k_B T <<$ the potential barrier ΔE in the excited state, the cluster of $m+1$ excited molecules survives as metastable state.

(3) $[\delta_e(m+1) + \delta_g(m)]/2 \geq \omega_0 > \delta_g(m)$:

Here the distorted electronic excited state at the 0-th molecule is stabilized below the electronic ground state. Under this condition, the distorted phase can thermally extend beyond the size m without additional optical pumping. The energy relation is drawn in Fig.2(3).

(4) $\delta_g(m) \geq \omega_0$:

This inequality means that not only the distorted electronic excited state at $q_0 = \sqrt{S}/\varpi$ but the undistorted one at $q_0 = 0$ are below the lowest state of the electronic

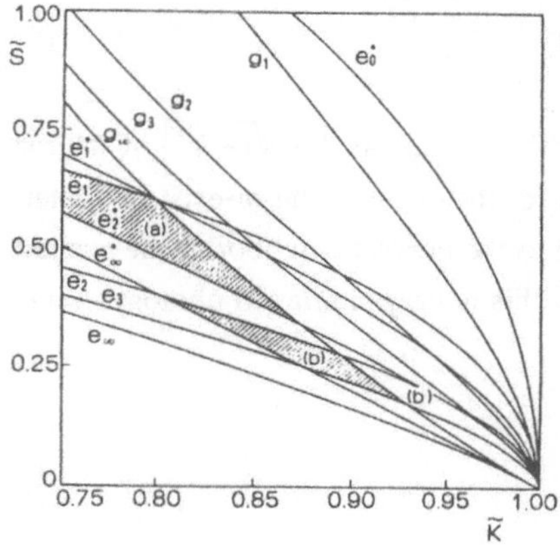

Figure 3. The phase diagram over $\tilde{S} \equiv S/\omega_0$ and K. e_m, $e_m^{\bullet}$ and g_m curves denote the stability borders of m-molecules cluster displaced through electronic excitation. e_m: the metastability, $e_m^{\bullet}$: the absolute stability, and g_m: the spontaneous extention of the m-displaced molecules cluster.

ground state at the 0-th and $(m+1)$-th molecules. Once this condition is satisfied, the distorted cluster size increases without any supply of optical and thermal energy.

Note here that these conditions depend not only on two parameters S/ω_0 and K but also on the size m of the distorted cluster of optically excited molecules. We introduce three critical sizes m_{c1}, m_{c2} and m_{c3} of the distorted cluster by

$$\omega_0 = \delta_e(m_{c1}+1), \tag{10a}$$

$$\omega_0 = \left[\delta_e(m_{c2}+1) + \delta_g(m_{c2})\right]/2, \tag{10b}$$

$$\omega_0 = \delta_g(m_{c3}). \tag{10c}$$

Such a short cluster as $m < m_{c1}$ disappears mainly through radiative decay. Once the size m becomes larger than m_{c1}, the cluster becomes metastable for $m_{c1} \le m < m_{c2}$ and absolutely stable $m_{c2} \le m < m_{c3}$. The metastable clusters decay thermally, reduce their size and finally disappear at such high temperature as $k_B T > \Delta E$. On the other hand, the excited cluster with $m_{c2} \le m < m_{c3}$ expands thermally and such a large cluster as $m \ge m_{c3}$ expands to the crystal size without any optical or thermal energy supply. Figure 3 shows these three stability conditions for each $m: e_m(\omega_0 = \delta_e(m))$, $g_m(\omega_0 = \delta_g(m))$ and $e_m^{\bullet}(\omega_0 = [\delta_e(m+1) + \delta_g(m)]/2)$, on the parameter plane $\tilde{S} \equiv S/\omega_0$ and K.

We can also show that the excitation far from the cluster is attracted to the nearest neighbor of the cluster of m excited molecules at $l = 1, 2, \cdots, m$. We denote the excitation energy at the site $-n$ outside the cluster by

$$E(-n,m) = \omega_0 - \delta_e(-n, m+1) + \frac{S}{2\overline{\omega}^2}. \tag{11}$$

Then it is evaluated as

$$\delta_e(-n,m+1) = \delta_e(m+1)\left(\frac{1-\sqrt{1-K^2}}{K}\right)^n, \tag{12}$$

where $\delta_e(m+1) \equiv \delta_e(0,m+1)$ in agreement with Eq.(9). As $\left\{1-\sqrt{1-K^2}\right\}/K < 1$ for $K < 1$, the excitation energy $E(-n,m)$ is smaller for the closer to the m-excited cluster. Therefore the electronic excitation is attracted to the nearest neighbor of the excited cluster through the finite excitation transfer. This is very similar to photosynthetic system.

4. Photo-polymerization and -isomerization

Several kinds of photo-induced structure changes are understood on the phase-diagram of Fig.3.

4.1 Photo-polymerization of diacetylene crystal is triggered by ultraviolet light irradiation and is accelated [7]. The interaction mode relevant to the photo-polymerization of diacetylene crystal is such a linear combination of rotations of the excited monomer and its neighbors as polymerization is induced on a sheet of diacetylene crystal. The oligomer corresponds to the lowest-energy configuration in the electronic excited state in the present model. From the observed experiment [7]: (1) Photo-excited monomer decays out; (2) Only when two nearest neighboring molecules are photo-excited, a dimer is formed; (3) This dimer is metastable and thermal reaction works even at 100°K producing trimer, tetramer, hexamer, ⋯, successively in an order of one hour, and finally the polymer crystal is formed after a few days; (4) These oligomers are confirmed to be metastable by ESR and absorption spectrum. From these observations, we can derive the following inequalities: (1) $\omega_0 > \delta_e(1)$, (2) $\omega_0 < \delta_e(2)$, (3) $\left[\delta_e(3)+\delta_g(2)\right]/2 > \omega_0 > \delta_g(2)$ and (4) $\omega_0 > \delta_g(\infty)$. The corresponding phase region is marked by (a) in Fig.3.

4.2 Photo-induced isomerization was observed in many kinds of polydiacetylene crystals, e.g., TCDU, ETCD and PDA-(10-8), -(12-8) and -(14-8) [2, 3]. The so called acetylene (A) and the butatriene (B) types correspond to the electronic ground and excited states, respectively, in the present model. (1) The yield of this A-to-B transition increases more strongly than by the third power of the irradiated light intensity [9]. This means $\delta_e(m \geq 3) > \omega_0 > \delta_e(2)$. (2) The B-phase is absolutely stable as it extends over the crystal after annealing ETCD and PDA-(12-8), and the fine particles of PDA-(12-8) become occupied by the B-phase after laser irradiation at room temperature. Then we have $\omega_0 < \left[\delta_e(\infty)+\delta_g(\infty)\right]/2$. (3) When we know whether the isomerization is induced spontaneously after partial irradiation or is thermally accelated, we can have another inequality $\omega_0 > \delta_g(\infty)$ or $\omega_0 < \delta_g(\infty)$. The former and latter cases correspond, respectively, to the region (b′) and (b) in Fig.(3).

5. Partial color change

We have found it possible to induce partially the structure change for the materials satisfying $\delta_e(\infty) < \omega_0 < \delta_g(\infty)$. To investigate the interaction among clusters of distorted molecules and their dynamics, the following numerical simulation is performed: First a one-dimensional chain of size $N = 8192$ is chosen and the excited molecules are randomly distributed with a given probability P at $t = 0$. Second we calculate the lattice relaxation energy at each site and we allow the radiative decay if possible. Third we repeat the second procedure for the new configuration of the excited and ground states. Thus we trace the time evolution of this system like a deterministic cellular automaton. Here one time step is the radiative decay time, and the thermal effect and the excitation transfer are neglected.

The final number of the excited molecules N_e divided by the total number N is obtained as a function of ω_0 relative to $\delta_e(\infty)$ and $\delta_g(\infty)$. We find [3] that $N_e/N = 1$ for $\omega_0 < \delta_g(\infty)$ while $N_e/N = 0$ for $\omega_0 > \delta_e(\infty)$. For $\delta_g(\infty) < \omega_0 < \delta_e(\infty)$, N_e/N takes any value between 0 and 1 depending upon the initial excitaion degree $P \equiv N_e(t=0)/N$. This dependence is drawn in Fig.4 for the case of $K = 0.7$. These phenomena were already observed in PDA (14-8). The change of N_e/N against P is very sharp for small P in the case $\omega_0 = \delta_g(\infty)$. These curves are symmetric with respect to the point $P = \frac{1}{2}$ and $N_e/N = \frac{1}{2}$. This is because the ground and excited states are symmetric against the lattice distortion as long as the model of assembly of two-levels is applied. This property of partial coloring is very useful as materials for optical-neural networks because these are endowed with the learning and teaching abilities [10].

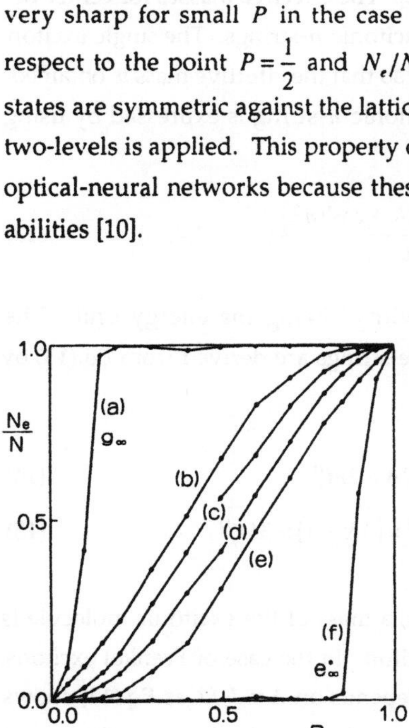

Figure 4 The final number N_e of the distorted molecules through electronic excitation divided by N as a function of the probability of the initial excitation (=proportional to the optical intensity). $K = 0.7$, $\tilde{S} = 1.03$ (a), 0.80 (b), 0.66 (c), 0.55 (d), 0.48 (e) and 0.42 (f).

6. Excitonic *n*-strings

Electronic structures of charge transfer excitons were calculated by using the Frenkel exciton model taking a donor-acceptor pair as a unit cell [11, 12] and compared with experiments and detailed calculations [13]. In evaluating the electronic structure of excitonic n-strings, the symmetry breaking, practically accompanied with lattice distortions, was used implicitly. Donor electron may transfer to acceptors on the left and right hand sides so that a single charge-transfer exciton can be described as a linear combination of these. When two excitons are created from neighboring donors, two acceptors are chosen on the same side of those donors because attractive dipolar interaction is most effectively used. Of course, we can make linear combination of these two states to make the excitonic 2-string but the dipolar interaction is much larger than tunneling frequency. Therefore we may consider excitonic n-strings to have the induced static dipolemoment. For the moment in this Section, we derive the electronic structure and nonlinear optical responses due to excitonic n-strings without taking account of the lattice distortion explicitly.

The electronic structures and their eigen-energies are derived by diagonalyzing the Hamiltonian (1) with $Q_l = 0$. Analytical result for infinite system $N = \infty$ and numerical result for $N = 14$ are derived in [11, 12]. The effective masses for center-of-mass motion were derived analytically for the excitonic n-strings. The single exciton has the dispersion relation $E^{(1)}(K) = \omega_0 - 2J\cos K$ so that the effective mass is obtained as $M_f^{(1)} = 1/(2J)$. The dispersion relation of excitonic n-string is expressed by using the Bethe ansatz as

$$E_s^{(n)} = n(\overline{\omega}_0 - 2\Delta) - \frac{2\sinh\lambda\left[(-1)^{n-1}\cos K + \cosh n\lambda\right]}{\sinh n\lambda} \quad (n = 2,3,4,\cdots), \tag{13}$$

where $\overline{\omega}_0 \equiv \omega_0/J$, $\Delta \equiv J_z/J$, and $-\cosh\lambda = \Delta$ with J being the energy unit. The effective masses $M_s^{(2)}$ and $M_s^{(3)}$ of two- and three-strings are derived from Eq.(13) by putting $n = 2$ and $n = 3$:

$$M_s^{(2)} = J_z/J^2 = M_f^{(1)} \times 2\Delta > 2M_f^{(1)}, \tag{14}$$

$$M_s^{(3)} = \left(4J_z^2 - J^2\right)/2J^3 = M_f^{(1)} \times \left(4\Delta^2 - 1\right) > 3M_f^{(1)}, \tag{15}$$

both for $\Delta > 1$. In the case of Wannier excitons, a mass of the excitonic molecule is just twice as large as that of a free Wannier exciton. In the case of Frenkel excitons, however, the mass of the excitonic two-string depends on $\Delta \equiv J_z/J$ as Eq.(14) shows and is heavier than twice the free exciton mass for the case $\Delta > 1$, where the bound state is possible at $K = 0$. The three-string turns out to have a mass that is much heavier than three times of $M_f^{(1)}$. These results come from the complicated propagation mechanism of n-strings; for the n-string, n steps are necessary in order to propagate as a whole by a single site. Here the n-string must be dissolved virtually and successively into $(n-m)$- and m-strings in the intermediate states $(1 \le m \le n-1)$, in each of the n steps. As a result, the effective mass of the excitonic n-string increases rapidly with n; the wave function of the n-string in real space tends to be strongly localized with increasing n.

7. Exciton-mediated carrier transport

When the excitonic n-string is formed, its effective mass increases exponentially with n and well localized only through electronic origin. Once elementary excitations are localized, they become to interact more strongly with lattice vibrations than delocalized ones. Basing upon these considerations and the results obtained in Sections 2 to 6, we speculate the exciton-mediated carrier transport in strongly pumped Anthracene-PMDA.

First as mentioned in Sections 2 and 3, distorted clusters on n excited molecules are formed only through electron-phonon interactions. Second, we have found excitonic n-strings, i.e., bound states of n excitations only through electronic interactions. At the same time, the larger strings have been shown to become much heavier. This induces the much stronger exciton phonon interactions. Therefore once the cluster size n becomes larger than a critical size, we may easily expect that excitonic n-strings coupled with the lattice distortion become metastably or absolutely stable in the sense that these can not radiatively decay. These correspond to the energy diagrams in Fig.2(2) and 2(3), respectively. In the present case of anathracene-PMDA, the excited states are ionic because of the charge-transfer exciton while the ground state is made of neutral donor-acceptor pairs. These excited clusters are considered as metastable or stable ionic states embedded in neutral matrix. This is a locally induced Neutral-Inonic (NI) transition. Recently theoretical paper was published along this line [14]. Within a single cluster, each induced static dipolemoment is aligned in the same direction, but the positively and negatively polarized clusters are equally populated without applied electric field. These clusters survive even after switching off the pump field for the cluster size n larger than the critical size m_{c1} or m_{c2} introduced in Section 3. We may expect the electric polarization to be induced under application of static electric field because the relative number of positively and negatively polarized clusters is changed under the static field.

It is also interesting to optically pump the neutral crystal under application of static field. Then the number of positively polarized clusters is larger than the opposite one and the effective net polarization is expected to survive after switching off the pump light. When we excite the charge-transfer exciton in the crystal with the net effective polarization, then these charge-transfer excitons are more easily trapped so as to increase the net polarization of ionic clusters. Because the cluster sizes or the number of clusters, say, in the positively polarized direction than others so that the optically excited charge-transfer excitons are more strongly attracted to the majority clusters. Some of these speculations will play some roles in testing the present model on Anthracene-PMDA and others.

References

[1] see, e.g., chap.4 of Excitonic Processes in Solids, by M. Ueta, H. Kanzaki, K. Kobayashi, Y. Toyozawa and E. Hanamura (Springer-Verlag, 1986).

[2] E. Hanamura and N. Nagaosa, J. Phys. Soc. Jpn. **56**, 2080 (1987).

[3] T. Nagai, N. Nagaosa and E. Hanamura, J. Luminescence **38**, 314 (1987).

[4] Y. Tokura, T. Kanetake, K. Ishikawa and T. Koda, Synthetic Metals **18**, 407 (1987)

[5] S. Koshihara, Y. Tokura, K. Takeda and T. Koda, Phys. Rev. Lett. **68**, 1148 (1992).

[6] H. Bässler, H. Sixl and V. Enkelmann, Polydiacetylenes (Advances in Polymer Science **63**) ed. H. -J. Cantow (Springer-Verlag, 1984).

[7] H. Sixl, in Electronic Properties of Polymers and Related Compounds (Springer Series in Solid State Sciences **63**) ed. H. Kuzmany, M. Mehring and S. Roth (Springer-Verlag, 1985) p. 240.

[8] M. Hasegawa, Chem. Rev. **83**, 507 (1983).

[9] T. Kanetake, Doctral thesis (University of Tokyo, 1990).

[10] E. Hanamura and N. Nagaosa, in Polymers in Information Storage Technology, ed by K. L. Mittal (Plenum, 1989) p. 67.

[11] H. Ezaki et al., Solid State Commun. **88**, 211 (1993).

[12] H. Ezaki, T. Tokihiro and E. Hanamura, Phys, Rev. **B50**, 10506 (1994).

[13] M. Kuwata-Gonokami et al., Nature **367**, 47 (1994).

[14] Y. Toyozawa, Solid State Commun. **84**, 255 (1992).

Mean-Field Theory of Photo-Induced Structural Phase Transitions

T. Ogawa[1] *and N. Nagaosa*[2]

[1]Department of Physics, Tohoku University, Aoba-ku, Sendai 980-77, Japan

[2]Department of Applied Physics, The University of Tokyo, Bunkyo-ku, Tokyo 113, Japan

Abstract. A photoinduced structure change is treated as a nonequilibrium phase transition to investigate its switching dynamics and threshold behavior. To this end, the mean-field theory of the photoinduced structural phase transition is developed to clarify how local microscopic structural distortions in each molecule lead cooperatively to a global macroscopic structure change in many-molecule systems. A three-dimensional long(infinite)-range intersite coupling between the local distortions plays a crucial role in inducing threshold behaviors to bring about the nonlinearity inhered in the system. Dependence of the switching dynamics on optical pumping and temperature is clarified, and the time-resolved emission spectra are also calculated with this model.

1. Introduction

Structural changes as a result of relaxation after an optical pumping have been studied in the fields of photochemistry. In condensed matters, particularly in solids, the interaction among local structural distortions brings about the nonlinearity and can give rise to a new cooperative phenomenon [1], which is photoinduced switchings between two macroscopically distinguishable states.

An essential question is how the local microscopic structural distortions induced by optical pumping lead cooperatively to a global macroscopic change of the total system. In some polydiacetylene (PDA) crystals, the photoinduced structure change (called the AB transition) and the threshold behavior as a function of the pumping intensity were clearly observed. Then we treat this photoinduced structure change as a photoinduced nonequilibrium phase transition, stressing its threshold behavior and nonlinearity. When the three-dimensional interaction among the local structural distortions is large enough and the energy difference between the two states (distorted and undistorted) is small, the nucleation process, which prevents the threshold phenomena, can be neglected. In such a case, we can employ the mean-field treatment to discuss the structure change from a viewpoint of a phase transition.

Main aim of this paper is to present a theory as simple as possible for describing the dynamics of the photoinduced structure changes, paying a special attention to their cooperative nonlinearity. Such a theory provides us a novel viewpoint of the cooperative transition dynamics, which are described within a framework of classical nonlinear dynamics with single degree of free-

Springer Series in Solid-State Sciences, Volume 124
Relaxations of Excited States and Photo-Induced Structural Phase Transitions
Ed.: K. Nasu

dom. In particular, the mean-field model of the photoinduced structural phase transitions is investigated to clarify the transition dynamics and its dependence on optical pumping and temperature. In the switching dynamics, both the optical- and thermal-transition processes are incorporated. A phase diagram in a material-parameter space is given for characterizing the photoinduced structural transition to discuss the respective roles and qualitative differences between the optical and thermal processes. Moreover, time-resolved emission spectra are calculated. We also comment on a close analogy to the kinetic Ising model, leading to a physical insight and deep understandings.

2. The Minimal Model for the Structural Transition

We start with a simple model which consists of the localized two electronic levels $|1\rangle$ and $|2\rangle$ at site i with the Franck-Condon energy $E_{\rm FC}$. The state $|2\rangle$ is coupled with the interaction mode Q_i (treated as a classical variable) relevant to the structural change with the lattice relaxation energy S. The interaction modes ate site i and j are coupled with the coupling constant K_{ij}, whose diagonal elements K_{ii} is zero. The system Hamiltonian is written in a dimensionless scale as

$$\mathcal{H} = \sum_i |2_i\rangle \left(1 - \tilde{S}\tilde{Q}_i\right) \langle 2_i| + \frac{\tilde{S}}{2}\sum_i \tilde{Q}_i^2 - \frac{\tilde{S}}{2}\sum_{ij} K_{ij}\tilde{Q}_i\tilde{Q}_j, \qquad (1)$$

with $|1_i\rangle\langle 1_i| + |2_i\rangle\langle 2_i| = 1$. Here $\tilde{S} \equiv S/E_{\rm FC}$ $(0 \le \tilde{S} \le 2)$ and $\tilde{Q}_i \equiv Q_i/\sqrt{S}$. We here note that in this model, a species of elementary excitations and final (absolutely stable) structures at thermal equilibrium need not to be specified.

The relaxation of Q_i to the thermal equilibrium in each parabola finishes within the time of an order 10^{-13} sec. Therefore we assume that Q_i is always distributed around the bottom of one of the parabolas with the distribution functions:

$$F_1(\tilde{Q}_i) = (2\pi\tilde{T})^{-1/2}\exp[-(\tilde{Q}_i - \tilde{\Delta}_i)^2/2\tilde{T}], \qquad (2)$$

$$F_2(\tilde{Q}_i) = (2\pi\tilde{T})^{-1/2}\exp[-(\tilde{Q}_i - \tilde{\Delta}_i - 1)^2/2\tilde{T}], \qquad (3)$$

where $\tilde{T} \equiv k_{\rm B}T/S$ and $\tilde{\Delta}_i \equiv \sum_{j\neq i} K_{ij}\tilde{Q}_j$.

There are two elementary processes in the transition between $|1\rangle$ and $|2\rangle$ states: optical and thermal processes. The transition probability $P_{ab}^{\rm optical}$ from $a = |2_i\rangle$ state to $b = |1_i\rangle$ state of the ith site via the optical transition is given as

$$P_{ab}^{\rm optical} = \int d\tilde{Q}_i\, F_a(\tilde{Q}_i)\left[P_{ab}^{\rm down}(\tilde{Q}_i)\Theta(1 - \tilde{S}\tilde{Q}_i) + P_{ab}^{\rm up}(\tilde{Q}_i)\Theta(\tilde{S}\tilde{Q}_i - 1)\right], \qquad (4)$$

where $P_{ab}^{\rm down}(\tilde{Q}) \propto \tau_r^{-1}[1+f_B(\delta E)]|\delta E|^3$ and $P_{ab}^{\rm up}(\tilde{Q}) \propto \tau_r^{-1} f_B(-\delta E)|\delta E|^3$ with $\delta E \equiv E_a(Q) - E_b(Q)$, $f_B(\epsilon) = (e^{\epsilon/k_{\rm B}T} - 1)^{-1}$, and the spontaneous-emission lifetime τ_r. Only the vertical transition with respect to $\tilde{Q}_i$ is possible (the Franck-Condon principle). The unit step $\Theta(x)$ specifies which of $|2\rangle$ and $|1\rangle$ is lower in energy for each $\tilde{Q}_i$. For simplicity, we neglect the off-diagonal elements between the two states, which induce the hybridization of two branches near the crossing point.

On the other hand, the thermal transition rates are assumed to be the attempt frequency τ_t^{-1} times the activation factor:

$$P_{ab}^{\text{thermal}} = \tau_t^{-1} \exp[-(E_c - E_a^*)/\tilde{S}\tilde{T}], \tag{5}$$

where E_c and E_a^* (E_b^*) are the energy of the crossing point of the two parabolas and the energy of the bottom of the a (b) parabola, respectively, i.e., $E_c = (2\tilde{S})^{-1} - \tilde{\Delta}_i$, $E_1^* = -\tilde{S}\tilde{\Delta}_i^2/2$, and $E_2^* = 1 - \tilde{S}(1+\tilde{\Delta}_i)^2/2$.

In strictly one-dimensional systems, quantum fluctuation and the nucleation processes play crucial roles in low-temperature properties, leading to *no* threshold behavior and *no* phase transitions [2]. Here we access the photoinduced structural transition from a viewpoint of higher-dimensional features [3]. Then we employ the mean-field approximation to the mode-mode coupling term. The influence from the surrounding sites to the transition probabilities at the ith site is represented by the average values. Thus the intersite interaction is described by the mean displacement $\langle Q \rangle$, which is i-independent, and then the model is reduced to a single-site problem:

$$\mathcal{H}_{\text{MF}} = |2\rangle \left[1 - \tilde{S}(1 + K\langle\tilde{Q}\rangle)\tilde{Q}\right] \langle 2| + \frac{\tilde{S}}{2}\tilde{Q}^2. \tag{6}$$

Here $K \equiv \sum_j K_{ij}$ $(0 \leq K \leq 1)$ and $\tilde{\Delta} \equiv K\langle\tilde{Q}\rangle = \frac{K}{1-K} n_2$. States of the total system is specified only by n_2, which is the ratio of the number of $|2\rangle$ sites to the total number of sites $(0 \leq n_2 \leq 1)$. This treatment is exact for ∞-dimensional systems. Validity of the mean-field approximation will be discussed afterwards.

3. Phase Diagram at Low Temperature

Hereafter, assuming that all sites of the system are initially in $|1\rangle$ state [i.e., $n_2(t=0) = 0$], dynamics of the photoinduced structure change is described by the temporal evolution of $n_2(t)$. This obeys the rate equation of motion:

$$\frac{d}{dt} n_2(t) = [1 - n_2(t)]\, I_{\text{pump}}(t) - \frac{dU(n_2)}{dn_2}, \tag{7}$$

where $I_{\text{pump}}(t) > 0$ is the optical pumping (proportional to the creation rate of the number of $|2\rangle$ sites) and $U(n_2)$ is an effective potential for the n_2 dynamics, i.e,

$$U(n_2) = -\int dn_2 \left\{ [P_{12}^{\text{thermal}} + P_{12}^{\text{optical}}][1 - n_2(t)] - [P_{21}^{\text{thermal}} + P_{21}^{\text{optical}}] n_2(t) \right\}. \tag{8}$$

The collective nature of the system is built-in through the dependence of the transition probabilities [hence the potential $U(n_2)$] on n_2. After an optical pumping, the system is considered to transit to another state if $n_2 \to 1$, while

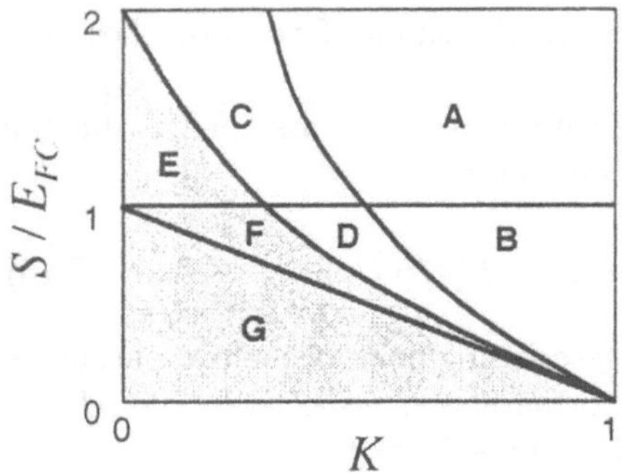

Fig. 1: Phase diagram of the system in terms of the photoinduced structural transition is given on the parameter plane of $(K, \tilde{S})$ at $\tilde{T} \ll 1$. The systems only in regions A–D can exhibit the structural phase transition under an appropriate optical pumping. In regions A and B (B and D) the spontaneous emission can be observed in the relaxation process to $|2\rangle$ ($|1\rangle$) state after an optical pumping. In dark region (E–G), the system is not switchable even for strong optical pumping.

otherwise ($n_2 \to 0$) the system fails to switch and goes back to the initial state.

We here consider the case of low temperature $\tilde{T} \ll 1$, which is experimentally relevant. When the solution of $dU(n_2)/dn_2 = 0$ lies between 0 and 1, the system can yield the photoinduced structural transition under an appropriate optical pumping. At $\tilde{T} = 0$, this switching condition is

$$2\frac{1-K}{1+K} < \tilde{S} < 2, \tag{9}$$

as shown in Fig. 1. Otherwise, the system never shows the structural switching irrespective of the pumping.

The switching region in Fig. 1 is further classified in terms of the optical stability. Whether the spontaneous emission is possible or not depends on the relative relation of the two parabolas. When

$$n_L \equiv \frac{(1-K)(1-\tilde{S})}{K\tilde{S}} < n_2(t) < \frac{1-K}{K\tilde{S}} \equiv n_U, \tag{10}$$

the system is optically stable against the spontaneous emission, that is, the system yields no light emission at low temperature. Then the relative relations of n_L, n_U, and 0, 1 leads to the qualitative difference in the dynamics of the structural transition. The relations $0 < n_U$ and $n_L < n_U$ always hold, and there are three boundaries corresponding to $n_L = 0$, $n_L = 1$, and $n_U = 1$. As a result, the parameter plane $(K, \tilde{S})$ is divided into seven regions, as shown in Fig. 1.

4. Numerical Results

Here we carry out numerical integrations of Eq. (7) to discuss the effect of pumping and the time-resolved emission spectra. The optical pumping is assumed to be the simplest rectangular shape as $I_{\text{pump}}(t) = I_p$ only for $0 \le t \le \Delta t_p$ and otherwise $I_{\text{pump}}(t) = 0$. We note that our present model cannot describe how the structural change dynamics depends on the frequency of pumping light, which is fixed to be more than E_{FC} in this paper.

Temporal dynamics of the transition process is found to be sensitive to the pump pulse duration (Δt_p) and the pump pulse intensity (I_p). Figure 2 shows the switching time distribution on the plane of I_p and Δt_p. There exists a threshold pulse intensity I_p^{th}, which is approximately evaluated as

$$I_p^{\text{th}} \sim \left(\frac{1}{\tau_t} + \frac{1}{\tau_r}\right) \frac{1-K}{K} \frac{1-\tilde{S}}{\tilde{S}}. \qquad (11)$$

Numerically-evaluated values of I_p^{th} are well described by the above approximate formula. Dependence of I_p^{th} and the switching time t_{switch} on K, $\tilde{S}$, and $\tilde{T}$ will be reported in detail elsewhere.

Next we shall discuss the time-resolved emission spectrum $W_{21}(\omega)$ $[W_{12}(\omega)]$ in the radiative $|2\rangle \to |1\rangle$ $[|1\rangle \to |2\rangle]$ decay process, which is given as

$$\begin{aligned} W_{ab}(\omega) \;\propto\; & F_a\left([E_{\text{FC}} \mp \hbar\omega]S^{-1/2}\right) P_{ab}^{\text{down}}\left([E_{\text{FC}} \mp \hbar\omega]S^{-1/2}\right) \\ & - F_a\left([E_{\text{FC}} \pm \hbar\omega]S^{-1/2}\right) P_{ab}^{\text{up}}\left([E_{\text{FC}} \pm \hbar\omega]S^{-1/2}\right), \qquad (12) \end{aligned}$$

with the upper (lower) sign for the $|2\rangle \to |1\rangle$ ($|1\rangle \to |2\rangle$) case. At low temperature, the $|2\rangle \to |1\rangle$ spontaneous emission is possible when $n_2(t) < n_L$, while

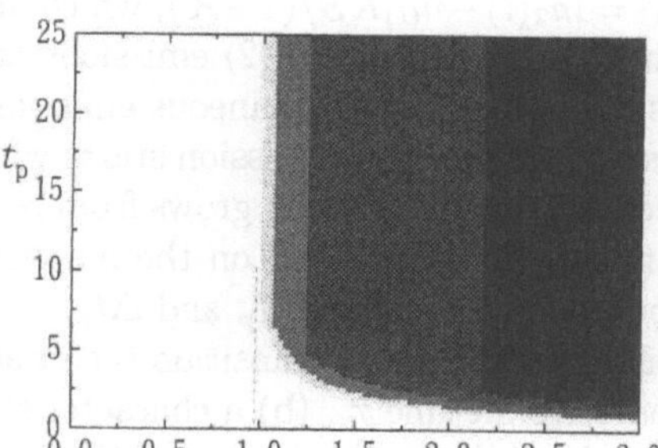

Fig. 2: The switching-time distribution on the plane of I_p and Δt_p. Material parameters are chosen to be $K = 0.6$, $\tilde{S} = 0.6$ (region D), $\tau_t/\tau_r = 10^{-2}$, and $\tilde{T} = 0.05$. In dark and black regions, the system exhibits the structural transition with the switching time $t_{\text{switch}} = \tau_t \sim 6\tau_t$. In a white region, the system cannot switch to the other state. There exists a threshold pumping intensity $I_p^{\text{th}} \sim 0.96$ (denoted as a broken line).

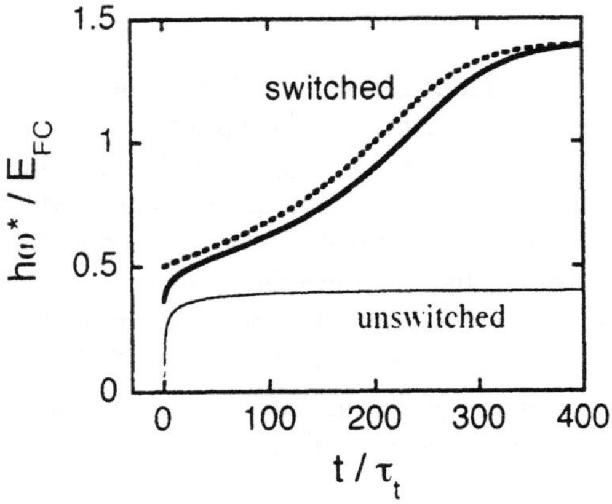

Fig. 3: Temporal evolution of the central frequency of the spontaneous emission spectra is drawn for $K = 0.8$, $\tilde{S} = 0.6$ (region B), $\tau_t/\tau_r = 10^{-2}$, and $\tilde{T} = 0.05$. The time origin ($t = 0$) is chosen to be the time at the end of pumping. The duration time of a pumping pulse is fixed at $\Delta t_p = 30\tau_t$. When $I_p = 0.7$ (thick broken line) and $I_p = 0.2156$ (thick solid line), the system yields the structural switching. On the other hand, the system goes back to the initial state with no structural transition when $I_p = 0.215$ (thin broken line) and $I_p = 0.1$ (thin solid line), which are almost overlapped. For $I_p = 0.7$, 0.2156, and 0.1, there is no time delay of the onset of emission after the end of pump pulse. But for $I_p = 0.215$, there is a time delay and the central frequency starts from zero. In this figure, the time delay is so small that it cannot be seen.

the $|1\rangle \to |2\rangle$ spontaneous emission when $n_2(t) > n_U$. The central frequency of the Gaussian emission spectrum is dependent on time as $\hbar\omega_{21}^*(t)/E_{\rm FC} = [n_L - n_2(t)]K\tilde{S}/(1-K)$ and $\hbar\omega_{12}^*(t)/E_{\rm FC} = [n_2(t) - n_U]K\tilde{S}/(1-K)$, which are shown in Fig. 3 for region B, where both $|2\rangle \to |1\rangle$ and $|1\rangle \to |2\rangle$ emissions are possible. In general, there are two types of onset of the spontaneous emission; (i) just after the optical pumping finishes, the spontaneous emission starts with a *finite* frequency, and (ii) the central frequency of the emission grows from zero *after a time delay* from the finish of pumping. These depend on the material parameters, K and $\tilde{S}$, and also on the pumping pulse shape, I_p and Δt_p.

The temporal dynamics of the photoinduced structural transition is characterized by several time scales: (a) the radiative lifetime τ_r, (b) a characteristic time of thermal transition $\tau_t \exp[(\text{barrier height})/k_{\rm B}T]$, (c) characteristic times of optical pumping I_p^{-1} and Δt_p, (d) the appearance time of a critical droplet τ_m (the nucleation process is neglected in the mean-field model assuming large τ_m), and (e) the observation time in experiments. We can expect a variety of transition dynamics by controlling these time scales; temperature controls the time scales (a) and (d), choice of materials affects (a), (b), and (d), and the optical pumping changes (c). For the design of photoswitching and photomemory materials, the controllability of these time scales is of crucial importance.

5. Discussion

5.1. Relation to the Kinetic Ising Model

Here we make comments on the relation of our model to the Ising model. Integrating out the interaction modes Q_i, the partition function of the model Eq. (1) is $Z \propto \mathrm{Tr}_{\{\sigma\}} \exp[-\mathcal{H}_{\mathrm{eff}}(\{\sigma\})/k_\mathrm{B}T]$, where the effective Hamiltonian is given by

$$\mathcal{H}_{\mathrm{eff}}(\{\sigma\}) = -h\sum_i \sigma_i - \frac{1}{2}\sum_{i\neq j} J_{ij}\sigma_i\sigma_j, \tag{13}$$

with the "magnetic field" h and the "exchange" J_{ij}:

$$h = \tilde{S}/(2-2K) - 1, \tag{14}$$

$$J_{ij} = (\tilde{S}/N)\sum_{\vec{k}} \omega_{\vec{k}}^{-2} \cos[\vec{k}\cdot(\vec{R}_i - \vec{R}_j)]. \tag{15}$$

Here N is the total number of sites, $\vec{R}_i$ is the lattice vector of the ith site, and $\omega_{\vec{k}}^2 = 1 - \sum_j K_{ij} \exp[i\vec{k}\cdot(\vec{R}_i - \vec{R}_j)]$.

The mean-field approximation to the Ising model is equivalent to replacing J_{ij} by the infinite ranged constant $N^{-1}J \equiv N^{-1}SK/(1-K)$. The transition temperature in the absence of the field h is given by $\tilde{T}_c = J/4 = K/(4-4K)$. Below this temperature, the free energy $\mathcal{F}(M)$ as a function of the magnetization $M = \langle\sigma_i\rangle = n_2 - \frac{1}{2}$ has double minima. The switching phenomena are expected only in the case of double minima in $\mathcal{F}(M) - hM$. The condition for the switching Eq. (9) is reduced to $|h| < \frac{1}{2}J$, which is nothing but the stability condition of the uniform metastable state against the single spin flip. Introducing the stochastic dynamics to the Ising model (kinetic Ising model), we can discuss the nonequilibrium dynamics of the system.

The close relation between our model and the Ising model can induce us to study the photoinduced structural transition with randomness and/or finite-range anisotropic couplings in terms of the spin glass and the neural networks.

5.2. Dynamics at Finite Temperature

The switching condition at general temperature is analytically given as $G_-(K) < \tilde{S} < G_+(K)$ for $4\tilde{T}/(1+4\tilde{T}) < K < 1$ with

$$G_\pm(K) = \left[\frac{1}{2} + \frac{K}{1-K} g_\pm\left(\tilde{T}\frac{1-K}{K}\right)\right]^{-1}, \tag{16}$$

$$g_\pm(x) = \frac{2x}{1 \pm \sqrt{1-4x}} + x\ln\left[\frac{1-2x\pm\sqrt{1-4x}}{2x}\right]. \tag{17}$$

The switching region in the phase diagram on the $(K, \tilde{S})$ plane shrinks as $\tilde{T}$ increases. Only the switching region B with large K and small $\tilde{S}$ survives the thermal fluctuations. However, the switching becomes harder because the threshold intensity I_p^{th} is very large there. We also note that at finite temperature, the optical transition process is always possible due to the thermal broadening of the distribution functions $F_n(\tilde{Q}_i)$ around the bottom of the potential parabolas. Therefore the criteria by n_L and n_U become ambiguous, i.e., qualitative difference among regions A–D tends to vanish. Moreover, the relative importance of the thermal transitions increases drastically, and hence the switching and relaxation times become shorter as $\tilde{T}$ increases.

We then conclude that in order to obtain the structural switching phenomena with reasonable pumping intensity, one should keep the system temperature low compared with the lattice relaxation energy.

5.3. Concluding Remarks

The mean-field model does not include the nucleation process, i.e., the fluctuation. Here we shall mention the validity of the mean-field treatment. In our model, the energy difference between the two states is $|h|$ and the coupling energy between the sites is J. With the aid of the continuum approximation [4], the critical droplet size is evaluated as $L_{\rm cr} \sim J/|h|$. When $L_{\rm cr} < 1$, the nucleation process plays a crucial role in the dynamics, and the threshold behavior vanishes due to the fluctuation. On the other hand, the mean-field treatment is valid only when $L_{\rm cr} \gg 1$, i.e., near the diagonal line $\tilde{S} \simeq 2(1-K)$ in the phase diagram. Our model is applicable to the materials with large interchain/interlayer couplings, where the phase-transition-type switchings are observed.

In conclusion, we have proposed a simple mean-field model to describe the photoinduced structural change as a nonequilibrium phase transition. We have classified the transition dynamics on the $(K, \tilde{S})$ plane, and clarified the threshold behavior, time scales, and the time-resolved spontaneous emission spectra. Temperature dependence of the dynamics has also been discussed. Although our model is too simple for full understanding of experimental results, it offers a novel viewpoint and many new findings and can be one of the starting points for exploring guiding principles in searching for and designing the new materials in terms of photomemory and photochromism. There are many important extensions of our model to be carried out in the future.

Acknowledgements

The authors would like to thank Prof. T. Luty, Prof. V. Emelianov, Prof. E. Hanamura, Prof. K. Nasu, and Dr. S. Abe for stimulating discussion. One of the authors (T.O.) thanks K. Tatani for his help in numerical computations.

References

[1] *Optical Properties of Low-Dimensional Materials*, edited by T. Ogawa and Y. Kanemitsu (World Scientific, Singapore, 1995).
[2] E. Hanamura and N. Nagaosa, J. Phys. Soc. Jpn. **56**, 2080 (1987).
[3] N. Nagaosa and T. Ogawa, Phys. Rev. B **39**, 4472 (1989).
[4] J. D. Gunton and M. Droz, *Introduction to the Theory of Metastable and Unstable States* (Springer-Verlag, Berlin, 1983).

Part II

Optical Excitations and Photo-Induced Phenomena in $BaBiO_3$, La_2CuO_4 and Other Perovskites

Spectroscopic Study of a 3d Peierls Insulator in $BaBiO_3$

H. Uwe

Institute of Applied Physics, University of Tsukuba,
Tennoudai 1-1-1, Tsukuba-shi, Ibaraki, 305, Japan

Abstract. For $BaPb_xBi_{1-x}O_3$, we have found a long-tail absorption with a square-root spectral shape which is consistent with a picture of the indirect allowed exciton. The exciton consists of an electron and a hole at either side of an indirect gap formed by the three-dimensional CDW. The indirect-gap energy deduced from the analyses nearly coincides with twice of the activation energy of the electronic transport. The exciton binding energy is found to be comparable with the band gap for a nearly stoichiometric $BaBiO_3$, indicating that possible Bose condensation of thermal excitons might be related to the enormous density of states which contributes to transport phenomena. In $BaPb_xBi_{1-x}O_3$, the deduced indirect gap energy once increases and then decreases with the Pb substitution, indicating formation of the Pb 6s band above the band gap with increasing the band width to merge into the lower Bi-6s band at the metal insulator transition.

1. Introduction

A perovskite-type oxide of $BaBiO_3$(BBO), a parent material of superconducting $BaPb_x$-$Bi_{1-x}O_3$(BPB)[1] and $Ba_{1-x}K_xBiO_3$(BKB)[2], is a semiconductor with a superlattice distortion of frozen-in breathing mode at R-point.[3] Cubic BBO would be a metal with a half-filled band. However, infrared observation of an optical-phonon of anti-parallel motion of Bi ions at R-point indicates formation of an insulator with the charge density wave(CDW).[4.5] In fact, the infrared activity for the anti-parallel motion of adjacent Bi ions cannot be realized without differently-charged states of Bi ions in the crystal with the center of symmetry. In addition, in an optical reflection measurement, a sharp optical conductivity peak at 1.9 eV was observed, which was inferred to correspond to the transition over the Peierls gap,[6] since a similar sharp optical conductivity peak was observed in a one-dimensional CDW insulator of $K_{0.3}MoO_3$.[7]

However, an experiment of electronic transport exhibited much smaller activation energy of 0.24eV than the visible CDW gap of 1.9 eV.[8] This indicates that the simple picture of the CDW insulator should not be applied to $BaBiO_3$. As the experiments which investigate the electronic structure, X-ray-absorption near-edge-structure(XANE) from Bi(*2p*) state was performed to exhibit a bump above the edge which is absent in Bi_2O_3 and through those of $Ba_{1-x}K_xBiO_3$.[9] This gives evidence for the Bi(6s) hole states just above the chemical potential. Since XANE from O(1s) state exhibits O(2p) hole state, on the other hand, O(2p)-Bi(6s) hybridized states will be formed and correspond to the conduction band. Recent photo-emission study combined with O(1s) XANE indicates a gap structure

Springer Series in Solid-State Sciences, Volume 124
Relaxations of Excited States
and Photo-Induced Structural Phase Transitions
Ed.: K. Nasu

between the split hybridized states of Bi(6s) and O(2p) with a much smaller energy than the visible CDW-gap energy.[10]

Recently, we investigated a mid-infrared absorption for single crystals of $BaBiO_3$, in which the spectra can be analyzed with a picture of the indirect allowed excitons consisting of electrons and holes presumably near L($\pi/2a$, $\pi/2a$, $\pi/2a$)- and W(π/a, $\pi/2a$,0)-points in the Brillouin zone(BZ), respectively.[11] The deduced indirect-gap energy E_g was found to coincide nearly with twice of the activation energy of the transport properties. Therefore, $BaBiO_3$ can be viewed as a semiconductor having an electronic structure with the gap at the surface of the 3-dimensional BZ which results from the superlattice structure with the R-point frozen-in phonon.

In this paper, we review experimental studies for the mid-infrared absorption and reflection for exciton in $BaBiO_3$ and also report on recent experiments on $BaPb_xBi_{1-x}O_3$(x= 0-0.2), in order to see how the indirect gap and the exciton characteristics change as approaching to the insulator-metal transition. Finally, we discuss on the problem that the density of states which contributes to the transport is comparable to the density of Bi atoms, being indicated by the Hall effect measurement.

2. Sample preparation and experimental procedure

Single crystals of $BaPb_xBi_{1-x}O_3$(nominal x=0, 0.04, 0.08, 0.12, 0.16) were grown from solutions of BaO_2, Bi_2O_3 and Pb_3O_4. Electron probe microanalysis indicated that the ratio of Ba to Bi is 1.025 for the present $BaBiO_3$ and is 1.05 for the previous sample of [11]. Exciton absorption spectra were found to be different between those. A large flat surface of the crystal was normal to the pseudo cubic <100> direction. The crystal was first optically polished on the large surface to which the reflectivity spectra were measured from 500 to 5000 cm^{-1} by the Fourier transform infrared spectrometer(FTIR). Then, the samples were made into thin plates and optically polished with thickness of 100μm and the infrared transmission spectra were measured. The absorption coefficient α and the refractive index were deduced self-consistently from the reflectivity and the transmissivity.

3. Absorption spectra of excitons in BaBiO3

Figure 1 shows spectra of absorption coefficient α of BaBiO3 with the ratio of Ba to Bi of 1.05 at 30K. Prominent features below 2000cm^{-1} are found to be overtones of LO phonons ($2LO_2$=820cm^{-1}, $2LO_1$=1144cm^{-1}, $3LO_1$=1729cm^{-1}) judged from the fundamental frequencies[4,5] of LO_1=572cm^{-1} and LO_2=410cm^{-1} [11]. These overtone spectra well correspond to the resonance Raman spectra,[12] in which the overtone phonons should not be assigned to ones due to the breathing mode of the oxygen octahedra but to the LO phonons. The overtone is found to be observed because of the unharmonic nature of the LO phonons not of the strong excited-electron-phonon interaction, since we observe those in infrared spectra.

In addition, above ca 1000cm^{-1}, one finds a temperature-dependent broad band which increases the intensity with elevating temperature. The band at 30K has a frequency

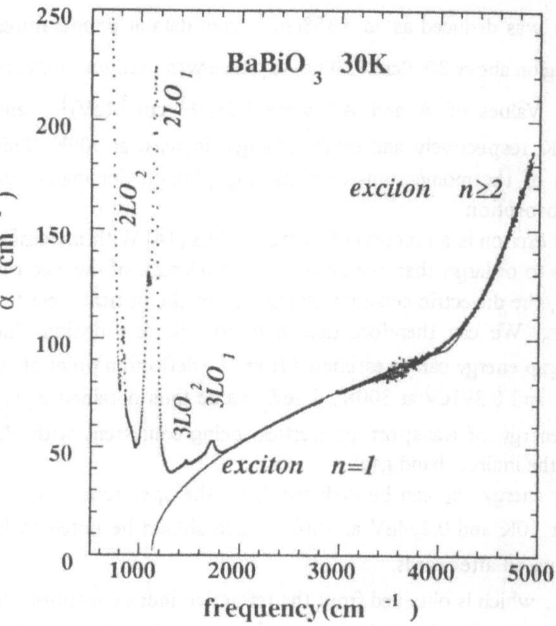

Fig.1 Absorption coefficient α of $BaBiO_3$ at 30K. Prominent features below 2000cm^{-1} are overtones of LO phonons. The solid line is a fitted curve of the indirect-exciton absorption.

dependence of an upward-convex square-root behavior below 4000cm^{-1}, followed by an abrupt increase. These are characteristics of absorption spectra α by indirect allowed excitons[13] between the indirect energy gap E_g, in which, for the principal quantum number of n_{ex}=1 and the binding energy

$$E_b = \mu e^4 / 2\varepsilon_1^2 \hbar^2, \tag{1}$$

$$\alpha = A(1+n_p)(\hbar\omega - E_{1+})^{1/2} + An_p(\hbar\omega - E_{1-})^{1/2} \tag{2}$$

where $E_{1\pm} = E_g - E_b \pm \hbar\omega_p$ and $n_p = 1/[exp(\hbar\omega_p / kT) - 1]$.

For $n_{ex} \geq 2$ and the binding energy

$$E_b' = \varepsilon_1^2 E_b / 4\varepsilon_2^2, \tag{3}$$

$$\alpha = A^*(1+n_p)(\hbar\omega - E_{2+})^{3/2} + A^* n_p(\hbar\omega - E_{2-})^{3/2}, \tag{4}$$

where $E_{2\pm} = E_g - E_b' \pm \hbar\omega_p$, and the dielectric constant, ε_1 and ε_2 for excitons of n_{ex}=1 and 2, is a function of exciton radius.[14]

A fitted absorption curve due to the excitons of n_{ex}=1 and 2 is shown in Fig1. Here, frequency of momentum-compensating phonon for infrared absorption by the indirect exciton with a finite wave vector was deduced as ω_p=358cm^{-1} from data at temperatures higher than 200K in frequency region above 3000cm^{-1}. The ω_p value was used for analyses of data at lower temperatures. Values of A and A* were 1.3x10^2 cm^{-1}/(eV)$^{1/2}$ and 1.7×10^{3}cm^{-1}/(eV)$^{3/2}$ below 200K, respectively and exhibited slight increase at 300K. This indicates that thermal excitation of the momentum-compensating phonon dominates the temperature dependence of the absorption.

The dielectric screening for exciton is a function of exciton radius.[14] With increasing the radius from the atomic radius to or larger than the de Broglie wavelength of the electron or hole with LO-phonon energy, the dielectric constant changes from the optical dielectric constant ε_∞ to the static one ε_0. We can therefore take a reasonable assumption that $\varepsilon_1^2/4\varepsilon_2^2 << 1$. Then, the indirect-gap energy can be estimated from the deflection point of α and ω_p as E_g=0.459eV at 30K and 0.391eV at 300K. The E_g value thus obtained equals nearly twice of the activation energy of transport properties, being consistent with the picture of a semiconductor with the indirect band gap.

Now, the exciton binding energy E_b can be deduced from the spectrum of n_{ex}=1 using E_g and ω_p as 0.360eV at 30K and 0.254eV at 300K which should be noted to be comparable with E_g, being discussed afterwards.

From values of E_b and ε_∞ which is obtained from the refractive-index spectrum, the reduced mass μ of the exciton and the effective Bohr radius a_H^* are evaluated as 0.325m_0 and 5.69Å at 30K and 0.319m_0 and 6.83Å at 300K, respectively.

Since the CDW gap opens three-dimensionally at the BZ boundary with the breathing-mode distortion, the minimum energy of the conduction band at the L-point(π/2a, π/2a, π/2a) in the pseudo-cubic notation lies which is nearest to the Γ-point(0,0,0), while the maximum energy of the valence band lies at the W-point(π/a, π/2a, 0) which is farthest from the Γ-point. Therefore, the indirect transition with the smallest energy gap seems to be possible between the L and W points, and the transition is allowed, since the wave function of charge-transferred state has opposite parity to that of the CDW state. The observed sharp optical conductivity at 1.9eV seems to originate from the fact that the vertical transition has nearly constant energy between the CDW and charge transferred states on the BZ boundary surfaces of hexagons centered at the L-point.

Mattheiss and Hamann[15] and Liechtenstein et al.[16] calculated the band structure by taking account of the breathing-mode distortion and the rotation of the oxygen octahedra. Both of the results exhibit that the indirect-gap energy is much smaller than the direct-gap one in qualitative agreement with the present experiment. Recently, Chowdhury[17] calculated the direct and indirect exciton absorption by using an extended Peierls-Hubbard model. The obtained spectra are found to well correspond to the conductivity peak at 1.9 eV and the mid-infrared absorption.

4. Excitons in $BaPb_xBi_{1-x}O_3$

Recently, we measured spectra of exciton absorption in $BaPb_xBi_{1-x}O_3$ in the semiconducting region.[18] The spectra can be decomposed into absorptions for n=1 and n≥2 and multi-phonon one. In Fig.2, the deduced values of E_g and E_b, and the phonon frequency ω_p, are

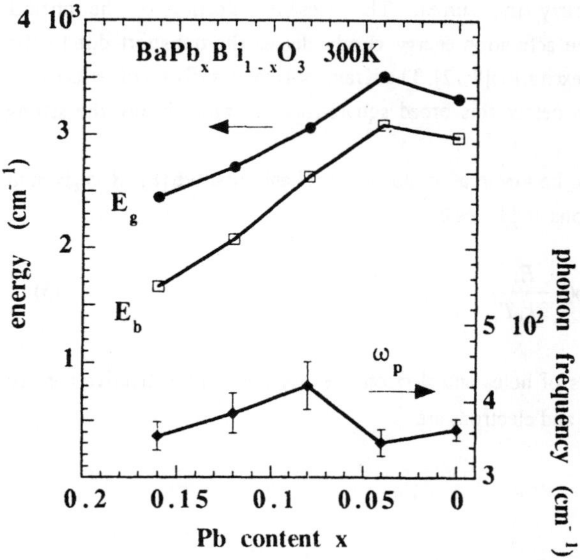

Fig.2 Values of the band-gap energy E_g, the exciton binding energy E_b, and the phonon frequency ω_p for the photon absorption as functions of nominal Pb content x.

plotted as functions of nominal Pb content x. Here, for the sample of $BaBiO_3$, the ratio of Ba to Bi is 1.025. We find that E_g nearly coincides with one for the ratio of 1.05, but E_b is considerably larger than the previous one.

With increasing Pb, E_g increases at first and then decreases. This behavior can be understood as follows; with substitution of Pb for Bi, the widths of two split Bi bands decrease at first with decreasing the Bi concentration to broaden the gap width, but the width of the Pb band which is located above the bottom of the upper Bi band increases afterwards and the optical transition from the top of the lower Bi band to the bottom of the Pb band determines the gap energy which decreases with x. The virtual-crystal model might be difficult for the interpretation of the first increase of the gap energy with x. The metal-insulator transition in BPB will be the result of the merge of the Pb band into the lower Bi band. This picture was first proposed by Yoshioka and Fukuyama[19], and is discussed in connection with observation of the pseudo gap in the optical reflection spectra.[12]

Since the exciton binding energy E_b is comparable with E_g, we could expect the condensation of the exciton with decreasing E_g. However, the refractive index increases with decreasing E_g, so that E_b decreases. This behavior was also observed in the less stoichiometric sample of the Ba to Bi ratio of 1.05, in which, with increasing temperature, E_g decreases while E_b decreases more rapidly. [11]

5. Possible Bose-condensation of thermal indirect excitons

The activation energy of the electronic transport, 0.24eV,[8] is found close to half of the presently-found indirect gap, and can be interpreted by a picture that both of thermally-excited holes and electrons carry the current. The present detection of the optical absorption corresponding to the activation energy clearly denies the transport due to the optical-forbidden two-electron excitation[8,12]. The Kramers-Kronig analysis of reflectivity as in [12] might be difficult to detect this broad square-root spectrum below the strong CDW-gap absorption.

The Hall coefficient R is, however, anomalously small and positive.[8] R is given in case of equal densities of electrons and holes by

$$\frac{1}{eR}=\frac{\mu_p+\mu_n}{\mu_p-\mu_n}\sqrt{N_pN_n}\exp(-\frac{E_g}{2k_BT}), \tag{5}$$

where μ_p and μ_n are mobility's of holes and electrons, respectively. The effective density of states N_p and N_n for holes and electrons are

$$N_p=2g_W(\frac{m_pk_BT}{2\pi\hbar^2})^{\frac{3}{2}}, \tag{6}$$

$$N_n=2g_L(\frac{m_nk_BT}{2\pi\hbar^2})^{\frac{3}{2}} \tag{7}$$

respectively, where $g_W=6$ and $g_L=4$ are multiplicity numbers of valleys at W and L points in the BZ and m_p and m_n are effective masses at those points. Since the zone-boundary surface at L point is normal to Γ-L direction, the constant-energy surface centered at L point might be more anisotropic than at W point. Therefore, the effective mass at L might be larger than at W, which might explain the observed positive R value, although effects of either band splitting or polaron would change the circumstance. The factors of g_W and g_L play a role to decrease R by one order of magnitude. However, $\mu_p/\mu_n=1.01$ should be concluded, on assuming $m_n=m_p=2m_0$, to interpret the observed small R. Otherwise, those masses would be much larger, ca $20m_0$, improbable because of polaron effect at dc or lower frequencies than LO-phonon one, contrary to the case of rapid motion of exciton.

As shown in the previous section, the more stoichiometric $BaBiO_3$ exhibits the comparable energy of the binding energy of the exciton with E_g, indicating possible Bose condensation of excitons due to the thermal or spontaneous excitations, in which the transition temperature might be higher than a few thousands K for the density corresponding to the Bi density. The E_b value itself is not so large, but the E_g one is small and the refractive index is small, resulting in the possible condensation. In this case, the dissociation of the exciton contributes to the carrier generation with dissociation energy of E_g and with a quite high density of states, a half of the Bi density. This will give the small Hall coefficient as observed. We have recently observed a piezoelectric mechanical oscillation under the RF electric field in $BaBiO_3$,[20] suggesting the broken symmetry due to the condensation followed by a lock-in transition from the incommensurate to homogeneous electronic phase.

6. Summary

We have discussed the indirect allowed exciton which consists of an electron and a hole at either side of an indirect gap formed by the three-dimensional CDW. The indirect-gap energy deduced from the analyses nearly coincides with twice of the activation energy of the electronic transport. The exciton binding energy is found to be comparable with the band gap for the more stoichiometric BBO, indicating possible Bose condensation of thermal excitons. In BPB, the deduced indirect gap energy once increases and then decreases with the Pb substitution, indicating formation of the Pb 6s band above the band gap with increasing the band width to merge into the lower Bi-6s band at the metal insulator transition.

Acknowledgment

We thank Prof. K. Nasu for providing a chance to present these results and information on recent theoretical calculations and Prof. H. Sumi for helpful discussions and the Cryogenic Center for use of the FTIR apparatus.

References

1. A. W. Sleight, J.L. Gillson, and P. E. Bierstedt, Solid State Commun. **17** , 27 (1975).
2. L. F. Mattheiss, E. M. Gyorgy, D. W. Johnston Jr., Phys. Rev. B **37**, 3745 (1988), R. J. Cava, B. Batlogg, J. J. Krajewski, R. Farrow, L. W, Rupp Jr., A. E. White, K. Short, W. F. Peck, T. Kometani, Nature **332**, 814 (1988).
3. D. E. Cox, A. W. Sleight, Acta. Crystallogr, B **35**, 1 (1979).
4. S. Uchida, S. Tajima, A. Msaki, S. Sugai, K. Kitazawa, S. Tanaka , J. Phys. Soc. Jpn. **54**, 4395 (1985).
5. A. Inoue, A. Iyo, H. Uwe, T. Sakudo, Y. Tanaka, M. Tokumoto, Advances in Superconductivity IV, Proc. 4th Int. Symp. Supercon. (ISS '91), 139.
6. S. Tajima, S. Uchida, A. Msaki, H. Takagi, K. Kitazawa, S. Tanaka, Phys. Rev. B **32**, 6302 (1985).
7. G. Travaglini, P. Wachter, Phys. Rev. B **30**, 1971 (1984).
8. H. Takagi, S. Uchida, S. Tajima, K. Kitazawa, S. Tanaka, Proc. 18th Int. Conf. Phys. Semiconductors, Stockholm, 1851 (1986).
9. S. M. Heald, D. DiMardio, M. Croft, M. S. Hegde, S. Li. Greenblatt, Phys. Rev. B **40**, 8828 (1989).
10. H. Namatame, A. Fujimori, H. Takagi, S. Uchida, F. M, F. de Groot, J. C. Fuggle, Phys. Rev. B **48**, 16917 (1993-I).
11. H. Uwe and K. Tachibana, Advances in Superconductivity VII, Proc. 7th Int. Symp. Supercon.(ISS '94) 165.
12. S. Uchida, K. Kitazawa, S. Tanaka, Phase Transitions **8**, 95 (1987).
13. R. J. Elliott, Phys. Rev. **108**, 1384 (1957).

14. H. Haken, J. Phys. Chem. Solids **8**, 166 (1959).
15. L. F. Mattheiss, D. R. Hamann, Phys. Rev. B **28**, 4227 (1983).
16. A. L. Liechtenstein, I. I. Mazin, C. O. Rodriguez, O. Jepsen, O. K. Andersen, M. Methfessel, Phys. Rev. B **44**, 5388 (1991-II).
17. A. Z. Chowdhury, Thesis, The graduate University for Adv. Studies (1996).
18. H. Uwe and Y. Yamamoto, Czechoslovak J. of Phys. **46** S5, 2689 (1996).
19. D. Yoshioka and H. Fukuyama, J. Phys. Soc. Jpn. **54**, 2996 (1985).
20. M. Nao and H. Uwe, presented at the Spring meeting of the Phys. Soc. Jpn, 1996.

Evidence of Polaronic Deformations in High T_c Materials Observed by Photo-Induced IR Absorption and Photo-Induced XAS

G. Ruani and C. Taliani

Istituto di Spettroscopia Molecolare C.N.R., via Gobetti, 101, I-40129 Bologna, Italy

Abstract. Doping the semiconducting parent compound of the high temperature superconducting systems by laser illumination is used to investigate possible charge-phonon interaction in the normal state. In this paper we present an analysis of the lattice distortions induced by photogenerated carriers in different families of cuprate perovskites like $YBa_2Cu_3O_{6+x}$, La_2CuO_4, and Nd_2CuO_4. Comparison of photoinduced (PA) features with doping induced modes and with Raman scattering measurements of semiconducting and metallic systems reveals strong similarities between most of the observed PA vibrational mode and those (Raman or IR) of the superconducting material. In order to investigate the microscopic photoinduced structural and electronic variation in $ErBa_2Cu_3O_{6+x}$ we have also performed photoinduced X-ray absorption spectroscopy (P-XAS). The results indicate that a contraction of the in plane copper (Cu(2)) and apical oxygen (O(4)) bond occurs and also that the photoinjected holes responsible of the distortion are situated on O(4) with p_z symmetry. A comparison of our findings with possible theoretical scenarios is also presented.

1. Introduction

The normal state properties of high T_c superconductors as well as the microscopic mechanism of superconductivity in this materials is not yet understood. Electronic correlation effect, strong charge-lattice interactions, antiferromagnetic coupling and strong anharmonicity of oxygen atoms that are all present in cuprates have been invoked by theoreticians to understand the physics of these materials, and superconductivity itself.

The study of high critical temperature superconductors by means of low energy optical spectroscopy (Raman scattering, reflectance and photoinduced IR absorption) has played a significant role in characterising materials and in investigating low energy excitations as phonons, electronic or magnetic transitions and their interactions. In this view photoinduced absorption provides information on these quasi-particles and their interaction at very low doping concentration regime.

The observation of photogenerated polaronic distortion associated to new electronic transitions in HT_c semiconducting parent compounds have been reported by many authors [1-9]. The PA differential spectrum shows new vibrational modes associated with local distortions, and also bleachings of modes typical of the undistorted structure. A centrosymmetric system undergoes the mutual exclusion selection rule between IR and Raman modes; the photoinduced local lattice distortion in the undoped compound induce a reduction of symmetry, destroying the inversion symmetry and providing dipole moment in the distorted cell, as a result a moderate IR activity of formally forbidden Raman modes can be induced allowing their observation in the PA spectrum. IRAV modes can also derive from the change in energy of IR active modes of the undistorted lattice due to the charge-lattice interaction. This will show a bleaching of the peak corresponding to bare phonon (ω_0) and the appearance of a peak of the lattice vibration at the distorted site (ω_p) [10]. The observation of bleachings in the PA spectrum can also be associated with a screening effect associated with the photogeneration of mobile carriers. In this paper we report also, for the first time, photoinduced X-ray absorption measurements in a semiconducting system ($ErBa_2Cu_3O_{6.15}$)

2. Experimental

2.1. Photoinduced absorption (PA)

Pressed pellets of the powdered ceramic materials with CsI (at 1% weight concentration) are investigated by transmission measurements by using a modified FT-IR interferometer in vacuum (mod. Bruker IFS 113v) with a Ge/Cu liquid He cooled detector (0.03 - 0.5 eV). Measurements are carried out at low temperature in a flow liquid helium cryostat. Carriers are photogenerated in the semiconducting materials by laser light. The spectra reported in this paper have been obtained by using a c.w. Ar^+ laser at 2.54 eV. Typical intensity of $-\Delta T/T$ (T = transmittance) observed by a photoinduced experiment at low temperature (about 10 K) in these materials is of the order of 1/1000 or smaller, this requires high sensitivity of the experimental set-up and accuracy in performing the experiment.

The PA signal can be of the same order of magnitude of the changes induced in the spectrum by heating of the sample. In order to avoid misinterpretation of the photoinduced absorption spectrum features with those due only to heating effects, the PA spectrum is always compared with the thermomodulated spectrum where a hot source is substituted to the laser.

2.2. Photoinduced X-ray absorption (P-XAS) in $ErBa_2Cu_3O_{6+x}$

X-ray absorption spectra were measured by the dispersive spectrometer at LURE-DCI synchrotron radiation facility at Orsay using a Si(111) polichromator crystal with a Darwin width limited energy resolution of 1 eV at 9 KeV. This spectrometer, because of the great stability in X-ray optics, is a very performant instrument for experiences demanding great sensibility in the measurement of X-ray absorption spectra differences [11, 12]. P-XAS was carried out on pressed pellet of $ErBa_2Cu_3O_{6.15}$ in LiF. The dimension of the ceramic particles were of about 1 μ and we have used LiF powder for its low absorption coefficient both in the optical region for laser excitation and in the Cu-K edge energy region around 9 KeV. The $ErBa_2Cu_3O_{6.15}$ equivalent thickness was approximately 2 μ, considering an homogeneous distribution of the material in the LiF powder. Measurements were carried out at 30K using a helium closed cycle cryostat. Photoexcitation was obtained by a c.w. Ar^+ laser using the 2.54 eV line at different power density. The pellets was mounted at 45° respect to the X-ray beam axis. The optical line was built to drive the laser beam to both side of the pellets, to improve its volume actually irradiated. Spectra were obtained by using a similar technique as for PA measurements. The photoinduced X-ray absorption coefficient difference spectra $\Delta\mu$ is equivalent, in the limit of small variation of the absorption coefficient, to calculate $-\Delta T/T$ as normally performed in experiments of photoinduced IR absorption.

3 Photoinduced absorption

3.1. PA in La_2CuO_4

The PA spectrum of La_2CuO_4 shows (see Fig. 1) the photobleaching of two phonon modes at 380 and 690 cm^{-1} and the appearance of IRAV modes at 422, 463 and 570 cm^{-1}. plus a peak at 670 cm-1 that is thermoinduced. The PA spectrum is very similar to that was reported by Kim et al. [2]. Both the bleaching peaks are associated with two E_u oxygen modes: the peak at 380 cm^{-1} is assigned to the bending mode of the apical oxygen and the one at 690 corresponds to the in plane oxygen copper stretching mode [13]. A decrease in intensity of these modes are also observed in Sr doped samples because of the screening of free carriers in the CuO_2 plane [14]. The peak at 422 cm^{-1}, coincide with the A_g mode of the apical oxygen parallel to the Cu-O plane [15]. The 463 cm^{-1} peak is quite close in energy to the most intense IR doping induced features observed at low temperature[13] that has been interpreted as a polaronic derived structure[16]. The broad feature around 570 cm^{-1} do not correspond to any feature neither in the Raman nor in the doping induced spectrum. One can try to interpret this peak as the mode derived from the in plane E_u copper oxygen stretching mode in the site where the polaron is sitting. Exact diagonalisation calculations [10] show that the formation of small polarons in Cu-O planes is allowed for electron phonon coupling strength (λ) larger than λ_c. The presence of the polaron induce a change in energy of the phonon modes. In case of hole doping a softening is expected, on the other hand, electron doping will induce a hardening of the bare mode. So a large shift in energy of the vibrational mode with respect to the bare mode will correspond to a value of λ much larger than λ_c implying a very low mobility of the polarons.

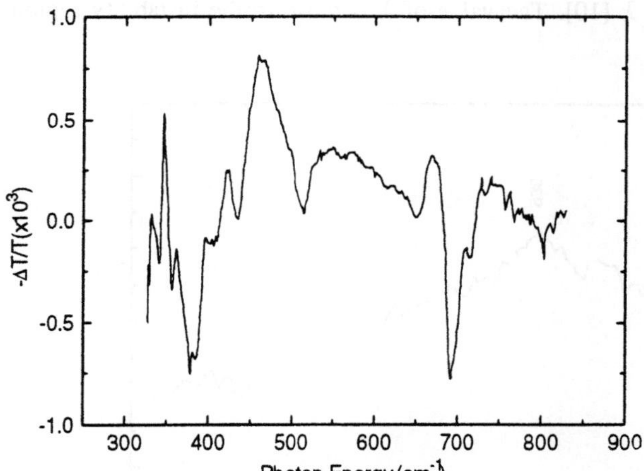

Fig.1 Photoinduced absorption spectrum of La_2CuO_4 performed at 10 K

3.2. PA in Nd_2CuO_4

The photoinduced spectra of Nd_2CuO_4 show several PA peaks in the spectral region below 600 cm^{-1}: a bleaching around 520 cm^{-1} and other peaks whose relative intensity strongly depend on laser power density and a broad peak with maximum around 930 cm^{-1}. (see Fig. 2)

The interpretation of the PA spectral features is not trivial. In fact, the infrared spectrum of Nd_2CuO_4 has phonons with different symmetry with about the same energy ($E_u \perp c$; $A_{2u} // c$) at about 515 cm^{-1} [17]; in analogy with the PA spectrum of La_2CuO_4 the 521 cm^{-1} bleaching can be associated to the E_u mode (in plane Cu-O stretching).

The appearance of new absorption band in this system cannot be attributed to any Raman mode that could become IR active because of the breaking of the local symmetry as in the case of $YBa_2Cu_3O_{6+x}$ or La_2CuO_4: in fact there is no Raman mode in the T' structure involving atoms close to the CuO_2 plane where the carriers can be induced [17]. By comparing the PA features with doping induced IRAV modes it is possible to observe a similarity with the weak peak at 382 cm^{-1} [14, 16]. Also the broad band with maximum at about 930 cm^{-1} almost coincide with the so called *d* band in electron doped system [16]. The structures in the low energy side of this band are reproducible and some of these structures are also present in doped samples and have been interpreted as overtones of local modes associated to polarons in CuO_2 planes [16].

The photogenerated electron hole pairs in Nd_2CuO_4 probably are accommodated both in the CuO_2 plane this mean that we will expect features associated to both electrons and holes. By attributing the PA peak at 461 cm^{-1} to local modes associated to an "hole" polaron that corresponds to the $E_u \perp c$-axis bleached phonon mode at 521 cm^{-1} (in plane Cu-O stretching) it is possible to estimate the electron-lattice coupling strength for hole doped Nd_2CuO_4 using the relation between the softening of this mode ω_0^2/ω_p^2 with λ [10]. The value of λ is close to the instability region around λ_c.

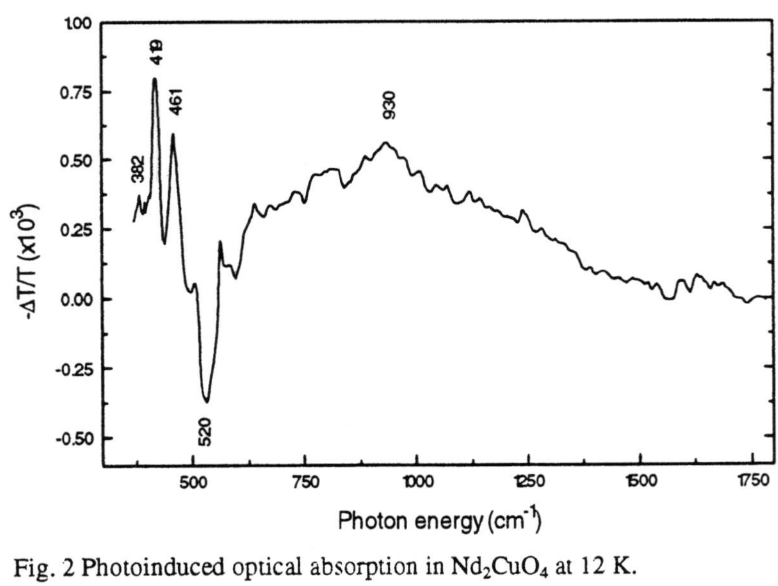

Fig. 2 Photoinduced optical absorption in Nd_2CuO_4 at 12 K.

The crystal structure of Nd_2CuO_4 is more complex than the T' structure were the CuO_2 planes are forming isolated elements. It has been shown that more than 10 % of the Cu atoms is bonded to an apical oxygen [18]. Moreover there are no available data of reflectance measurements // *c*-axis. These facts open more possible interpretations of PA data of the Nd_2CuO_4 compound.

3.3. PA in $YBa_2Cu_3{}^{16}O_{6+x}$ $YBa_2Cu_3{}^{18}O_{6+x}$.

The most relevant features in the photoinduced absorption of the semiconducting $YBa_2Cu_3O_{6+x}$ (Y-123) system in the region of the Cu-O vibrational modes are a broad peak around 510 cm^{-1} (phonon energy are referred to ^{16}O sample), two narrower bands at about 410 and 445 cm^{-1}, and a bleaching of the IR allowed mode at 610 cm^{-1} (Fig. 3).

As already discussed in earlier papers [8, 9, 19] the photoinduced modes at 510 and 445 cm^{-1} are very close in energy to two of the Raman active modes of the fully doped materials indicating that the local distortion induced by the photoinduced species are similar to those induced by chemical doping. The bleached mode corresponds to the E_u in plane Cu(2)-O(2) stretching; the oscillator strength reduction of this mode can be due either to the decrease of volume (the distorted area of the CuO_2 plane around the photoinduced carrier acquires a different dipole and a different oscillator energy) or to a screening of the mode due to 'free' carriers. The vibrational origin of the photoinduced features is confirmed by the observation of the isotope shift $\Delta\omega/\omega$ in the PA spectrum of a ^{18}O substituted sample [19]. All peaks (and bleaching) are red shifted (Fig. 3). Let us focus on the strongest features (the peak at 515 cm^{-1}). In Fig. 4 the photoinduced IR-absorption spectrum of the $YBa_2Cu_3O_{6.15}$ compound (c) is compared with the IR excited Raman spectra of fully-doped $YBa_2Cu_3O_{7-y}$ [20] (a) and low-doped $YBa_2Cu_3O_{6.15}$ [19, 21] (b). All the three spectra are characterised by a peak at about 500 cm^{-1}, indicating that the photogenerated polaron-like structures in semiconducting Y-123 (Fig. 4 c) as well as those resulting from local deformations around short-chains in the same material

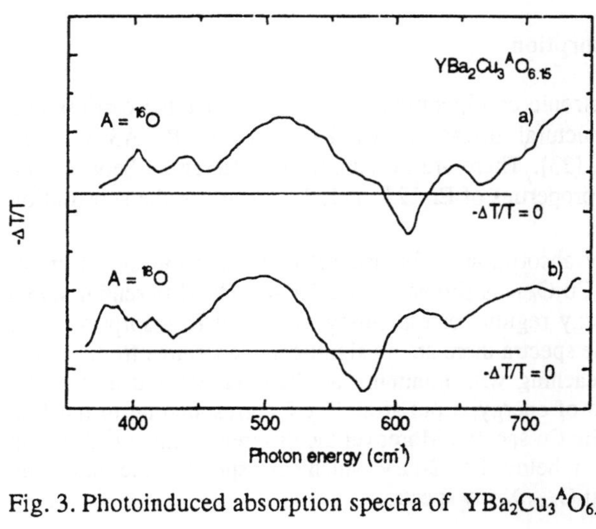

Fig. 3. Photoinduced absorption spectra of $YBa_2Cu_3{}^AO_{6.15}$: a) A=16, b) A=18.

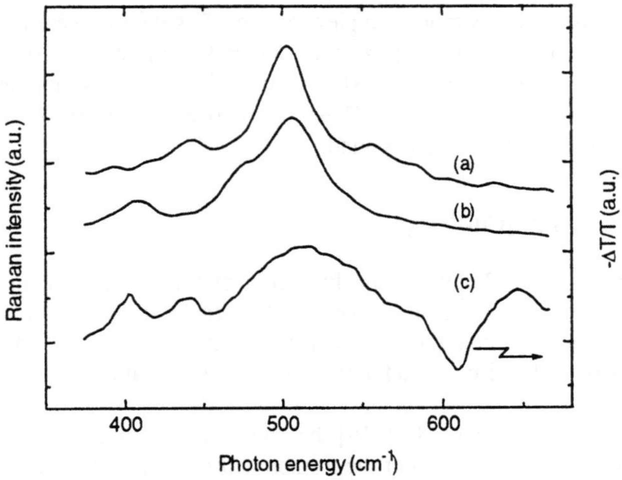

Fig. 4. Comparison between **a)** IR-excited Raman spectrum of single crystal $YBa_2Cu_3O_{7-y}$ in the x(zz)x̲ geometry; **b)** IR-excited Raman spectrum of polycrystalline $YBa_2Cu_3O_{6.15}$ and **c)** photoinduced IR-absorption spectrum of $YBa_2Cu_3O_{6.15}$. All three spectra are characterised by a peak around 500 cm^{-1}.

(Fig. 4 b) also exist in the high-oxygen doped system (Fig. 4 a). This indicates that both photoinduction and oxygen doping of semiconducting Y-123 result in the generation of polaronic states which mainly affects the apex oxygen position along the *c*-axes.

The peak at 510 cm^{-1} shows the smallest isotope shift (4.1 %) with respect to the other peaks in the PA spectrum and it is even smaller than the shift observed in the Raman spectra of the metallic samples [22]. This observation agrees with the anharmonic nature of the apical oxygen O(4) potential and suggests that the largest deformation of the lattice (polaron) is confined around the O(4) site.

4. Photoinduced X-ray absorption

In order to identify the electronic configuration of the photogenerated carriers and possibly the associated structural distortion, we have studied P-XAS of well characterised $ErBa_2Cu_3O_{6.15}$ [23]. There are no significant differences between the structure and the electronic properties of Er-123 Y-123 , except for the substitution of Er on Y.

The photoinduced X-ray absorption difference between laser on and laser off absorption spectra of $ErBa_2Cu_3O_{6.15}$ is shown in Fig. 5 [24]. The differential signal can be observed in the energy region corresponding to the Cu-K absorption edge while in the other parts of the spectra there are no significant departures from zero. It is possible to observe a bleaching with minimum at about 12 eV and a shoulder around 5 eV where the zero of energy is put at 8977 eV corresponding to the first inflection point of the metallic Cu spectra. Moreover the difference signal is different from zero in the energy region below 18 - 20 eV which correspond to the maximum of the absorption of $ErBa_2Cu_3O_{6.15}$ X-ray spectra.

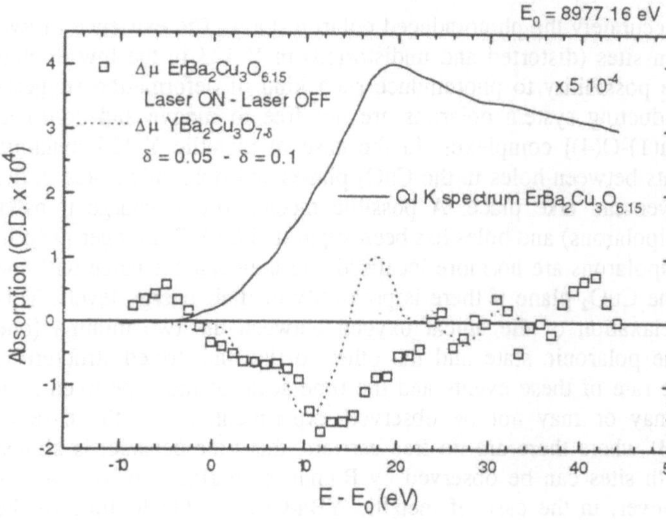

Fig. 5. XAS (full line) and P-XAS (open squares) spectra of $ErBa_2Cu_3O_{6.15}$. The dotted line correspond to the difference XAS spectrum of $YBa_2Cu_3O_{6.95}$ $YBa_2Cu_3O_{6.90}$ from ref. [26]

The interpretation of the photoinduced X-ray absorption difference may be found on comparison with data concerning the variation of the Cu-K absorption edge with oxygen vacancies doping keeping in mind that the photo-doping does not produce the stoichiometric variations of chemical doping.

A detailed study of oxygen doping dependence of Cu-K edge features in $YBa_2Cu_3O_{6+x}$ [25] shows that very similar difference spectra are observed in the evolution of doping from x = 0.8 to x = 0.95 (Fig. 6 of ref. [25]) which is the region where the system has the maximum value of T_c, above 90 K. In this regime the difference spectra show many similarities to the first derivative of the absorption for single crystal in *c*-axis orientation between 0 and 15 eV.

The peak around 12.5 eV that is observed in the difference spectra can be associated to the change of the Cu(2)-O(4) distance [25, 26]. The similarity of this feature with the one observed by P-XAS suggests that the photoinduced holes in Er-123 are accommodated in the $O(4)p_z$ orbitals inducing a contraction of the copper plane apical oxygen distance as occurs in the high doping regime.

5. Results

It has been shown that in all the semiconducting parent compounds of high T_c materials that have been investigated it is possible to photoinduce polaronic-like deformation in the Cu-O sublattice. In the case of Nd_2CuO_4 and La_2CuO_4 the interpretation of the spectra is not trivial: there are not sufficient data for an unambiguous identification of the type of distortions that are photogenerated. In the case of the Y-123 system, the large number of information we have allows to localise and describe more accurately the photoinduced polaron states. The existence of two frozen apical oxygen sites (distorted and undistorted) in Y-123 in the low doping regime [27] and the possibility to photoinduce such kind of deformation suggests that in the semiconducting system polarons are not free to moove and confined within the [O(4)-Cu(1)-O(4)] complexes. In the case of metallic Y-123 dynamic charge-transfer events between holes in the CuO_2 planes and polaronic states in the charge reservoir layer can take place. A possible mechanism of charge transfer between polarons (bipolarons) and holes has been suggested by J. Ranninger [28]; in the metallic states bipolarons are no more localised and they can exchange with the mobile carriers in the CuO_2 plane if there is proximity of their energy levels. This corresponds to a relaxation of the apical oxygen between the two minima (one corresponding to the polaronic state and the other to the undistorted structure) . Depending upon the rate of these events and the time scale of the experiment, the two distinct sites may or may not be observed experimentally. In the case of $YBa_2Cu_3O_{6+x}$ ($x<0.3$) where there are no free carriers, this time constant is almost infinite and thus both sites can be observed by Raman scattering and by c.w. PA measurements. However, in the case of metallic $YBa_2Cu_3O_{7-y}$, the lifetime of the distorted states become too short and it is possible to observe only the average value between the two sites.

A second possible scenario, consistent with our findings, is the one suggested by Bianconi et al. [29]. Polarons in the doped system are forming stripes inducing a corrugation of the CuO_2 plane [29]; this structure is derived by EXAFS analysis of BISCO and LASCO where a bimodal distribution of apex oxygen-copper plane is found. The stripes, formed by polarons (bipolarons), are insulating forming a superlattice of Josephson coupled quantum wells (quasi one dimensional metal) leading to an increase of the density of state at the Fermi energy and consequently of T_c.

Acknowledgements

We own special thanks to all colleagues who have collaborated with us in the original research reported here: Dr. Zamboni, Dr. A.J. Pal, Dr. F. Baldi, Dr. M. Missori and Dr. A. Bianconi. We also acknowledge the technical assistance of Mrs. P. Mei, T. Bonfiglioli and A. Martiniello.

References

[1] A. J. Epstein *et al.*, in *High Tc Superconductors*, Eds. A. Bianconi and A. Marcelli (Pergamon Press, Oxford, 1989), p. 87.
[2] Y. H. Kim *et al.*, Phys. Rev. B **36**, 7252 (1987).
[3] Y. H. Kim *et al.*, Phys. Rev. **36**, 7252 (1987).
[4] Y. H. Kim, S.-W. Cheong, and Z. Fisk, Phys. Rev. Letters **67**, 2227 (1991).
[5] D. Mihailovic *et al.*, Phys. Rev. B **42**, 7989 (1990).
[6] D. Mihailovic and C. M. Foster, Solid State Commun. **74**, 753 (1990).
[7] D. Mihailovic *et al.*, Phys. Rev. B **44**, 237 (1991).
[8] C. Taliani *et al.*, Solid State Commun. **66**, 487 (1988).

[9] C. Taliani *et al.*, in *Electronic Properties of High Tc Superconductors and Related Compounds*, Eds. H. Kuzmany, M. Mehiring and J. Fink (Springer-Verlag, Berlin-Heidelberg, 1990), p. 280.
[10] J. Lorenzana and A. Dobry, Phys. Rev. B **50**, 16094 (1994).
[11] E. Dartyge, *et al.*, Nucl. Inst. Meth. A **246**, 245 (1986).
[12] H. Tolentino, E. Dartyge, F. A., *et al.*, J. Appl. Cryst. **21**, 15 (1988).
[13] A. V. Bazhenov, *et al.*, Phys. Rev. B **40**, 4413 (1989).
[14] G. A. Thomas, *et al.*, Phys. Rev. B **47**, 11369 (1993).
[15] C. Thomsen and M. Cardona, in *Physical properties of high temperature superconductors I*, Ed. D. M. Ginsberg (World Sci., Singapore, 1989), p. 409.
[16] P. Calvani, *et al.*, Phys. Rev. B **53**, 2756 (1996).
[17] E. T. Heyen, *et al.*, Solid State Commun **74**, 1299 (1990).
[18] P. G. Radaelli, *et al.*, Phys. Rev. B **49**, 15322 (1994).
[19] G. Ruani, *et al.*, Physica C **185**, 963 (1991).
[20] G. Ruani, in *Phase Separation in Cuprate Superconductors*, Eds. K.A. Müller, E. Sigmund (Springer-Verlag, Berlin, 1993), p. 333.
[21] V. N. Denisov, Phys. Rev. B **48**, 16714 (1993).
[22] G. Ruani *et al.*, Physica C **226**, 101 (1994).
[23] P. G. Radaelli *et al.*, Phys. Rev. B **45**, 4923 (1992).
[24] G. Ruani, C. Taliani, M. Missori, A. Bianconi *et al.*, (in preparation).
[25] H. Tolentino *et al.*, Physica C **192**, 115 (1992).
[26] H. Tolentino *et al.*, in *Electronic properties of High Tc Superconductors and related compounds*, Eds. H. Kuzmany, M. Mehring and J. Fink (Springer-Verlag, Berlin, Heidelberg, 1990), Vol. 99, p. 176.
[27] G. Ruani, P. Guptasarma, and C. Taliani, Solid State Commun. **96**, 653 (1995).
[28] J. Ranninger, in *Anharmonic Properties of High Tc Cuprates*, Eds. D. Mihailovic, G. Ruani, E. Kaldis and K.A. Müller (World Sci., Singapore, 1995), p. 204 and refs. therein.
[29] A. Bianconi and M. Missori, J. Phys. I France **4**, 361 (1994).

Part III

Optical Excitations, Lattice Relaxations and Photo-Induced Phenomena in M-X Chains

Relaxation Processes of Photo-Excited States in Wide-Gap MX Chain

Y. Wada, N. Matsushita[1], *and M. Yamashita*[2]

National Institute for Research and Inorganic Materials, 1-1 Namiki, Tsukuba, Ibaraki 305 Japan
[1]College of Arts and Sciences, University of Tokyo, 3-8-1 Komaba, Meguro, Tokyo 153, Japan
[2]Graduate School of Human Informatics, Nagoya University, Furochou, Chikusa, Nagoya 464-01, Japan

Abstract. Time evolution in msec region, temperature dependence, and excitation photon energy dependence of the photoinduced absorption spectra of wide gap MX-chain complexes have been measured. It has been found that the photoinduced absorption bands are caused by three states: the photo-generated neutral soliton pair, the polaron pair, and a state the origin of which is not clear. Neutral soliton pairs are most efficiently generated from the charge transfer excitons. Polaron pairs are generated only from the higher excited states. It is found that the generation of the self-trapped excitons and that of the neutral soliton pairs compete with each other in the relaxation processes of the charge transfer excitons.

1. Introduction

Recently, grate many studies have been made on the relaxation processes of the photo excited states in one-dimensional electron systems. The halogen-bridged mixed valence metal complexes (MX-chain complexes) are considered to be prototype materials for the studies of the relaxation processes of the phtoexcited states in one-dimensional electron systems [1]. In MX-chain complexes metal (M=Pt, Pd, and Ni) and halogen (X=Cl, Br, and I) ions form one-dimensional linear chain as M^{4+}-X^--M^{2+}-X^--M^{4+}-X^--M^{2+}-X^-. The metal ions are surrounded by the ligand molecules such as ethylenediamine (en) and cyclohexanediamine (chxn). In Pt- and Pd complexes, the halogen ion distortions from the central position of the neighboring metal ions cause charge density waves (CDW's) with periods of twice the metal-metal distances. Since the CDW ground states have twofold degeneracy as in polyacetylene, the investigation of the relaxation processes of the photoexcited states in these materials has gained considerable interest from a fundamental point of view [1-16]. One interesting feature of the MX-chain is that the physical properties, such as the energies of CDW band gaps and magnitudes of the halogen ion distortions can be changed in a wide range by the substitution of the metal, the halogen, the counter ions, and the ligand molecules [1,3]. Additionally, single crystals can be easily prepared.

There are characteristic differences between the relaxation processes of the photoexcited states in MX-chain complexes with small CDW band gaps (E_{CT}<2.0eV) [9] and those in MX-chain complexes with wide CDW band gaps (E_{CT}>2.5eV) [5]. Here, E_{CT} is the peak energy of the charge transfer (CT) exciton absorption band. In a small gap MX-chain complex the CT-excitons relax nonradiatively to the ground states. Polarons and solitons are generated from the higher excited states [9]. In a wide gap MX-complex a self-trapped exciton (STE) luminescence band [2,3] and two or three photo-induced absorption (PA) bands [5] are observed when the complex is irradiated within the CT-exciton absorption band. It has been found that the relaxed states which cause these PA-bands have S=1/2 spins. The PA-bands have been assigned to polaron absorption bands [5,16] or neutral soliton absorption bands [7]. The assignments are still controversial.

Springer Series in Solid-State Sciences, Volume 124
Relaxations of Excited States and Photo-Induced Structural Phase Transitions
Ed.: K. Nasu

In wide gap MX-chain complexes little attention has been paid to the dependence of the relaxation processes of the photo excited states on the excitation photon energy. In most of the experimental studies of the relaxation processes of the wide gap MX-chain complex the photon energies of excitation light are within the CT-exciton absorption bands. In such cases, only the relaxation processes of the CT-excitons were studied. However, the binding energies of the CT-excitons in wide gap MX-complexes are quite large: larger than 0.5eV [8]. It may be quite hard for charge separated states to be generated from CT-excitons. Therefore, in order to elucidate all over features of the relaxation processes of the photoexcited states in wide gap MX-chain complexes, it is necessary to investigate the relaxation processes of higher excited states. It has been found in $[Pt(en)_2][Pt(en)_2Br_2](SO_4)_2.6H_2O$ (Pt-Br-SO_4) that the feature of the PA-spectrum changes depending on the excitation photon energy [11]. In order to elucidate the details of the PA-bands, time evolution in msec region, temperature dependence, and excitation photon energy dependence of the PA-spectra have been measured.

Generation of the STE's is one of the important relaxation processes of the CT-excitons in wide gap MX-chain complexes. Since there is no potential barrier between free excitons and STE's in one-dimensional electron system, CT-exciton (free exciton) is expected to efficiently relax to STE's. However, little investigation has been made on the relaxation processes of STE's. In order to investigate the relaxation processes of STE's, lifetimes and their temperature dependence of the STE luminescences in several MX-chain complexes have been measured.

2. Experiment

In the time resolved PA spectra measurement samples are irradiated by a XeCl excimer laser or by a dye laser pumped by the excimer laser. A W-lamp was used as light source of the probe light. The probe light was passed through samples, polarizer, and a monochromator and then detected by a photomultiplier or InGaAs photodiodes. The time evolution of the PA was recorded by a digital storage oscilloscope. The time resolution of the system is about 10μsec.

In the time integrated PA-spectra measurements 500W size high pressure Hg lamp was used as an excitation light source. The light was passed through a monochromator and a polarizer and was chopped (0.5Hz). The samples were irradiated by the light. The probe light was chopped (470Hz) and detected by a photomultiplier, by a Ge photodiode, or by a PbS photodetector. The AC component of the output signal of these detectors are detected by a lock-in amplifier which is synchronized with the chopping frequency of the probe light. The AC component of the output signal of the lock-in amplifier is detected by another lock-in amplifier which is synchronized with the chopping frequency of the excitation light.

In the time resolved luminescence measurements the samples were excited by a dye laser synchronously pumped by a cw mode-locked Ar ion laser. Excitation light in the UV region was obtained by the frequency doubling of the dye laser pulse. The duration of the pulse was about 5psec. The luminescence was passed through a spectromator and then detected by a streakcamera. The time resolution of the system is about 30psec.

3. Results

Fig. 1 shows the polarized luminescence spectrum and the polarized component of the imaginary part of the dielectric constant (ε_2) of Pt-Br-SO_4 parallel to the chain axis at 2K. Imaginary part of the dielectric constant has been obtained from the reflection spectrum by Kramers-Kronig transformation. The broad strong absorption band at 2.67eV is a CT-exciton absorption band [3,8]. The broad luminescence band is caused by the radiative recombination of the STE's [2,3]. Such STE luminescence bands are commonly seen in wide gap MX-chain complexes [3].

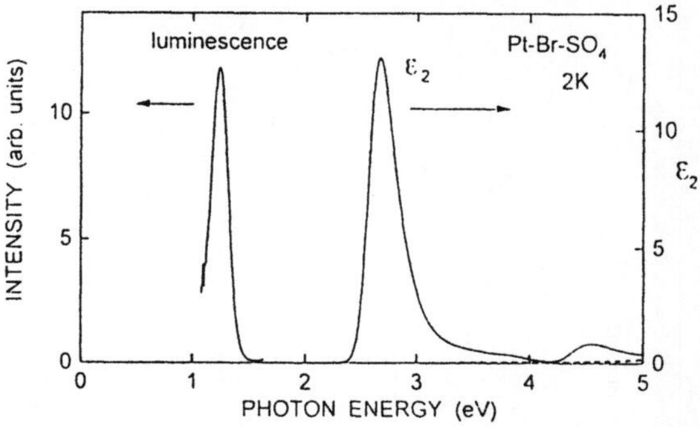

Fig. 1. Polarized luminescence and and ε_2 spectra parallel to the chain axis of Pt-Br-SO_4.

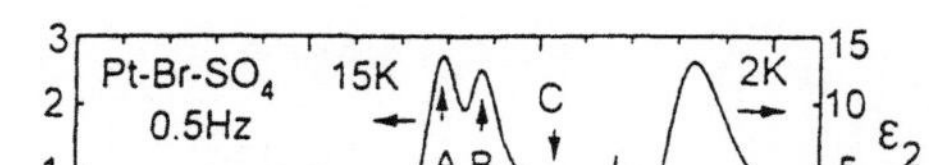

Fig. 2. Excitation photon energy dependence of PA-spectra of Pt-Br-SO_4.

Fig. 2 shows the polarized PA spectra of Pt-Br-SO_4 parallel to the chain axis for different excitation photon energies (E_{ex}'s). Here, excitation light is polarized parallel to the chain axis (P_{ex}//chain). All the PA bands shown in the figure are polarized parallel to the chain axis. When the excitation photon energies are within the CT-exciton absorption band, three PA-bands marked by A, B, and C are observed and no large change of the PA spectra has been observed. Such PA-bands are commonly observed in wide gap MX-chain complexes when these complexes are irradiated within the CT-exciton absorption bands. On the other hand, new PA bands marked by D, E, F, and G appear in the case of higher energy excitation. This shows that there are characteristic differences between the relaxation processes of CT-excitons and those of the higher excited states.

Fig. 3. shows the excitation spectra of the A-, D-, E-bands, STE luminescence at 2K, and photo current at room temperature of Pt-Br-SO_4. Here, the excitation light is polarized

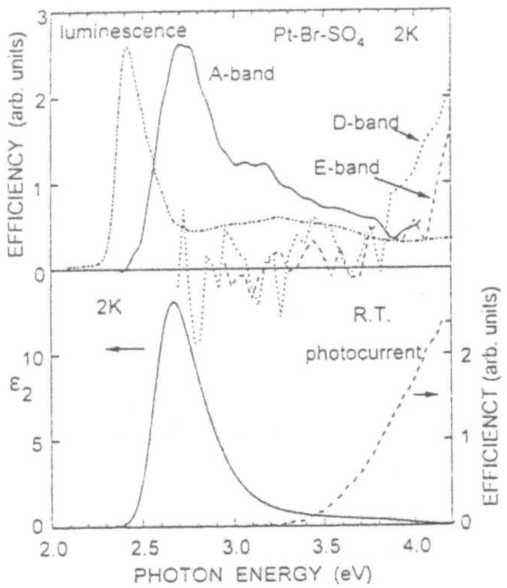

Fig. 3. Excitation spectra of the A-, D-, E-band, STE luminescence, and photocurrent in Pt-Br-SO4.

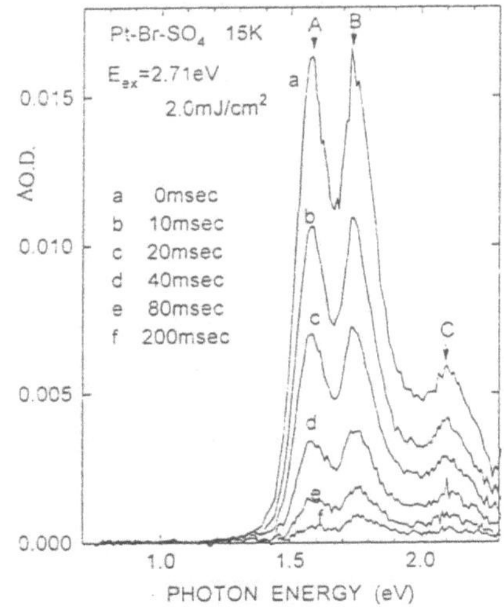

Fig. 4. Time evolution of PA-spectra of Pt-Br-SO4 for Eex=2.71eV.

parallel to the chain axis. The polarized component of ε_2 parallel to the chain axis is also shown in the figure. The A-band is most efficiently generated by the light with photon energy close to the peak energy of the CT-exciton absorption band. On the other hand, the D- and E-bands are only generated by the light with much higher photon energy. It is found that the maximum of excitation spectrum of the STE luminescence is located at the tail of the CT-exciton absorption band. The efficiency of the STE luminescence decreases steeply above the energy of the maximum. On the contrary, the efficiency of the A-band rises steeply from the same energy. This shows that the generation of STE's and that of the states which cause the A-band compete with each other in the relaxation processes of CT-excitons.

The time resolved polarized PA spectra (P_{ex}//chain) parallel to the chain axis of Pt-Br-SO_4 at 15 K for E_{ex}=2.71eV are shown in Fig. 4. The PA-bands decay non-exponentially. No change of the intensity ratio among the A-, B-, and C-bands was detected during the period shown in the figure. The decay of these PA-bands are much faster than that reported in [Pt(en)$_2$][Pt(en)$_2$Cl$_2$](ClO_4)$_4$ (Pt-Cl-ClO_4, E_{CT}=2.72eV). In the complex, decay of the A- and B-bands is not detected at 77K [5]. However, in Pt-Cl-SO_4 no time evolution measurement in msec or μsec region was made. Time evolution of the A- and B-bands of several wide gap MX-chain complexes including Pt-Cl-ClO_4 have also been measured at 15K. In all complexes, The intensity of the A- and B-bands become less than 1/e within a second. Temperature dependence of the PA-spectra (P_{ex}//chain) of Pt-Br-SO_4 for E_{ex}=2.85eV has been measured. It is seen that above 35K, the intensities of the A-, B-, and C-bands decreases steeply as the increase of the temperature. No change of the intensity ratio among the A-, B-, and C-bands was detected. Above 65K, these PA-bands can not be detected. Since the A-, B-, and C-bands show the same time evolution and the same temperature dependence, these bands are considered to be caused by the same relaxed state.

The time resolved polarized PA spectra (P_{ex}//chain) parallel to the chain axis of Pt-Br-SO_4 at 15K for E_{ex}=4.03eV are shown in Fig. 5. As seen in the figure, all PA bands decay non-exponentially. It is found that the D-and G-bands show quite similar decay. No change of the

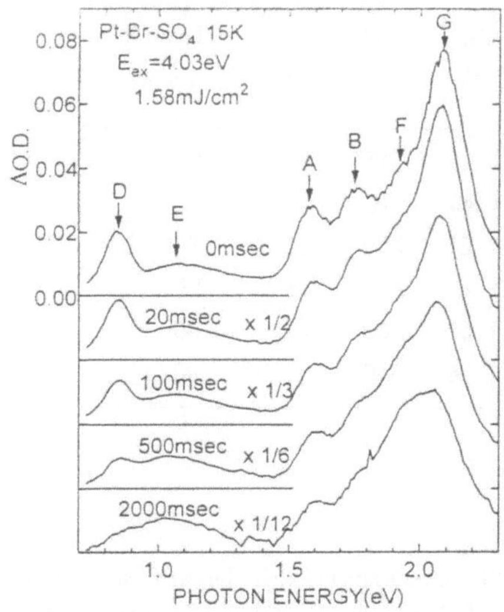

Fig. 5. Time evolution of the PA spectra of Pt-Br-SO4 for Eex=4.03eV.

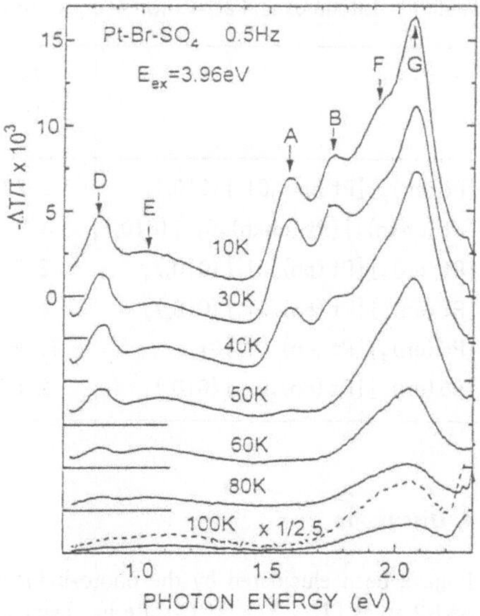

Fig. 6. Temperature dependence of the PA-spectra of Pt-Br-SO_4 for E_{ex}=3.96eV.

intensity ratio between the D- and G-bands was detected over the period shown here. Two seconds after the excitation, PA-bands other than the E- and F-bands are very week. No change of the intensity ratio between the E- and F-bands was detected over the period from 0sec to 10sec.

Fig. 6 shows temperature dependence of the polarized PA-spectra (P_{ex}//chain) parallel to the chain axis for E_{ex}=3.96eV. Below 30K, the PA-bands show no temperature dependence. Above 40 K, intensities of the PA-bands decreases as increasing the temperature. The intensities of the A-, and B-bands become very week above 50K. The D- and G-bands disappear above 100K. Only the E- and F-bands can be observed above 100K. No change was found in the intensity ratio between the D- and G-bands and in that between the E- and F-bands in the temperature range shown in the figure. According to these experimental result, it is concluded that the D- and G-bands are caused by the same relaxed state (DG-state). It is also concluded that E- and F-bands are caused by the same relaxed state (EF-state). Time resolved PA-spectra for a excitation photon energy of 4.03eV of several wide gap complexes including Pt-Cl-ClO_4 have been measured. PA-bands which correspond to the D-, E-, F- and G- bands are commonly seen in these complexes. The time evolutions of theses bands are qualitatively the same as those observed in Pt-Br-SO_4. In some complexes the PA-bands which correspond to the D- and G-bands are separated into two bands.

The STE luminescence of the wide gap complexes measured show exponential decay. The lifetimes of the complexes are temperature independent at low temperature. Above several tens of K, the lifetimes decrease as increasing temperature. The temperature dependence can be explained by the superposition of a temperature independent processes and a thermally activated processes. The lifetimes (τ's) at 2K and activation energies (ε's) are listed in table I. In MX-complexes with metal ion alternation (MXM'X-complexes) do not have twofold degeneracy in the ground states (TDGS). As seen in table I, the activation energies of the STE's of the complex with TDGS is a little smaller than that of the complexes without TDGS.

Table I. Lifetimes and activation energies of the STE luminescence.

	E_{CT} (eV)	E_{STE} (eV)	τ (2K) (psec)	ε (meV)	Twofold degeneracy
$[Pd(en)_2][Pt(en)_2Cl_2](ClO_4)_4$	3.22	1.65	45	43	NO
$[Pt(chxn)_2][Pt(chxn)_2Cl_2](ClO_4)_4$	3.2	1.49	70	17	YES
$[Pt(pn)_2][Pt(pn)_2Cl_2](ClO_4)_4$	2.85	1.26	230	28	YES
$[Pt(en)_2][Pt(en)_2Cl_2](ClO_4)_4$	2.72	1.22	230	15	YES
$[Pd(en)_2][Pt(en)_2Cl_2]Cl_4$	2.65	1.55	280	32	NO
$[Pd(en)_2][Pt(en)_2Br_2](ClO_4)_4$	2.59	1.56	440	48	NO

4. Discussion

It have been elucidated by the photo-induced ESR measurements that the ABC-states have S=1/2 spins [7]. The ABC-state has been assigned to a photo-generated neutral soliton pair [7] or to a photo-generated polaron pair [5,16] since these states have S=1/2 spin. As seen in Fig. 3, the ABC-states are most efficiently generated from the CT-excitons. In wide gap MX-chain complexes, the electron-hole biding energies of excitons are very large [8]. In such cases it is considered to be very hard for charged states such as polaron pairs to be generated from the CT-excitons. The excitation spectra of the A-band is different from that of the photo-current at room temperature. In small gap MX-chain complexes, polarons are not generated from the CT-excitons [9]. In these complexes the electron hole binding energy in the CT-exciton is smaller [8]. Theoretical studies predict that polaron pairs are only generated from the higher excited states [13(b)]. According to these facts, it is hard for the ABC-state to be assigned to a polaron pair. Theoretical studies predict that neutral solitons cause PA-bands located near the center of the band gap [14,15]. The energies of the A-and B-bands are 59% and 65% of E_{CT}, respectively. If the binding energies of the CT-exciton is taken into account, the A- and B-bands are located closer to the center of the band gap. Theoretical studies predict that a neutral soliton absorption band appears at an energy higher than the center of the band gap in a wide gap MX-chain complex [14,15]. Therefore, we conclude that the ABC-state is a neutral soliton pairs.

As seen if Fig. 3, the generation of the STE's and that of neutral soliton pairs compete with each other in the relaxation processes of the CT-excitons. CT-excitons with larger lattice vibration energy seem to efficiently relax to neutral soliton pair. At the efficiency maximum energy of the STE luminescence, the generation efficiency of the neutral soliton pair is very small. Theoretical studies predicts that the CT-excitons relax to the STE's immediately after the photogeneration [1(c),13(a)]. Theoretical studies also predict that the potential barrier between the STE and the soliton pairs is very small and that STE's are efficiently relax to soliton pairs and to the ground state [13(a)]. However the experimental result shows that the generation of the neutral soliton pairs from the STE's in the thermal equilibrium is not the dominant process in the generation processes of the neutral soliton pairs from CT-excitons at low temperature. The generation of the neutral soliton pairs may be completed before the STE's reach the thermal equilibrium at low temperature. As seen in table I, the life times of the STE's are too short for the radiative lifetimes. This have been attributed due to the dominant nonradiative annihilation processes of the STE's [12]. This is consistent with the theoretical prediction mentioned above. According to the experimental results mentioned above, the STE's in the thermal equilibrium is considered to be dominantly relax to the ground state at low temperature. Generation efficiency of the soliton pairs from the STE's in the thermal equilibrium may be small at low temperature. As seen in table I, there is no definite difference between the life times of the STE's at low temperature in MX-chain complexes with TDGS and those in MXM'X-complex in which separated soliton pairs are unstable. This is consistent with the dominant nonradiative recombination processes of the STE's in the thermal equilibrium to the ground state. The instability of the soliton pairs may make the activation energies of the STE's in the MXM'X-complex larger than those of the MX-chain complex with TDGS.

Theoretical studies predict that a polaron causes a pair of PA bands in the higher energy side and in the lower energy side of the center of the band gap [14,15]. In small gap MX-chain complexes, a pair of broad PA-bands caused by photo-generated polaron pairs are commonly observed. Both the DG-state and the EF-state cause a pair of PA-band in higher and lower energy side of the center of the band gap. Both DG- and EF-states are generated only from the higher energy excited states. Theoretical studies predict that a photo-generated charged soliton [14,15] and a bipolaron [13(c)] cause PA-bands near the center of the band gap. None of the D-, G-, E-, and F- bands are located near the center of the band gap. Therefore, either DG-state or EF-state can not be considered to be a charged soliton pair or a bipolaron pair. As seen Fig. 6, D-and G-band are hardly detected above 100K. Therefore EF-state may be the photocarrier at room temperature. The pair of E- and F-bands is quite similar to the polaron absorption bands which are commonly observed in small gap MX-complexes [9]. Therefore, the EF-state is considered to be a photogenerated polaron pair. At present the details of the DG-state is not clear. One possible candidate may be another kinds of polaron with different lattice distortion.

Excitation light power density dependence of the time evolution of the A-, D-, and E-bands has been investigate at 15K. In the range from 0.035mJ/cm^2 to 35mJ/cm^2, no change of the decay rate of the A-band was observed. In the range from 0.36mJ/cm^2 to 11mJ/cm^2, no change of the decay rate of D- and E-bands was also observed. These results indicate that photogenerated neutral solitons, polarons, and DG-states are dominantly annihilated by the geminate recombination processes.

The quantum efficiencies of the generation of the neutral soliton pairs and that of the polaron pairs can be estimated from the intensity of the PA-bands if the oscillator strengths of the neutral soliton absorption bands and those of the polaron absorption bands are known. Since these values are not known, the values obtained by the theoretical study [14] were used. The integrated oscillator strengths of both neutral soliton pair and polaron pairs are assumed to be 20. Then, the quantum efficiency of the neutral soliton pair generation for Eex=2.71eV and that of the polaron pair generation for E_{ex}=4.03eV in Pt-Br-SO_4 at 15K are $3x10^{-3}$ and $2x10^{-2}$, respectively. These values are considerably small. The time resolution of the measurement system is 10μsec. One possible origin of such small quantum efficiencies may be the much faster geminate recombination processes within the time resolution.

5. Summary

From the CT-excitons STE's and neutral soliton pairs which cause A-, B-, and C-bands are generated. It has been found that the generation of STE's and that of neutral soliton pairs compete with each other. Nonradiative relaxation processes to the ground state is considered to be dominant in the relaxation processes of the STE's in the thermal equilibrium at low temperature. From the higher excited states polaron pairs which cause E- and F-bands and DG-states which cause D- and G-bands are generated in addition to STE's and neutral soliton pairs. The quantum efficiency of the neutral soliton pair generation for E_{ex}=2.71eV and that of the polaron pair generation for E_{ex}=4.03eV in Pt-Br-SO_4 are $3x10^{-3}$ and $2x10^{-2}$, respectively.

References

[1] (a) K. Nasu, J. Phys. Soc. Jpn. **52**, 3865 (1983); (b) **53**, 302 (1984); (c) **53**, 427 (1984).

[2] H. Tanino and K. Kobayashi, J. Phys. Soc. Jpn. **52**, 1446 (1983).

[3] Y. Wada, T. Mitani, M. Yamashita, and T. Koda, J. Phys. Soc. Jpn. **54**, 3143 (1985).

[4[N. Matsushita, N. Kojima, T. Ban, and I. Tsujikawa, J. Phys. Soc. **56**, 3808 (1987).

[5] S. Kurita, M. Haruki, and K. Miyagawa, J. Phys. Soc. Jpn. **57**, 1789 (1988).

[6] M. Haruki and S. Kurita, Phys. Rev. B**39**, 5706 (1989).

[7] N. Kuroda, M. Sakai, M. Suezawa, Y. Nishina, and K. Sumino, J. Phys. Soc. Jpn. **59**, 3049 (1990).

[8] Y. Wada and M. Yamashita, Phys. Rev. B**42**, 7396 (1990).

[9] H. Okamoto, T. Mitani, K. Toriumi, and M. Yamashita, Phys. Rev. Lett. **69**, 2248 (1992).

[10] F. H. Long, S. P. Love, B. I. Swanson, and R. H. Mckenzie, Phys. Rev. Lett. **71**, 762 (1993).

[11] Y. Wada, N. Matsushita, and M. Yamashita, Mol. Cryst. Liq. Cryst., **256**, 891 (1994).

[12] Y. Wada, U. Lemmer, E. O. Goebel, M. Yamashita, and K. Toriumi, Phys. Rev. B**52**, 8276 (1995).

[13] (a) A. Mishima and K. Nasu, Phys. Rev. B**39**, 5758 (1989); (b) **39**, 5763. (1989); (c) **40**, 5593 (1989).

[14] Y. Tagawa and N. Suzuki, J. Phys. Soc. Jpn. 59, 4074 (1990).

[15] K. Iwano and K. Nasu, J. Phys. Soc. Jpn. **61**, 1380 (1992).

[16] J. Tinka Gammel, A. Saxena, I. Batistic, A. R. Bishop, and S. R. Phillpot, Phys. Rev. B**45**, 6408 (1992).

Solitons, Polarons and Excitons in MX Chains

H. Okamoto

Research Institute for Scientific Measurements, Tohoku University, Katahira, Aoba-ku, Sendai 980, Japan

Abstract. We characterized the photoinduced gap states from photoinduced absorption (PA) spectroscopy in the halogen(X)–bridged metal(M) complexes having degenerate CDW state, $[Pt(en)_2][Pt(en)_2X_2](ClO_4)_4$ (X=Br, I) and those having non–degenerate CDW state, $[Pd(en)_2][Pt(en)_2X_2](ClO_4)_4$ (X=Br, I). In the former complexes, either spin–solitons or charged–solitons were observed depending on the gap energies. In the latter complexes, photogenerations of solitons are suppressed. We have also investigated the excitation profiles of the luminescence from the self–trapped exciton, which are compared with those of the PA signals in the two types of complexes. The results demonstrate that the luminescence process strongly competes with the dissociation to spin-soliton pairs followed by nonradiative recombinations.

1. Introduction

Elementary excitations associated with lattice instability in the 1–D half–filled electronic systems have attracted much attention from both theoretical and experimental point of views. It has been revealed that the concepts of solitons [1] and other nonlinear excitations are essential for understanding the optical and magnetic properties as well as the transport properties of the 1–D materials such as conjugated polymers and charge-transfer (CT) complexes. Subsequently, the halogen(X)–bridged metal(M) complexes (or equivalently the MX compounds) have provided a unique opportunity to study the instability of the 1–D Peierls–Hubbard system, because of the remarkable tunability of the electronic states [2,3].

In the MX compounds, the displacement of the bridging halogens, that is, the amplitude of the CDW states can be controlled by replacements of the metals (M=Pt, Pd, Ni) and the bridging halogens (X=Cl, Br, I), and the ligand molecules and the counter anions surrounding the 1–D chains [2,3]. In addition, the nondegenerate CDW states can be obtained in the hetero–metal complexes in which Pt and Pd ions arrange alternatively [4]. Such a control of the CDW states in the MX compounds allows us to make an advanced study of dynamics of the nonlinear excitations such as solitons and polarons.

In this paper, first, the chemical control of CDW is reviewed. After that, photoinduced gap states are characterized from photoinduced absorption (PA) spectroscopy on the homo–metal complexes, $[Pt(en)_2][Pt(en)_2X_2](ClO_4)_4$ and $[Pt(chxn)_2][Pt(chxn)_2X_2]X_4$ (X=Br, I), and the hetero–metal complexes, $[Pd(en)_2][Pt(en)_2X_2](ClO_4)_4$ (X=Br, I). Here, (en) and (chxn) denote ethylenediamine and cyclohexanediamine, respectively. Moreover, we report the excitation profiles of the PA signals, which are compared with those of luminescence from the self–trapped exciton (STE). On the basis of the comparison, we will discuss the relaxation processes of the photoexcited states in MX chains.

Springer Series in Solid-State Sciences, Volume 124
Relaxations of Excited States and Photo-Induced Structural Phase Transitions
Ed.: K. Nasu

2. Experimental

The PA measurements were made by using an FTIR spectrometer (Nicolet system 800 and Jasco IR–8000) equipped with an InSb or an MCT detector and a 25cm monochromator (Jasco CT–25) with a PbS cell or a photomultiplier depending on the energy regions. As an excitation light, an Ar–laser or a second harmonics of a Ti–Sapphire laser was used. In the measurements of the excitation–spectra of the PA signals and luminescence, a 650W halogen–tungsten lamp or a 75W Xe lamp monochromized by a f=10 cm monochromator (Jasco CT–10) was utilized. The single crystals used were synthesized in the same way as reported in ref.2.

3. Control of the CDW states

The MX compounds are represented as $[M^{3-\rho}A_2][M^{3+\rho}A_2X_2]Y_4$ $(0\leq\rho<1)$, where A and Y are the ligand and the counter anion, respectively. In the MX compounds, hydrogen(H)–bonds between the amino–groups of the ligands (A) and the counter anions (Y) support the 1–D MX chains, as schematically illustrated in Fig.1. The 1–D electronic state is composed of d_{z^2} orbitals of metals and p_z orbitals of bridging halogens. In the Pt or Pd complexes, the mono–valence state expressed as

$$--X^- --- M^{3+} --- X^- --- M^{3+} --- X^- --- M^{3+} --- X^- --- M^{3+} --- X^- ---$$

is unstable due to the electron–lattice (e–l) interaction and stabilized into the mixed–valence (CDW) state shown below.

$$- - X^- - M^{3+\rho} - X^- - - - M^{3-\rho} - - - X^- - M^{3+\rho} - X^- - - - M^{3+\rho} - - - X^-$$

In Table 1, abbreviations of the MX compounds discussed in this paper are listed. In the following, we will use them instead of the rational formula for convenience.

By replacing the constituent elements, we can obtain a variety of complexes having different physical parameters; e.g. the amplitude of the bridging–halogen distortion Δ and the CT exciton energy E_{CT}. The lowest excitation corresponding to the optical gap is the CT exciton transition expressed as $(M^{2+}(d^8), M^{4+}(d^6) \rightarrow M^{3+}(d^7), M^{3+}(d^7))$. In Fig.2, E_{CT} of the Pt complexes with (en) and (chxn) for the ligands are plotted against the distortion parameter d which is defined as 2Δ divided by the metal–to–metal distance L [5].

The two polymorphisms of the Br–bridged Pt complexes, $[Pt(en)_2][Pt(en)_2Br_2](ClO_4)_2$ –(I) and (II) are characterized by the different space group $P2_1/a$ [6] and C2/m [5], respectively. In these two complexes, the conformations of the (en) molecules are slightly different from each other, resulting in the difference of L and Δ (or d).

In the complexes with halogen for counter anions, the H–bonds between the amino groups of ligands and counter halogen ions are strengthened and then 2–D tight H–bond network is formed [3,7]. As a result, interchain interaction is enhanced and a direction of each halogen displacement is two–dimensionally ordered as shown below [3,7].

$$\begin{array}{l} - - X^- - M^{3+\rho} - X^- - - - M^{3-\rho} - - - X^- - M^{3+\rho} - X^- - - - M^{3+\rho} - - - X^- \\ - - X^- - M^{3+\rho} - X^- - - - M^{3-\rho} - - - X^- - M^{3+\rho} - X^- - - - M^{3+\rho} - - - X^- \\ - - X^- - M^{3+\rho} - X^- - - - M^{3-\rho} - - - X^- - M^{3+\rho} - X^- - - - M^{3+\rho} - - - X^- \end{array}$$

(Fig.1 shows the complex having a 2–D ordering of CDW.) In contrast, in the complex–

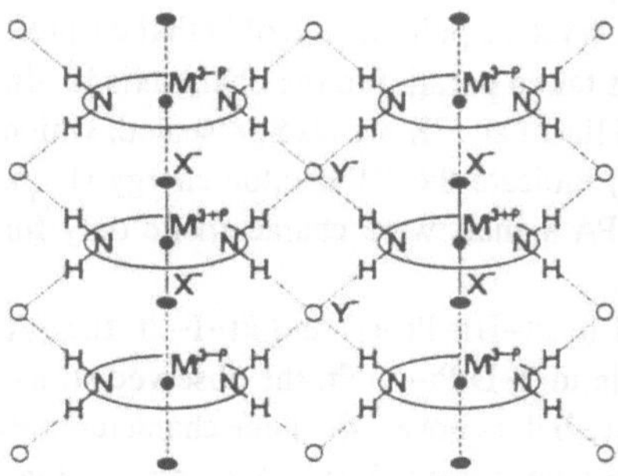

Fig. 1 Schematic structure of the halogen–bridged metal complex. The dotted lines show H–bonds.

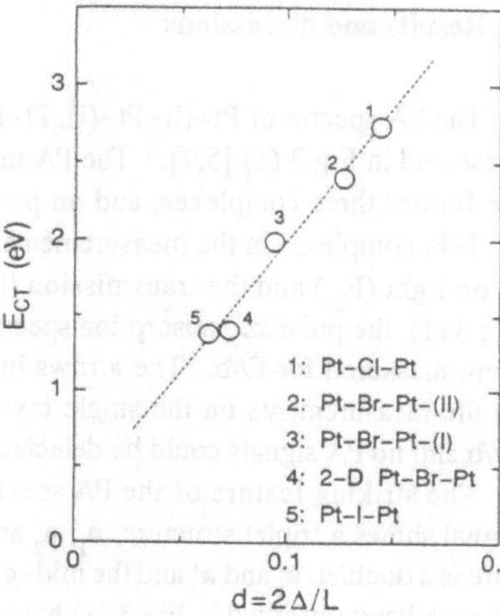

Fig. 2 The CT exciton energy E_{CT} as a function of the distortion parameter d in the Pt complexes (see Table 1).

Table 1. Abbreviations of the MX compounds.

	Abbreviations
$[Pt(en)_2][Pt(en)_2Cl_2](ClO_4)_4$	Pt–Cl–Pt
$[Pt(en)_2][Pt(en)_2Br_2](ClO_4)_4$–(II)	Pt–Br–Pt–(II)
$[Pt(en)_2][Pt(en)_2Br_2](ClO_4)_4$–(I)	Pt–Br–Pt–(I)
$[Pt(chxn)_2][Pt(chxn)_2Br_2]Br_4$	2–D Pt–Br–Pt
$[Pt(en)_2][Pt(en)_2I_2](ClO_4)_4$	Pt–I–Pt
$[Pt(chxn)_2][Pt(chxn)_2I_2]I_4$	2–D Pt–I–Pt
$[Pd(en)_2][Pt(en)_2Br_2](ClO_4)_4$	Pt–Br–Pd
$[Pd(en)_2][Pt(en)_2I_2](ClO_4)_4$	Pt–I–Pd

es with ClO_4 for counter anions, the interchain interaction through H–bonds is weak, and therefore, an ordering of CDW does not exist. The 2–D ordering of CDW suppresses the photogeneration of solitons as presented in the next section.

Another interesting way to control the electronic states is to utilize the metal alternation effect. By substituting a half of Pt ions by Pd ions, we can obtain the hetero–metal complexes where Pt and Pd ions are arranged alternatively. Since, the energy level of 4d orbitals in Pd ions is lower as compared with that of 5d orbitals in Pt ions, the CDW state is composed of $Pd^{3-\rho}$ ions and $Pt^{3+\rho}$ ions as shown below.

$$- - X^- - Pt^{3+\rho} - X^- - - - Pd^{3-\rho} - - - X^- - Pt^{3+\rho} - X^- - - - Pd^{3+\rho} - - - X^-$$

In this case, the CDW ground state is nondegenerate, where it is expected that solitons will not be produced. The interrelation between the homo–metal complexes and the hetero–metal complexes is analogous to that between the trans–$(CH)_x$ and cis–$(CH)_x$.

4. Results and discussions

The PA spectra of Pt–Br–Pt–(I), Pt–I–Pt, 2–D Pt–Br–Pt and 2–D Pt–I–Pt at 77K are presented in Fig.3 (1) [5,7]. The PA measurements are performed on single crystals for the former three complexes, and on powder samples dispersed in KI pellet for the 2–D Pt–I–Pt complex. In the measurements on single crystals, polarization of both the irradiation light (E_{ex}) and the transmission light (E) is taken parallel to the chain axis b. In Fig.3 (1), the polarized absorption spectra (dotted lines) at 77K are also presented, which were measured for E//b. The arrows in Fig.3 (1) indicate the CT exciton energy (E_{CT}). In the measurements on the single crystals, the PA signals were characteristic only for E//b and no PA signals could be detected for E⊥b.

The striking feature of the PA spectra is that in Pt–Br–Pt–(I) and Pt–I–Pt, the PA signal shows a triplet structure, $\mathbf{a}_1$, $\mathbf{a}_2$ and **b**, while in 2–D Pt–X–Pt, the observed structure is a doublet, $\mathbf{a}'_1$ and $\mathbf{a}'_2$ and the mid–gap structure **b** does not exist. Time characteristics of each band observed in Fig.3 (1) have been also investigated [5,7]. According to that, the band **b** decays in a different manner compared with the doublet, $\mathbf{a}_1$ and $\mathbf{a}_2$. The components of the doublet, $\mathbf{a}_1$ and $\mathbf{a}_2$ (or $\mathbf{a}'_1$ and $\mathbf{a}'_2$) show the same decay characteristic. Namely, the optically excited states include two different types of photo–products associated with the doublet $\mathbf{a}_1$, $\mathbf{a}_2$ ($\mathbf{a}'_1$, $\mathbf{a}'_2$) and the mid–gap structure **b**.

In the 1–D Peierls–Hubbard system, several localized levels associated with the lattice instability are considered to exist in an intragap region, as illustrated in Fig.4; namely,

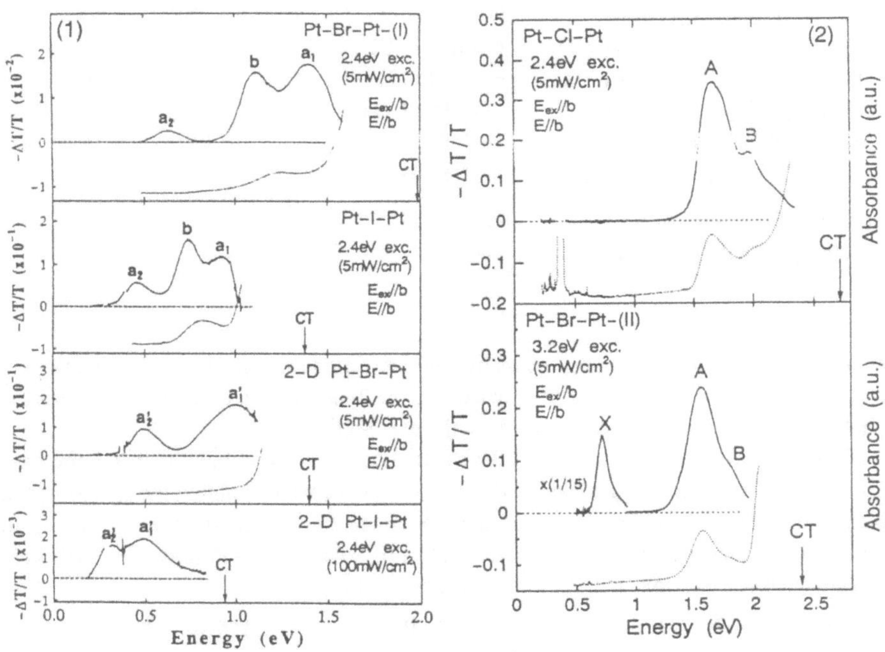

Fig. 3 PA spectra $-\Delta T/T$ at 77 K; (1) Pt–Br–Pt–(I), Pt–I–Pt, 2–D Pt–Br–Pt, 2–D Pt–I–Pt and (2) Pt–Cl–Pt, Pt–Br–Pt–(II). Dotted lines show polarized absorption spectra with E//b.

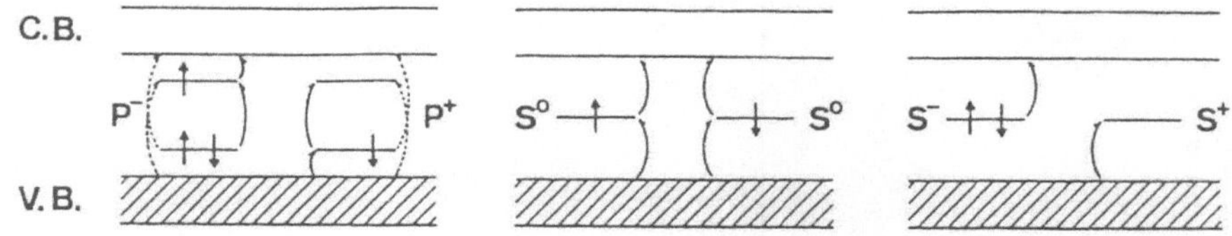

Fig. 4 Energy level diagrams of polarons (P^-, P^+), spin-solitons (S^0), and charged-solitons (S^-, S^+). Possible optical transitions are indicated by arrows.

polarons (P^- or P^+), and spin- and charged-solitons (S^0 and S^- or S^+) [8-10]. Dynamical properties of these particles are substantially determined by the correlation between the electron-lattice interaction (S) and the transfer energy (T) [8,9]. For S/T<1, it has been theoretically demonstrated that two absorption bands arise for a polaron and one mid-gap band for either of a spin- or charged-soliton. Possible transitions are indicated by the solid arrows. (The oscillator strengths of the higher transitions indicated by the dashed lines are suggested to be extremely weak.) Comparing the energy positions of the PA bands in Fig.3 (1) with the theoretical predictions, it is natural to assign the doublet $\mathbf{a}_1$, $\mathbf{a}_2$ ($\mathbf{a}'_1$, $\mathbf{a}'_2$) to polarons and the **b** band to spin- or charged-solitons [5,7].

These assignments are reasonable, taking account of the interchain interaction. Under the 2-D phase ordering of CDW, formation of a dissociated soliton pair requires a large scale of misalignment of CDW phases between the neighboring chains, which leads to excess energy. Accordingly, in 2-D Pt-X-Pt, soliton formation would be suppressed.

Here, we will discuss whether the origin of the **b** band is spin-solitons or charged-solitons. As for this, ESR measurements give important informations. In the ESR measurements on as-grown samples of Pt-I-Pt and Pt-Br-Pt-(I), a finite amount of paramagnetic spins (ca.10^{-4} per site) was detected [11]. In these two complexes, a weak absorption band was observed at nearly half of the gap energies as shown by the dotted lines in Fig.3 (1). However, in 2-D Pt-Br-Pt, neither such a defect-related absorption band nor paramagnetic spin has been detected. Taking account of the difference in the interchain interactions, they have been attributed to spin-solitons confined by some crystal imperfections. Considering the energy difference between the defect-induced absorption band and the PA band **b**, the PA band **b** is attributable to charged-solitons. The energy difference between the absorption of spin-solitons and that of charged-solitons is probably due to the on-site Coulomb repulsion energy **U**, similarly to the case of trans-$(CH)_x$. In trans-$(CH)_x$, it has been found that the absorption energy of spin-solitons (ca. 1.35 eV) is considerably larger than that of charged-solitons (ca. 0.5 eV) [12].

To obtain informations about the photogeneration process of the gap states, we measured the excitation profiles of the PA signals (I_{PA}) on Pt-Br-Pt-(I) and Pt-I-Pt, which are plotted in Fig.5. Since it has been ascertained that the annihilation process of the observed gap states could be explained by the bimolecular recombination model [5,7], the values of I_{PA} were normalized by using the relation $I_{PA} \propto N_{ex}^{1/2}$. N_{ex} is the photon number per unit volume, which has been evaluated by considering the power and the absorption coefficient (the absorption depth) of the excitation lights [13]. As seen, the CT exciton does not contribute to generation of charged-solitons and polarons in both complexes. This is considered mainly due to the strong excitonic effect on the photo-excited elec-

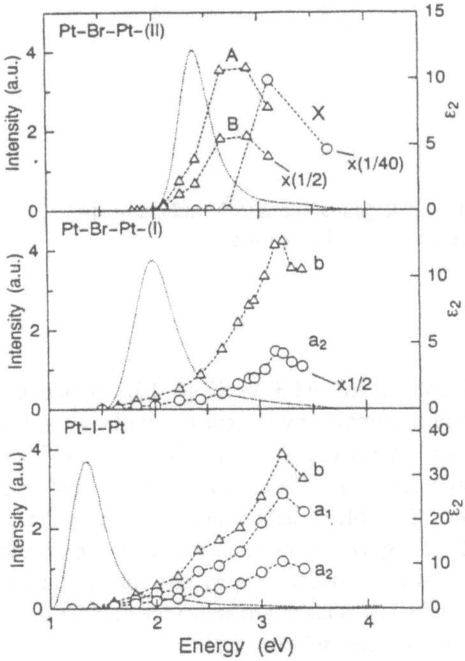

Fig. 5 Excitation profiles of the PA signals at 77K. Dotted lines show the ε_2 spectra.

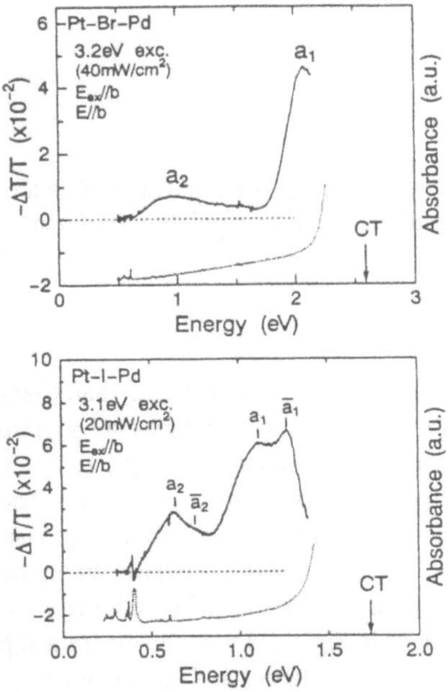

Fig. 6 PA spectra $-\Delta T/T$ at 77 K of Pt–Br–Pd and Pt–I–Pd. Dotted lines show polarized absorption spectra with E//b.

tron–hole pairs which is induced by the nearest neighbor Coulomb interaction V. In fact, the excitation energy dependence of the PA signals in Pt–I–Pt is almost the same as that of the photo–conductivities along the chain axis b reported by Haruki et al. [14], suggesting that a free electron–hole pair is able to contribute to the formation of charged–particles, that is, polarons and/or charged–solitons.

Let's proceed to discussions about the PA spectra of the MX compounds with larger optical gaps. In Fig.3 (2), presented are the PA spectra of Pt–Cl–Pt and Pt–Br–Pt–(II). Polarized absorption spectra near the band edge are also shown by the dotted lines. The feature of the PA spectrum of Pt–Cl–Pt shown in Fig.3 (2) is in good agreement with those previously reported [15]; there exists two absorption bands **A** and **B** just below the band edge in as–grown samples and they are enhanced by light irradiations. At 77 K, the photoinduced increases of the **A** and **B** bands never decay in Pt–Cl–Pt.

In Pt–Br–Pt–(II), the two PA bands are also observed at 1.55 eV and 1.79 eV. Spectral features of these bands are quite similar to those of the **A** and **B** bands in Pt–Cl–Pt, so that we labeled these bands as **A** and **B** in Fig.3 (2). In addition, at low energy side of the **A**, **B** bands, observed is a weak PA band which is labeled as **X** in Fig.3 (2). In Fig.5, the excitation profiles of the three PA bands observed in Pt–Br–Pt–(II) are presented. The **A** and **B** bands show the same behaviors, suggesting that these bands have the same origin, while the **X** band shows the different behavior, which is similar to the PA bands due to polarons observed in Pt–Br–Pt–(I) and Pt–I–Pt. It has also been found that the time dependences of the PA bands **A** and **X** were different from each other. (The PA bands **A,B** of Pt–Br–Pt–(II) decay even at 77 K, in contrast with the PA bands **A,B** of Pt–Cl–Pt.) These results indicate that photoexcited states include the two types of gap states characterized by the **A,B** bands and the **X** band in Pt–Br–Pt–(II).

Here, we will discuss the origin of the bands **A** and **X**. The energy of the band **X** is small, which is about 30% of E_{CT}. Considering such a small energy, the **X** band might be attributed to the lower energy transition of polarons, which is the $\mathbf{a}_2$ band discussed before. In Pt–Cl–Pt, it has been revealed from the photoinduced ESR result that the excited state responsible for the A band has spin s=1/2 [16]. Assuming that the **X** band is due to polarons, the **A** band is attributable to spin–solitons, since a charged–soliton has no spin. At the present stage, however, these assignments are so speculative. To determine the assignments of the PA signals more clearly, we investigated the metal alternation effects. We expect to obtain clear assignments of the PA bands by comparing the PA spectrum of the homo–metal complex with that of the hetero–metal complex, since photogenerations of solitons should be suppressed in the hetero–metal complex.

Fig.6 shows the polarized absorption and PA spectra for the hetero–metal complexes, Pt–Br–Pd and Pt–I–Pd. In both complexes, there are no prominent structures in the absorption spectra, similarly to the case of the 2–D Pt–Br–Pt complex. In Pt–Br–Pd, the two PA bands are observed. These two PA bands have been found to show the same time characteristics, suggesting that they have the same origin. Such a doublet structure is characteristic to the absorption bands of polarons. Therefore, we labeled these bands as $\mathbf{a}_1$ and $\mathbf{a}_2$. In the PA spectrum of Pt–I–Pd, four structures are observed. It has been ascertained that the four structures show the same decay characteristics, so that the same excited state is likely responsible for these structures. Taking account of the analogy to the PA spectra of 2–D Pt–Br–Pt and Pt–Br–Pd, these structures in Pt–I–Pd are also attributable to polarons. In Fig.6, therefore, these structures are labeled as $\mathbf{a}_1$, $\bar{\mathbf{a}}_1$, $\mathbf{a}_2$ and $\bar{\mathbf{a}}_2$. A small splitting between $\mathbf{a}_1$ ($\mathbf{a}_2$) and $\bar{\mathbf{a}}_1$ ($\bar{\mathbf{a}}_2$) is probably due to the electron–hole asymmetry. In the hetero–metal complex, electron polarons exist on Pt sites and hole polarons on Pd sites. Because of the difference in the on–site Coulomb repulsive energy **U**, the absorption energies for hole polarons will be slightly larger than those for electron polarons. The PA band labeled as $\mathbf{a}_2$ observed in Pt–Br–Pd is so broadened that it might include a doublet structure split due to the electron–hole asymmetry.

The strong midgap band **A** observed in the homo–metal complexes, Pt–Cl–Pt and Pt–Br–Pt–(II), is not detected in the hetero–metal complex. This is a strong evidence to assign the **A** band to spin–solitons. As a result, the **X** band in Pt–Br–Pt–(II) has been reasonably assigned to polarons (the $\mathbf{a}_2$ band). The existence of the higher energy band for polarons, that is, the $\mathbf{a}_1$ band could not be discriminated in Pt–Br–Pt–(II). It might be masked by the strong **A** band. The reason why the absorptions of spin solitons are composed of the two bands **A** and **B** is not clear at the present stage. The theoretical study by K. Iwano et al. indicates that the absorption spectrum of a spin–soliton show a doublet structure around the half of E_{CT} in the presence of lattice fluctuations [10]. This might explain the existence of the two PA bands **A** and **B**.

As discussed in detail above, in the Pt–X–Pt complexes having the large gap (E_{CT}>2 eV), spin–solitons are observed, while in the complexes having the small gap (E_{CT}<2 eV), charged–solitons are observed. Here, we will discuss the reason of the changes in the photogeneration of solitons. The formation energy of a charged soliton is mainly determined by the on–site Coulomb repulsive energy **U**. With increase of **T**, the magni–

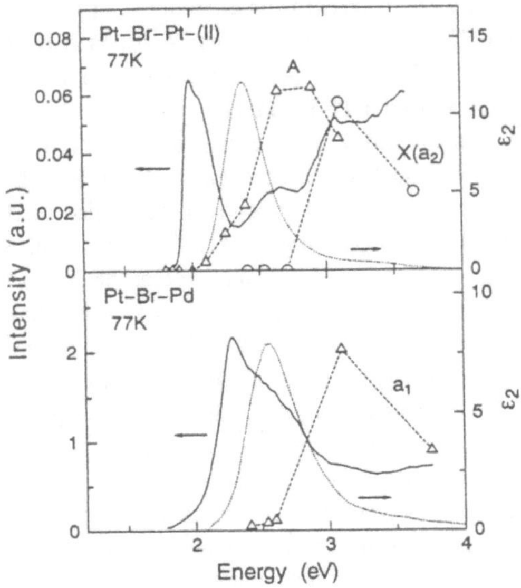

Fig. 7 Excitation profiles of the luminescence (solid lines) and the PA signals (circles and triangles) for Pt–Br–Pt–(II) and Pt–Br–Pd. Dotted lines show ε_2.

tudes of gap energies (E_{CT}) decrease. Simultaneously the effect of U is suppressed, so that the formation energy of charged–solitons will decrease relatively to that of spin–solitons. This may be a reason why the charged–solitons are observed in the small gap complexes and spin–solitons in the large gap complexes.

Next, I will discuss the relaxation process of the photoexcited states. It is interesting to clarify the interrelation between the luminescence process from STE and the dissociation process to soliton or polaron pairs followed by the nonradiative recombinations. In order to study such a competition in detail, we compared the excitation profiles of luminescence and PA signals for the two complexes, Pt–Br–Pt–(II) and Pt–Br–Pd. In Pt–Br–Pt–(II), both of spin–solitons and polarons are photogenerated, while in Pt–Br–Pd, only polarons are possible. From the comparison of the two complexes, we expect to discriminate the contributions of solitonic state on the relaxation process. In Fig.7, dotted lines show the ε_2 spectra, and solid lines show the excitation profiles of luminescence. There are two important differences between the two complexes. First, in the homo–metal complex, the luminescence decreases sharply around the CT exciton energy, while in the hetero–metal complex, such a sharp decrease of the luminescence intensity around the CT exciton energy is not observed. Second, the relative intensity of luminescence in the homo–metal complex is at least by one order smaller than that in the hetero–metal complex in the whole energy region. Such differences are attributable to the existence of photogeneration processes of solitons in the homo–metal complex.

We plotted the excitation profiles of the gap states in the same figure [13]. The triangles show spin–solitons and the circles show polarons. The sharp decrease of luminescence intensity corresponds to the increase of the photogeneration efficiency of spin–solitons in Pt–Br–Pt–(II). It suggests that the photogeneration process of solitons competes with the luminescence process. Since the decrease of luminescence and the increase of PA signals of polarons occur at almost the same energies (2.5–3 eV) in Pt–Br–Pd, the generation process of polarons might also compete with luminescence process.

Such relaxation processes have been theoretically studied by Mishima and Nasu [17]. According to that, there is only a small barrier between the CT exciton and the solitonic state, therefore solitons are photogenerated by the lowest CT exciton state. On the other hand, a large barrier exists between the CT exciton and the polaronic state, so that polarons cannot be generated from the CT excitons. Polarons are photogenerated only from the higher energy states corresponding to the electron–hole continuum. These theoretical expectations are qualitatively in agreement with our experimental results.

5. Summary

By focussing on the metal alternation effect as well as the interchain interaction, we characterized the PA bands of the several MX compounds. On the basis of the excitation profiles of the PA signals and the STE luminescence for Pt–Br–Pt–(II) and Pt–Br–Pd, we have deduced the relaxation process of the photo–excited states as follows. From high energy excited states corresponding to the electron–hole continuum, polaron pairs are generated. From CT excitons, the relaxation to STE and the dissociation to spin–soliton pairs compete with each other. When formations of spin–solitons are suppressed by the metal–alternation effect, luminescence is considerably enhanced.

This work has been made in collaborations with Y. Shimada (Tohoku Univ.), Prof. M. Yamashita (Nagoya Univ.), Prof. T. Mitani (JAIST) and Prof. Y. Oka (Tohoku Univ.).

References

1. W.P. Su, J.R. Schrieffer and A.J. Heeger, Phys. Rev. Lett. 42, 1698 (1979).
2. H. Okamoto, T. Mitani, K. Toriumi and M. Yamashita, Mater. Sci. Eng. 13, L9 (1992).
3. H. Okamoto, K. Toriumi, T. Mitani and M. Yamashita, Phys. Rev. B42, 10381 (1990).
4. Y. Wada, T. Mitani, K. Toriumi and M. Yamashita, J. Phys. Soc. Jpn. 58, 3013 (1989).
5. H. Okamoto, Y. Oka, T. Mitani, K. Toriumi and M. Yamashita, Mol. Cryst. Liq. Cryst. 256, 161 (1994).
6. K. Toriumi, M. Yamashita, S. Kurita, I. Murase, and T. Ito, Acta Cryst. B49, 497 (1993).
7. H. Okamoto, T. Mitani, K. Toriumi, and M. Yamashita, Phys. Rev. Lett. 69, 2248 (1992).
8. Y. Tagawa and N. Suzuki, Phys. Soc. Jpn. 59, 4074 (1990).
9. J.T. Gammel, A. Saxena, I. Batistic, A.R. Bishop, and S.R. Phillpot, Phys. Rev. B45, 6408 (1992).
10. K. Iwano and K. Nasu, J. Phys. Soc. Jpn. 61, 1380 (1992).
11. A. Kawamori, R. Aoki, and M. Yamashita, J. Phys. C18, 5487 (1985).
12. X. Wei, B.C. Hess, Z.V. Vardeny and F. Wudl, Phys. Rev. Lett. 68, 666 (1992).
13. We previously reported the excitation profiles of PA signals of Pt–I–Pt, Pt–Br–Pt–(I) and Pt–Br–Pt–(II) in ref.5. In ref.5, we didn't correct the data by taking account of the absorption coefficients. Therefore, the excitation profiles in ref.5 are different from those presented in this paper.
14. M. Haruki, M. Tanaka, and S. Kurita, Synth. Met. 19, 901 (1987).
15. S. Kurita, M. Haruki, and K. Miyagawa, J. Phys. Soc. Jpn. 57, 1789 (1988).
16. S. Kurita and M. Haruki, Synth. Met. 29, F129 (1989).
17. Mishima and K. Nasu, Phys. Rev. B39, 5758 (1989), ibid 39, 5763 (1989).

Lattice Relaxations Yielding Nonlinear Excitations in Photo-Excited MX Chains

K. Iwano

Photon Factory, National Laboratory for High Energy Physics,
Graduate University for Advanced Studies,
1-1 Ohu, Tsakuba 305, Japan email:iwano@nasuws1.kek.jp

Abstract. The natures of the photoinduced absorptions observed in halogen-bridged mixed-valent platinum complexes (so called MX chains) are still controversial. In order to shed the light there from the theoretical side, we investigate lattice relaxations that occur after photoexcitations and yield nonlinear excitations. First of all, we point out the possibility that there are various types of relaxation processes depending on the parameters of the system. Then we focus on several typical cases and show potential surfaces to illustrate how the relaxation proceeds to the destination. Lastly, we compare the calculated absorption spectra with those found in the measurements of photoinduced absorption and discuss their origins.

1. Introduction

The physics of one dimensional systems, particularly, the existence of nonlinear excitations such as solitons and polarons, has attracted much attention since the pioneering work by Su *el al.* [1]. In that work, the possibility of soliton excitations in the doubly degenerate ground states of polyacetylene was discussed in a mathematically well defined form for the first time. Following this work, many attempts to extend the theory have been performed because the original model was rather simple in the sense that the electron-electron interaction was not considered explicitly. However, it seems that those attempts were not so successful. The biggest shortcoming common to existing theories for polyacetylene is that they can not yet explain the nature of the absorption spectrum completely. When we discuss the photoexcitations of nonlinear excitations, the excited states in the Franck-Condon state correspond to the initial states of the relaxations. Thus, the present theories of polyacetylene have the disadvantage that even the starting point of relaxations is ambiguous.

While, MX chains give a good chance for study to overcome the above difficulty. Each material belonging to this group consists of one-dimensional chains and counter ions in between. In each chain, the platinum ions surrounded by ligands and the halogen ions are aligned alternately. This quasi one-dimensionality leads to the dimerization of the halogen sublattice and the mixed valency of platinum ions as Pt^{2+} and Pt^{4+}, namely, the charge density wave (CDW). In these materials, it is established that the absorption spectra are well explained in terms of one-dimensional charge-transfer (CT) excitons [2, 3, 4, 5]. Thus, we can say that, in order to treat electronic states, it is almost enough if we take

Springer Series in Solid-State Sciences, Volume 124
Relaxations of Excited States and Photo-Induced Structural Phase Transitions
Ed.: K. Nasu

into account up to two-body correlations of electron-hole pair, neglecting higher orders ones.

In these rather "simple" systems, many experiments have been performed to detect nonlinear excitations, because the two-fold degeneracy of the CDW ground states allows solitons to exist in addition to polarons. Based on the experimental data of photoinduced absorptions (PA), which are aimed for photoexcitation of nonlinear excitations, these CDW materials are separated into well defined two groups. The first group is the materials in which a rather small optical gap is observed. Their PA signals are well explained with an analogy to polyacetylene, assuming the photoexcited solitons and polarons [6]. While, in the second group, the materials show a rather wide gap, which means less itineracy of valence electrons. The observed PA signal has been interpreted as due to either solitons or polarons [7, 8, 9, 10], but it is still controversial.

In order to shed the light on this problem, here we try to obtain a whole picture of nonlinear excitations in MX chains. To do so, we make use of a one-dimensional single-band Peierls-Hubbard model, which has been successfully used to clarify their various optical properties. The details of the model and how to treat it is explained in the next section. In the third section, we present the energetics of various nonlinear excitations. The system parameters are changed rather freely, and it is pointed out that there can be various types of relaxations, depending on the parameters. Among those, we focus on several realistic cases in the forth section. The details of their relaxation paths are illustrated. In the fifth section, we reexamine the PA spectra theoretically and discuss the interpretations of observed ones. The last section is dedicated to conclusions.

2. Model and Method

We focus on a single chain of Pt's and halogens, and describe it by a one-dimensional single-band Peierls-Hubbard model as

$$H = -t_0 \sum_{l\sigma} \left(C^\dagger_{l\sigma} C_{l+1\sigma} + C^\dagger_{l+1\sigma} C_{l\sigma} \right) + U \sum_{l} n_{l\alpha} n_{l\beta} + V \sum_{l\sigma\sigma'} n_{l\sigma} n_{l+1\sigma'}$$
$$-S \sum_{l\sigma} (Q_{l+1} - Q_l) n_{l\sigma} + \frac{S}{2} \sum_{l} Q_l^2. \tag{1}$$

Here $C^\dagger_{l\sigma}$ and $C_{l\sigma}$ denote the creation and annihilation operator of an electron with spin σ in the d_{z^2} orbital of the lth metal atom, respectively, and Q_l is the dimensionless displacement along the chain of the lth halogen atom, as shown in Fig. 1. We have four parameters, t_0, U, V and S. They correspond to electron transfer energy, on-site and nearest-neighbor Coulomb energies and the site-diagonal electron-phonon interaction energy, respectively.

In the rest of this section, we briefly explain how to treat this model. The ground state is calculated using a Hartree-Fock (HF) approximation. There we follow a standard procedure, namely, the substitutions in U and V terms as

$$n_{l\sigma} n_{l'\sigma'} \rightarrow <n_{l\sigma}> n_{l'\sigma'} + n_{l\sigma} <n_{l'\sigma'}> - <n_{l\sigma}><n_{l'\sigma'}> \tag{2}$$

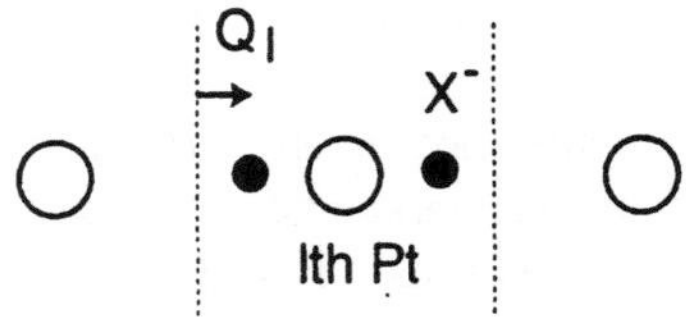

Fig. 1. A schematic picture of a part of the MX chain.

and

$$n_{l\sigma}n_{l+1\sigma} \to - < C^\dagger_{l\sigma}C_{l+1\sigma} > C^\dagger_{l+1\sigma}C_\sigma - C^\dagger_{l\sigma}C_{l+1\sigma} < C^\dagger_{l+1\sigma}C_\sigma > \\ + < C^\dagger_{l\sigma}C_{l+1\sigma} >< C^\dagger_{l+1\sigma}C_\sigma > . \quad (3)$$

While, the excited states are calculated by diagonalizing all the one-electron excitations with respect to the residual interaction beyond the HF approximation. The validity of restricting the states up to one-electron excitations is checked by the comparison with an RPA calculation [11]. The obtained energies of excitons are almost the same in those two calculations. The optical spectra are obtained by calculating the transition matrices of a current operator defined as

$$\hat{J} = iet_0 a \sum_{l\sigma} \left(C^\dagger_{l\sigma}C_{l+1\sigma} - C^\dagger_{l+1\sigma}C_{l\sigma} \right), \quad (4)$$

where a is the distance between neighboring two Pt's.

3. Energetics

In this section, we introduce possible nonlinear excitations in this system, particularly emphasizing on their creation energies. In Fig. 2, we show lattice patterns and electron distributions for the CDW ground state, a self-trapped exciton (STE), a neutral-soliton pair, a charged-soliton pair and a polaron-pair. To make the picture simple, the localized limit, namely, the zero transfer-energy case is assumed. Each well represents a d_{z^2} orbital of Pt, with its depth modified by the adjacent halogen ions. The top of the figure is the CDW ground state, in which deep and shallow wells are repeated with two or zero electron inside, respectively. Below that, the STE is shown to be localized around the center. In the case of neutral and charged solitons, the region between the two solitons is described by the other CDW state of the two-fold degeneracy.

When we think about the relaxations to these nonlinear excitations, the creation energy, which is defined as the energy difference between the CDW ground state and one pair of each excitation, is very important. In the localized limit, they are easily obtained as shown by Kuroda *et al.* [12]:

Table I. The creation energy of one pair of nonlinear excitations.

Excitation	STE	S_0S_0	S_+S_-	P_+P_-
Creation Energy	$5S$+$3V$-U	$6S$+$4V$-U	$4S$+$4V$	$6S$+$4V$-U

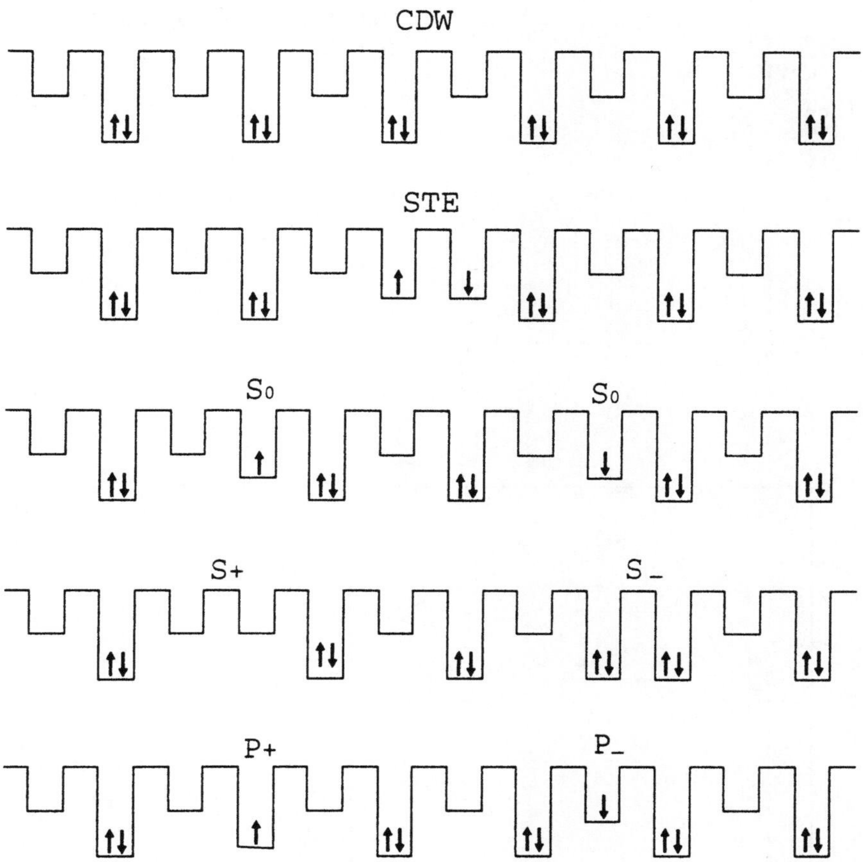

Fig. 2. Schematic pictures of the CDW ground state and various nonlinear excitations. S_o, $S_\pm$ and $P_\pm$ mean a neutral soliton, a charged soliton and a polaron, respectively.

In order to make the meaning transparent, we draw a phase diagram in Fig. 3. As is well known, above the boundary $U = 4S + 2V$, the spin density wave (SDW) is stable. The next boundary $U = 2S$ means the level crossing between the neutral-soliton pair and the charged-soliton pair. As is easily seen, when we increase U for fixed S, we pass though three phases: a phase where the charged-soliton pair is stable compared with the neutral-soliton pair (CS), and then a phase where neutral soliton becomes more stable, but the ground state is still CDW (NS), and finally SDW. This is quite natural because a neutral soliton, being nothing but an isolated spin, is regarded as a precursor to SDW.

Next, we consider finite transfer cases. The first example is located on the border line between the two regions of NS and CS in Fig. 3. Thus the energies of the neutral- and charged-soliton pairs happen to be the same in the localized limit. Moreover, the polaron-pair is always degenerate with the neutral-soliton pair in the same limit. Thus, the transfer being increased, these three states

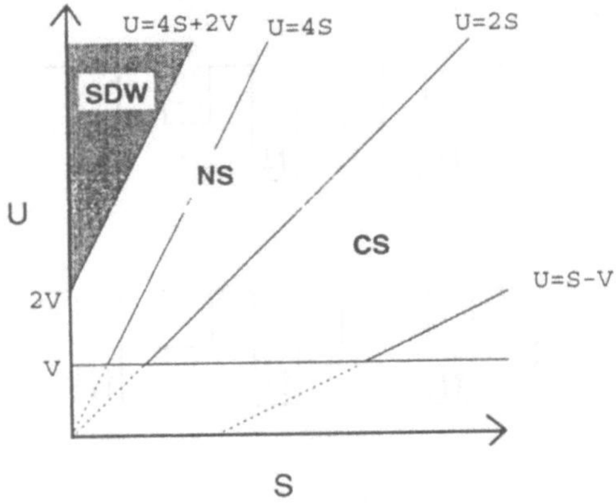

Fig. 3. A phase diagram at $t_0 = 0$.

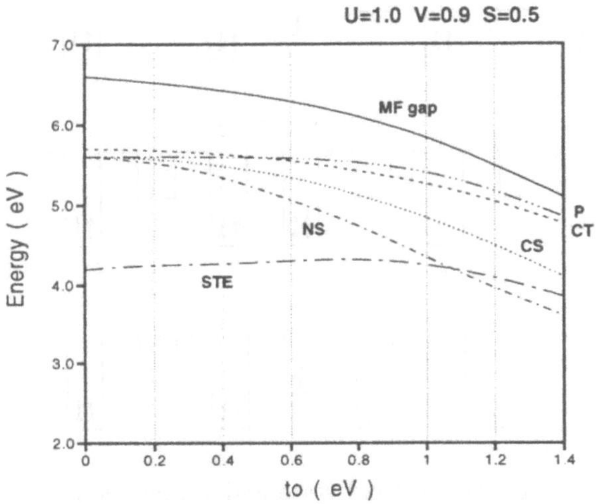

Fig. 4. The t_0 dependence of the creation energies of various nonlinear excitations. Here the other parameters are $U = 1.0$ eV, $V = 0.9$ eV and $S = 0.5$ eV. The abbreviations on the curves, i.e., MF gap, CT, STE, NS, CS and P mean the energy gap obtained by the HF calculation, the lowest CT exciton, a self-trapped exciton, a neutral-soliton pair, a charged-soliton pair and a polaron pair, respectively.

start from the same energy as shown in Fig. 4. In spite of this same starting energy, the neutral-soliton pair shows the steepest decrease and becomes lower than the STE in a region of relatively large transfer-energy. Here it should be noted that the STE energy does not change itself so much. The polaron energy also shows only a little change, and even becomes higher than the CT exciton energy in a large part of the region.

In the next figure, we show one more example. The overall tendency is the same as the previous one, except that the energy of a neutral-soliton pair is

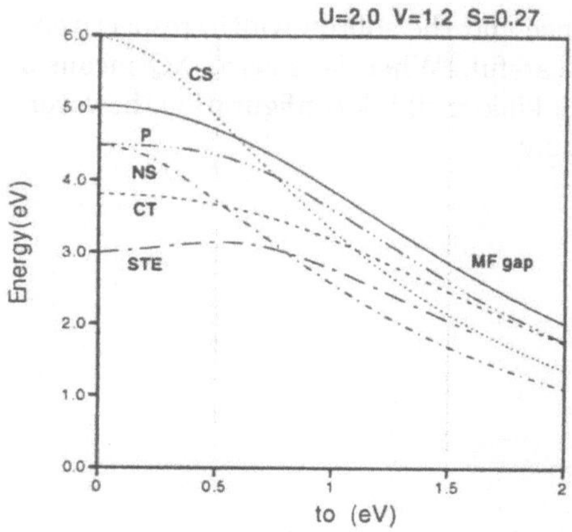

Fig. 5. Same as Fig.4, but $U = 2.0$ eV, $V = 1.2$ eV and $S = 0.27$ eV.

higher than the CT energy in a small transfer region. Thus the relaxation from the CT exciton to the neutral soliton is impossible in that region.

As a summary of this section, it is concluded that the creation energy and the relative stability of each excitation depends sensitively on the system parameters. Roughly speaking, when U is dominant over S, the neutral soliton is stable compared with the charged soliton.While, when vice versa, the latter is stable compared with the former in or near the localized limit. Even in this case, the former exceeds the latter in stability in an intermediate region, although this behavior is not explicitly mentioned here. As for the other excitations, what is important is that the energy of STE shows a smaller change even in increasing the transfer, which means that the energy of neutral-soliton pair and/or that of charged-soliton pair easily goes down below the STE energy . We can therefore say from the view point of energetics that the relaxations to neutral- and charged solitons from the STE become possible in a region of relatively large transfer.

4. Relaxation Paths

4.1. Soliton Channel (I)

In the previous section, we have particularly mentioned that the energy of a neutral-soliton pair goes down below the STE with a relatively large t_0, while the energy of a charged-soliton pair remains higher than that of a neutral-soliton pair, independent of t_0 when U is large enough compared with S. Then we calculate concrete potential surfaces for that situation, assuming an adiabatic approximation for the lattice movement. The lattice is deformed from the regularly dimerized pattern to a kink-antikink configuration through locally deformed patterns as STE. all of which are included in the expression as

$$Q_l = (-1)^l Q_0 \left[1 + \Delta Q \left(\tanh\theta((|l| - \frac{l_0}{2}) - 1\right)\right] \quad (5)$$

Here l_0 and θ is the soliton-soliton distance and the soliton width, respectively. As for the meaning of ΔQ, we must be careful. When l_0 is zero, ΔQ means a depth of the deformation. In the case of a kink-antikink configuration, however, it is not 2 but 1 as shown in the next figure.

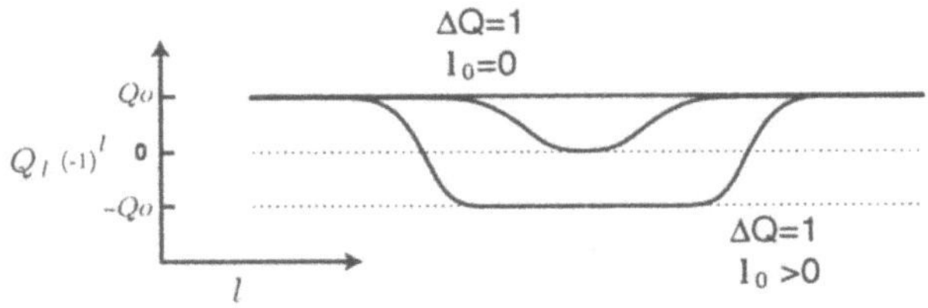

Fig. 6. A picture explaining a way of deforming the lattice.

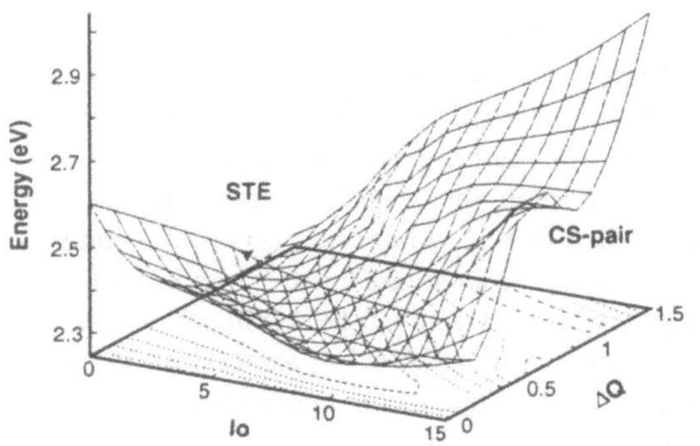

Fig. 7. The potential surface in the first excited state. The parameters are U=2.0 eV, V=1.2 eV, S=0.22 eV and t_0=1.3 eV.

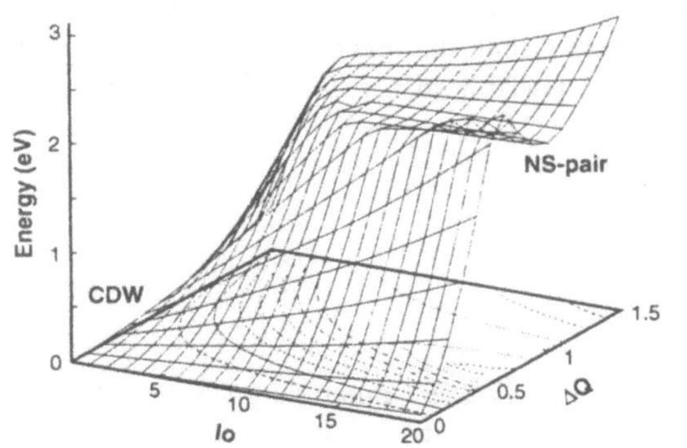

Fig. 8. The potential surface in the ground state. The parameters are the same as in Fig. 6.

In Figs. 7 and 8, we show potential surfaces in the first excited state and in the ground state, respectively. Each surface is drawn in respect to both l_0 and ΔQ, being optimized for θ in each state. In Fig. 7, the point at $l_0 = 0$ and $\Delta Q = 0$ corresponds to the singlet lowest CT exciton with parity odd in the regular dimerized lattice, which can relax directly to the STE along the curve on $l_0 = 0$ and $\Delta Q \neq 0$. The area around $l_0 \gg 1$ and $\Delta Q = 1$ corresponds to

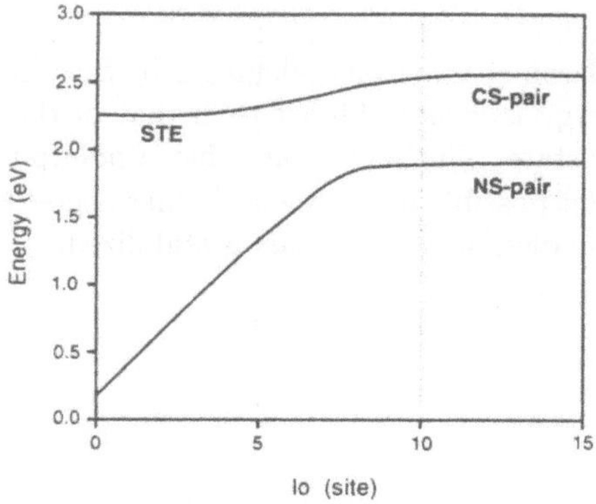

Fig. 9. The cross section at $\Delta Q = 1$ of the two surfaces in Figs. 7 and 8.

a charged-soliton pair, but it is higher than STE, consistent with the energy diagram in Fig. 5. In Fig. 8, on the other hand, the point at $l_0 = 0$ and $\Delta Q = 0$ corresponds to the CDW ground state. Again, the area around $l_0 \gg 1$ and $\Delta Q = 1$ is a soliton pair, but of a neutral soliton.

In order to look at both the figures at the same time, we show the cross section at $\Delta Q = 1$ in Fig. 9. As is mentioned before, the neutral-soliton pair is located below the STE. Thus, it is possible that the former is created as a result of nonadiabatic transition from the latter.

4.2. Soliton Channel (II)

As easily seen from the energy diagrams in Figs. 5 and 6, the energy of a neutral-soliton pair stays higher than that of the STE when t_0 is small. Even in this case, almost all the features are the same as in Fig. 9, except for the relative position of the STE and the neutral-soliton pair. What is important here is that the STE does no more relax to a completely separated neutral-soliton pair, and hence only a bound pair is a possible product.

4.3. Soliton Channel (III)

Let us return to the phase diagram in Fig. 3. As we already mentioned, a charged-soliton pair is energetically favored when $U < 2S$. Although this phase (CS) is somewhat invaded by the phase (NS) when t_0 is increased, it still exists. In this case, the ground state is connected to a charged-soliton pair, while the STE to a neutral-soliton pair, as discussed by Mishima *et al.* [13]. However, this case will be unrealistic because U is not considered to be so small. In order to prove this, we focus on the absorption spectra in the Franck-Condon state. For example, the most well known series of these materials are $[Pt(en)_2X_2][Pt(en)_2](ClO_4)_4$ with Cl, Br or I as X (hereafter abbreviated as PtX). Actually, in those spectra, all the CT peaks in the absorption spectra show a Lorentzian shape on the higher energy side, which fact means a strong exciton effect, i.e., a large V. Since U is larger than V in usual cases, we believe that U is also rather large.

4.4. Polaron Channel

An example of potential surfaces in a polaron channel was calculated by Suzuki *el al.* [14], in which the polaron-pair energy is located higher than that of the lowest CT exciton in the Franck-Condon state. This fact means that a polaron channel starting from the CT exciton is not possible unless we add other degrees of freedom to the deformation so that the distant polaron-pair is stabilized.

5. Absorption Spectrum

5.1. Neutral Soliton

So far, we have discussed relaxation channels that yield nonlinear excitations and shown that a neutral-soliton pair is the most likely candidate for the destination of the relaxation starting from the CT exciton. In this section, we assume the presence of a neutral soliton and calculate the absorption spectrum, which is directly related to PA. In Fig. 10, we plot a result in the case of U=2.0 eV, V=1.2 eV, S=0.22 eV and t_0=1.3 eV.

The larger absorption band around 2.6 eV is due to the CT exciton. This peak position is almost same as that of PtCl. The higher energy side of the peak is close to a Lorentzian, consistent with the observed shape. Its width at the half height 0.38 eV is only a little smaller than that of PtCl, nearly, 0.41eV. An extra band around 1.6 eV is due to the neutral soliton. This energy position is almost the same as the A peak in the PA spectrum [9], but B peak is not reproduced. As for the relative intensity between the main CT band and the gap absorption, it looks rather large in this figure. This is because we are assuming a rather high density of neutral solitons, that is, one soliton per 50 Pt atoms.

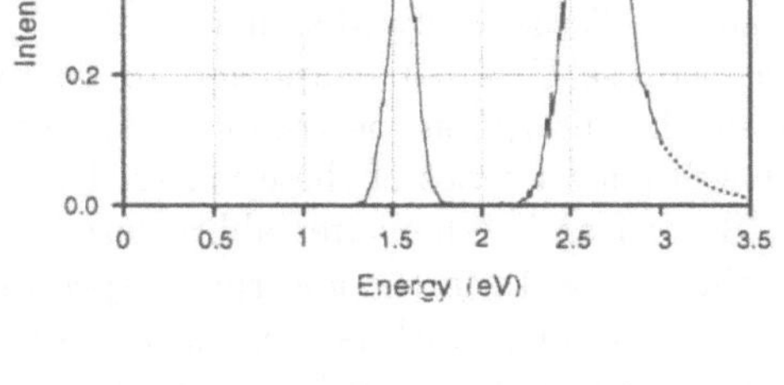

Fig. 10. The absorption spectrum in the presence of a single neutral soliton. The parameters are U=2.0 eV, V=1.2 eV, S=0.22eV and t_0=1.3 eV.

5.2. Polaron

In the rest of this section, we refer to the polaron case. The typical PA spectrum is shown in the previous paper [4]. There the used parameters are different from the present ones, but the spectral shape is essentially the same. Very roughly speaking, the difference in PA between a neutral soliton and a polaron is whether there exists a absorption band in the low energy region of the gap, namely, in the IR region, which band the soliton lacks. In the experiments, the existence of such a low-energy band was claimed once [15], but not reproduced by other measurements. We therefore do not make any assumption for the existence of this peak.

6. Conclusions

We have shown that, in the Peierls-Hubbard model that has a CDW ground state, there are various possibilities of relaxations. In the case of soliton channels, we have mentioned two possibilities: charged solitons or neutral solitons are created from the STE. It depends on the system parameters which is the case. In the case of polaron channels, on the other hand, it is not possible that polarons are created through the STE. They must be created from the photoexcitation with much higher energies than the lowest CT exciton.

As for the PA signal, the observed A band in PtCl and other similar materials is reproduced in the neutral-soliton case as in polaron case, while the B band is not clearly seen in the former case. To tell the truth, we have optically allowed levels at the energies corresponding to the B band, but they do not appear as an isolated peak because of very small intensity. However, it seems that the intensity of the B band is rather small also in the experimental spectra such as those in PtX (X=Cl and Br) [16]. Thus we think that it is a quantitative problem whether we obtain the B band as the second peak.

Summarizing the above results, a neutral-soliton pair is the most likely candidate for a product yielded by the photoexcitation into the CT band. However, the idea of a polaron-pair is not yet completely denied because of the "existence" of the low-energy PA band.

As future works, we are interested in time-dependent problems, particularly, the estimation of the STE life time. Moreover, in our present theory, only the paths to solitons through the STE are considered. Generally speaking, it also seems possible that solitons are created directly from the CT exciton. Such a problem will be very important in order to understand the whole dynamics in real materials.

References

[1] W. -P. Su, J. R. Schrieffer and A. J. Heeger, Phys. Rev. B**22** (1980) 2099.

[2] Y. Wada, T. Mitani, M. Yamashita and T. Koda, J. Phys. Soc. Jpn. **54** (1985) 3143.

[3] Y. Wada and M. Yamashita, Phys. Rev. B**42** (1990) 7398.

[4] K. Iwano and K. Nasu, J. Phys. Soc. Jpn. **61** (1992) 1380.

[5] K. Iwano and K. Nasu, J. Phys. Soc. Jpn. **62** (1993) 1778.

[6] H. Okamoto, K. Toriumi, K. Okaniwa, T. Mitani and M. Yamashita, Phys. Rev. Lett. **69** (1992) 2248.

[7] M. Tanaka, S. Kurita, T. Kojima and Y. Yamada, Chem. Phys. **91** (1984) 257.

[8] N. Kuroda, M. Sakai, Y. Nishina, M. Tanaka and S. Kurita, Phys. Rev. Lett. **58** (1987) 2122.

[9] S. Kurita, M. Haruki and K. Miyanaga, J. Phys. Soc. Jpn. **57** (1988) 1789.

[10] M. Haruki and S. Kurita, Phys. Rev. B**39** (1989) 5706.

[11] K. Nasu, J. Phys. Soc. Jpn. **53** (1984) 302.

[12] N. Kuroda, M. Kataoka and Y. Nishina, Phys. Rev. B**44** (1991) 13260.

[13] A. Mishima and K. Nasu, Phys. Rev. B**39** (1989) 5758.

[14] M. Suzuki and K. Nasu, Phys. Rev. B **45** (1992) 1605.

[15] R. J. Donohoe, S. A. Ekberg, C. D. Tait and B. I. Swanson, Solid State Commun. **71** (1989) 49.

[16] H. Okamoto, Y. Oka, T. Mitani, K. Totiumi and M. Yamashita, Mol. Cryst. Liq. Cryst. **256** (1994) 161.

Part IV

Optical Excitations, Lattice Relaxations and Photo-Induced Phenomena in Polymers

Domain-Wall Dynamics of Photo-Induced Phase Transition in Polydiacetylene Single Crystals

S. Koshihara, K. Akimoto, Y. Tokura[1]*, K. Takeda*[2]*, and T. Koda*[3]

Department of Applied Physics, Tokyo Institute of Technology, Tokyo 152, Japan

[1]Department of Physics, The University of Tokyo, Tokyo 113, Japan

[2]Tsukuba Research Laboratory, Japan Synthetic Rubber Co., Ltd., Tsukuba 305, Japan

[3]Department of Mathematical and Physical Sciences, Japan Women's University, Tokyo 112, Japan

Abstract. Domain-wall Dynamics of the photo-induced phase transition in polydiacetylene (PDA) single crystals has been investigated by time-resolved spectroscopy. It was found that primary process of the photo-induced phase transition is mostly completed within 50 ns. In addition, occurrence of a transient domain injection has been confirmed at temperatures far below and above T_c.

1. Introduction

Polydiacetylenes (PDAs), [$\nleftarrow$RC-C≡C-CR'$\nrightarrow_x$], are a family of representative π-conjugated polymers with various side-groups (R, R'), and one of unique properties in PDAs is a chromic phase transition which can be caused by various stimulations [1]. Most of the PDAs show two spectroscopically distinct phases: the gold colored phase and greenish one which are called the A- and B-phase, respectively. Among various PDAs, we have confirmed that single crystals of the alkyl-urethane substituted families (R=R'= $(CH_2)_4OCONH(CH_2)_{n-1}CH_3$, n=1-8; hereafter abbreviated as poly-4Un) undergo the first-order reversible A-B transition with change of temperature [2]. In Fig.1, we show a structure of poly-4Un.

Lately, we have discovered the reversibly (bi-directionally) photo-induced A-B transition in poly-4U3 single crystals [3]. The obtained results showed that the photo-carrier generation plays an important role in the driving process of the photo-induced phase transition. In this paper, we report on the overview of this unique photo-induced phenomenon in poly-4U3 single crystals and also on the dynamical aspects of it.

2. Characteristics of the Phase Transition in Poly-4U3 Crystals and Bi-Directional Phase Switching by Dichromatic Excitation

Single crystals and polycrystalline thin films of poly-4U3 show the first order phase transition (A-B transition) [2]. We can probe this transition by reflection (or absorption)

Springer Series in Solid-State Sciences, Volume 124
Relaxations of Excited States and Photo-Induced Structural Phase Transitions
Ed.: K. Nasu

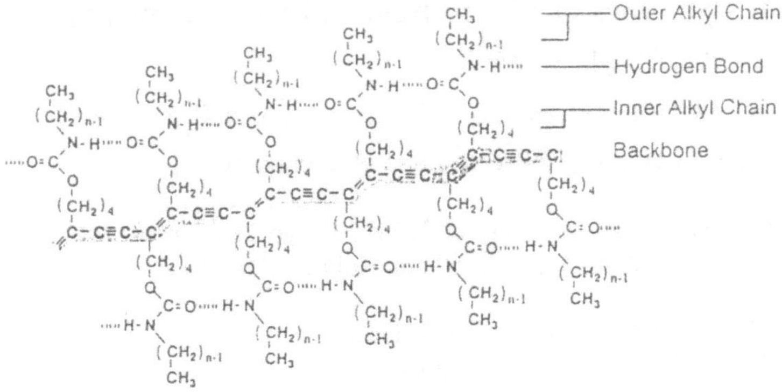

Fig.1: Structure of polydiacetylene poly-4Un.

and Raman spectroscopies. The reflection band which is observed at 1.95 eV in the A-phase (Fig.2(a); low-temperature phase) is due to the ^1B_u exciton (A-band) accompanied by its phonon (backbone vibrational mode) side band (A'-band). This excitonic reflection (or absorption) band shifted to ca. 2.35 eV (B-band) showing a similar phonon side band (B'-band) in the B-phase (Fig.2(b); high-temperature phase).

As observed in this figure, reflectance changes at 1.95 and 2.35 eV on the A-B transition are so large that the A-B transition is easily traceable by monitoring the reflectivity around these two photon energies. As an example, temperature dependence of the reflectivity at 1.95 eV is plotted in Fig.3. In the heating process (1->2->3), the system was kept in the A-phase below 400K, but showed an abrupt decrease of the reflectivity at 1.95 eV around 400 K in coincidence with the A-to-B transition. The A-B transition in poly-4U3 showed a wide hysteresis and the crystal was kept in the B-phase down to 360K in the cooling process (3->4->5) as shown in Fig.3.

Change in the π-electronic structure as observed upon the A-B transition is obviously due to the structural change of the carbon backbone, though the overall acetylene-type conformation seems to be maintained in the both phases [4]. The change of the bond structure can be probed by Raman spectroscopy [5]. The inset of Fig.2 shows the Raman spectra for the respective phases in the frequency region of the C=C symmetric stretching vibration on the carbon backbone. These show that the C=C stretching mode in the B-phase locates at ca. 50 cm^{-1} higher frequency than in the A-phase.

Recently we have found that a large photo-induced changes in reflectance and Raman spectra occur at a certain temperature in the middle of the hysteresis. Figure 4 shows the change in the reflectance spectrum of poly-4U3 single crystals induced by dichromatic excitation at 375K (points 2 and 4 of Fig.3). At the temperature point 2 (Fig.3), the crystal was kept in the A-phase as shown in Fig.4(a). After the excitation by a single shot of 2.58 eV (=$h\nu$) light pulse, the reflectance spectrum turned into the upper one of Fig.4(b). The observed shift of the reflectance maximum energy from 1.95 eV to 2.35 eV signals the occurrence of the phase transition from A to B. After the photo-induced change from the A- to B-phase, the same crystal was excited by a single shot of 3.18 eV (=hν') pulse (the second

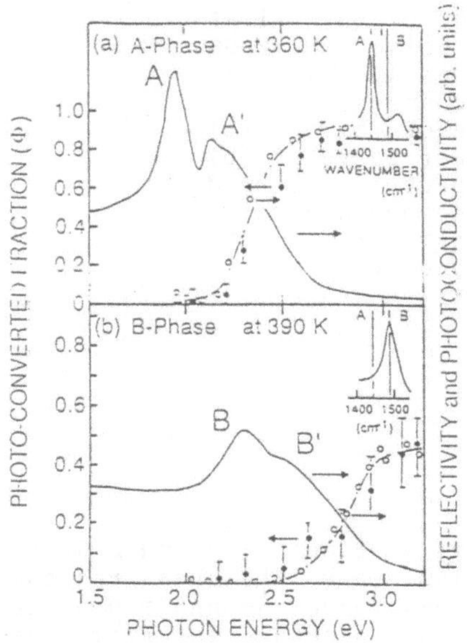

Fig.2: Reflection spectra in (a) the A-phase (observed at 360K) and (b) the B-phase (observed at 390K), and the excitation photon energy dependence (action spectra) of the photoconductivity (open circles). Dot-dashed-lines are the guide for the eye. Insets show the Raman spectra for the respective (A and B) phases in the frequency region of the C=C symmetric stretching mode. The photo-converted fraction (Φ) for the A-to-B and B-to-A directions are plotted as a function of the excitation photon energy by filled circles in (a) and (b), respectively.

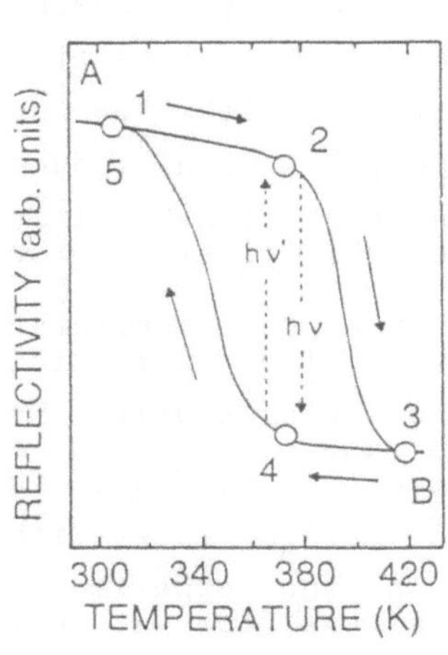

Fig.3: Temperature dependence of reflectivity of poly-4U3 single crystal observed at 1.95 eV. The crystal was heated up (1->2->3) and cooled down (3->4->5) in the order of the number. hν and hν' represent photon energies for excitation; hν=2.58 eV and hν'=3.18 eV.

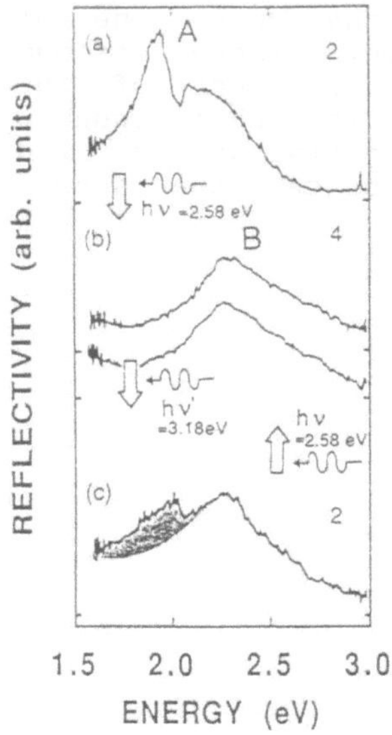

Fig.4: Reflection spectra before and after photo-excitation observed at 375 K (points 2 and 4 in Fig.3). (a) shows the reflection spectrum (the A-phase) before photo-excitation. After the irradiation by the first excitation pulse ($h\nu$=2.58 eV and 20 ns width), the spectrum changes to the upper curve of (b). (c) and the lower curve of (b) are the spectra observed after the second and third photo-excitations by $h\nu'$=3.18 and $h\nu$=2.58 eV, respectively.

excitation). Then the reflectivity between 1.5 and 2.0 eV increased and the new reflection band (hatched region) reminiscent of the A-phase region was formed around 1.9 eV as shown in Fig.4(c). This reflectance peak arising from the A-phase region could be completely extinguished by irradiation of the third excitation light pulse ($h\nu$=2.58 eV) (see the lower spectrum of Fig.4(b)). Observed spectral changes indicate that the bi-directional photo-switching between A- and B-phases occurs in poly-4U3 crystals as schematically shown in Fig.5

The efficiencies (Φ) of the photo-induced phase transition in A-to-B and B-to-A directions strongly depend on the excitation photon energy (see filled circles in Fig.2) and show quite parallel behavior to the excitation photon energy dependences of the photoconductivity (open circles in Fig.2). These results indicate that it is necessary to use a light pulse with an energy larger than the band gap energy ($E_g^{(A)}$ and $E_g^{(B)}$), for driving the photo-induced A-B transition in poly-4U3. For example, the photo-excitation with 2.58 eV light pulse, that is above $E_g^{(A)}$ but below $E_g^{(B)}$, can induce the A-to-B conversion, but not the B-to-A transition. This is the reason why the A-phase disappeared selectively after the third photo-excitation.

The conversion efficiencies also show the nonlinear dependences on the excitation intensity. The excitation photon density lower than 3x10^{18} photons/cm^3 cannot drive the phase transition. Hereafter, excitation intensity, i.e., a density of absorbed photons on the crystal surface, is abbreviated as cm^{-3}. The Φ reaches its optimum value with the photon density of 6x$10^{18}cm^{-3}$ [13]. In the experiments for Fig.6, the excitation intensity was always kept at ca.

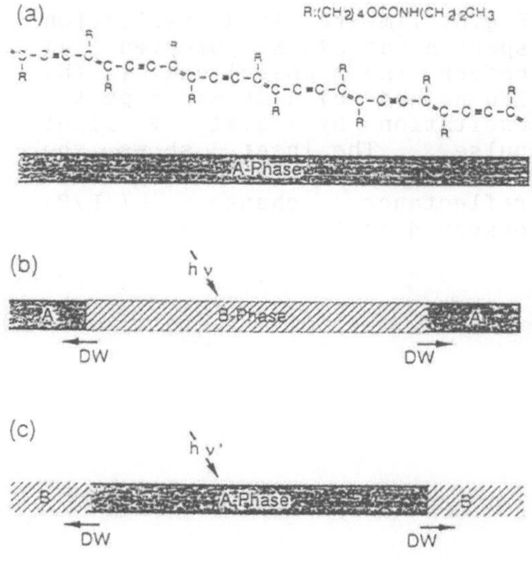

Fig.5: Schematic illustration of the bi-directional photo-induced phase switching in poly-4U3, (a) before photo-excitation, (b) A-to-B conversion by 2.58 eV (=hν)light pulse and (c) following B-to-A conversion by 3.18 eV (=hν')light pulse.

$6.5x10^{18}cm^{-3}$. It was also confirmed that the photo-excitation with the same intensity but below E_g^A causes no change in the spectra at any stage of the phase cycle.

3. Time-resolved Reflection Spectra in Case of the Permanent Phase Transition

The time-resolved study was carried out to measure the speed of the phase conversion and possibly detect the evolution process from the microscopic excited species into the macroscopic phase in the course of the photo-induced phase transition. Figure 6 is the time-resolved reflection spectra at various delay-times (Δt) after photo-excitation for the A-to-B transition. Crystals were kept at the temperature in the middle of the hysteresis region (375K: the points 2 and 4 in Fig.3). Photon energies of laser pulses were tuned to 2.58 eV and excitation intensities were set at $6.5x10^{18}$ cm^{-3}. The time-dependences of the reflectivity change ($\Delta R/R$) at photon energies 2.4 and 1.9 eV are shown in the inset. These photon energies, 1.9 eV and 2.4 eV, approximately correspond to the positions of the excitonic reflectance maxima in the A- and B-phases, respectively, and hence are suitable for probing the A-B transition.

Figures 6(b) and (c) show the reflection spectra at 150 ns and 2 ms after photo-excitation with 2.81 eV pulse, respectively. The reflectance maximum due to the A-phase

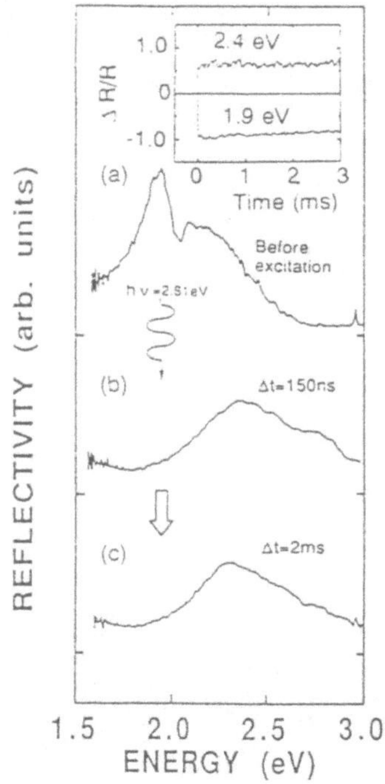

Fig.6: Time-resolved reflection spectra at 375 K observed (a) before (the A-phase) and at (b) 150 ns and (c) 2 ms after photo-excitation by a 2.81 eV light pulse. The inset shows the time-dependence of the reflectance change ($\Delta R/R$) observed at 1.9 and 2.4 eV.

seen in the original spectrum (Fig.6(a)) was extinguished immediately after excitation and a broad band characteristic of the B-phase was formed around 2.4 eV. The change in the reflection spectrum appears to be mostly completed within the period less than the resolution time (50 ns) of our experimental apparatus (see the inset), apart from a further slight red shift (about 0.04 eV) of the reflectance maximum position observed over a long period (Δt=2 ms). Thus, the photo-induced A-to-B phase transition occurs faster than 50 ns. The dynamics of the transition in the backward direction (B-to-A) was quite similar to that of the A-to-B conversion.

In addition to the fast primary process within 50 ns, a slow "red-shift" was observed in the temporal changes of the spectra. In previous study [6], it was reported that the thermally induced A-B transition in poly-4U3 crystals is accompanied not only by the structural changes in the carbon backbone but also with the configurational changes in side-groups. According to the ^{13}C NMR study by Tanaka et al. [6], a conformational change in the alkyl groups, especially a trans-gauche change in the inner-alkyl chains (see Fig.1), occurs upon the A-B transition. Therefore, the structural changes in both the backbone and side-groups are necessary for the whole process of the photo-induced A-B transition. The conformational change in the alkyl chains is likely to take much longer interval than the change in the backbone structure. On the basis of this speculation, we tentatively attribute the photo-converted band at a short delay time to the stressed backbones accompanied with the incommensurate configuration of the side-group.

4. Time-resolved Reflection Spectra in Case of the Transient Phase Change

By the time-resolved reflectance spectroscopy at the temperature points outside the hysteresis region (points 1(7); 300 K and 4; 420 K), we have confirmed the occurrence of the transient domain injection of the unstable phase (see Fig.7). At 300 K (low-temperature side out of the hysteresis loop), only the A-phase is stable and the crystal was kept in this phase before excitation as shown in the Fig.7(a). Within 50 ns after photo-excitation by 2.81 eV light pulse, the reflection band around 1.9 eV in the A-phase almost disappears and a new band around 2.5 eV whose spectral shape is characteristic of the B-phase appears (see Fig.7(b) Δt=100 ns). At Δt=100 μs, the reflectance maximum shifts to 2.3 eV and the 1.9 eV band (A-band) begins to reappear. At Δt=10 ms, the spectrum comes back to the same one as before photo-excitation. Obtained results clearly show that the transient domain injection as illustrated in Fig.8 occurs in poly-4U3 at 300 K.

The temporal shift of the photo-injected B-phase band from 2.5 to 2.3 eV may reflect the slower relaxation process of the side-group configuration than the change in the polymer backbone structure as discussed before.

A similar transient domain injection into the B-phase ground state has been observed at 420K (high-temperature side out of the hysteresis) by the irradiation of the 3.18 eV light pulse. For the unstable phase domain to be injected, additional energy is considered to be necessary. The lattice relaxation energy of the photo-excited species perhaps serves as effective fuel for the initial evolution of the local excited species into the macroscopic domain.

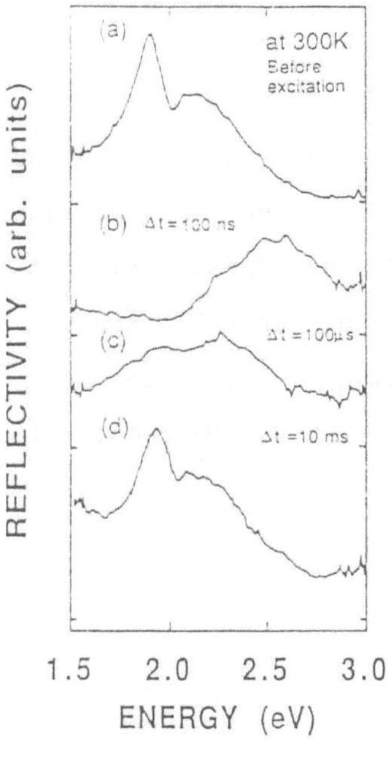

Fig.7: Photo-induced reflectance change at a temperature point below (300K) the transition temperature (T_c). (a) shows the spectrum of the A-phase before photo-excitation. The time-resolved reflectance spectra (b), (c), and (d) were observed at 100 ns, 100 μs, and 10 ms after photo-excitation by 2.81 eV (=hν) pulse, respectively.

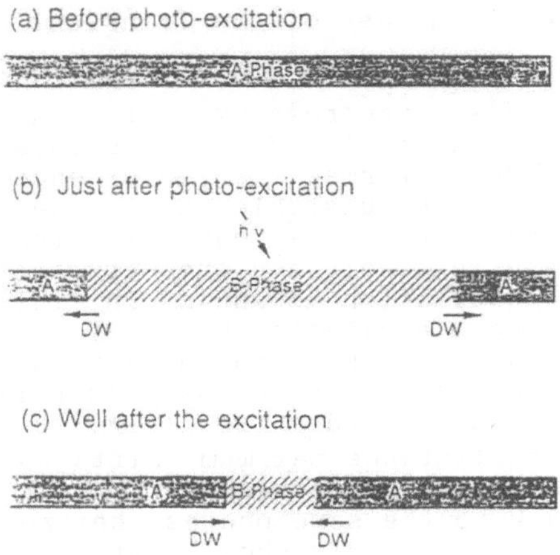

Fig.8: Schematic illustration of the transient domain injection observed at 300K.

Acknowledgment

We are much indebted to Mr. A. Itsubo, Mitsubishi Petrochemical Co. Ltd., for his help in the preparation of the samples. We also acknowledge fruitful discussion with Professor E. Hanamura, Professor N. Nagaosa and Professor T. Ogawa on the theoretical background of the photo-induced phase transition. This work was supported by a Grant-In-Aid for Scientific Research from the Ministry of Education, Science, Culture and Sports, Japan and also supported by Ciba-Geigy Foundation (Japan) for the Promotion of Science.

References

1) Polydiacetylenes, NATO ASI Series E, vol 102, ed. by D.Bloor and R.R.Chance, (Nijhoff, Dordrecht 1985).
2) S.Koshihara, Y.Tokura, K.Takeda, T.Koda and A.Kobayashi, J. Chem. Phys. 92 (1990) 7581.
3) S.Koshihara, Y.Tokura, K.Takeda and T.Koda, Phys. Rev. Lett. 68 (1992) 1148.
4) A.Kobayashi, H.Kobayashi, Y.Tokura, T.Kanetake and T.Koda, J. Chem. Phys. 87 (1987) 4962.
5) G.J.Exarhos, W.M.Risen Jr. and H.Baughman, J. Am. Chem. Soc. 98 (1976) 481.
6) H.Tanaka, M.A.Gomez, A.E.Tonelli and M.Thakur, Macromolecules 22 (1989) 1208.

Dynamics of Excitons in Polymer Solids

S. Abe

Electrotechnical Laboratory, 1-1-4 Umezono, Tsukuba 305, Japan

Abstract. Several important aspects of excitons in condensed conjugated polymer systems are discussed theoretically. The topics includes intrachain characteristics of a one-dimensional exciton, the formation of an exciton-polaron, the contribution of biexcitons, and interchain excitation and charge transfer processes.

1. Introduction

The optical properties of conjugated polymers have been studied intensively in recent years because of their potential applications for nonlinear optical materials, light emitting diodes, and other photonic devices. Through the studies it has been gradually recognized that excitons play important roles in this class of materials [1]. It is by now clear that excitons are primary photoexcitations in many conjugated polymers, and the photocreated excitons are subject to various dynamical processes, including intrachain lattice relaxation, biexciton formation, interchain excitation transfer and charge separation, and so on. Some of these topics associated with the dynamics of excitons are discussed in the present paper.

2. Basic Model

A theoretical description convenient for the discussion of excitons is the so-called configuration interaction (CI) method for a tight-binding model of π electrons. We consider the Pariser-Parr-Pople for a single chain of polyacetylene as a representative model of conjugated polymers [2]:

$$H = H_0 + H_{\text{int}}, \tag{1}$$

$$H_0 = -\left[t + (-1)^n \delta t\right]\left(C^\dagger_{n+1,s} C_{n,s} + C^\dagger_{n,s} C_{n+1,s}\right), \tag{2}$$

$$H_{\text{int}} = \sum_n U \rho_{n\uparrow} \rho_{n\downarrow} + \frac{1}{2} \sum_{n,s} \sum_{m(\neq n),s'} V_{nm} \rho_{ns} \rho_{ms'}, \tag{3}$$

Springer Series in Solid-State Sciences, Volume 124
Relaxations of Excited States and Photo-Induced Structural Phase Transitions
Ed.: K. Nasu

where $C^{\dagger}_{n,s}$ creates an electron at site n and spin s, and $\rho_{ns} = C^{\dagger}_{n,s} C_{n,s}$. In the one-electron part H_0, δt is the transfer energy modulation due to bond alternation. We use a parametrized form of electron-electron interaction: the on-site U and the long-range part $V_{nm} = Va / R_{nm}$ (Pople potential) or $V_{nm} = \left[(1/U)^2 + (R_{nm}/Va)^2\right]^{-1/2}$ (Ohno potential). The typical parameter ranges that we use mostly are $\delta t \sim 0.1t$ - $0.2t$, $U \sim 2t$ - $3t$, $V \sim t$ corresponding to intermediate interaction strength. Later we will add interchain interaction part to the Hamiltonian. We use the Hartree-Fock approximation for the ground state and the CI among single or double excitations for excited states. The latter is necessary to discuss biexciton states [3].

3. Intrachain Exciton

3.1 Exciton Size

The most characteristic feature of an one-dimensional exciton is its size, i.e., the average electron-hole distance. In ordinary semiconductors a Wannier exciton is a weakly bound electron-hole pair analogous to a hydrogen atom. Its size is determined by the effective Bohr radius given by $a_B^* = \hbar\varepsilon / e^2\mu^*$ using the dielectric constant ε and the reduced effective masses μ^* of the electron-hole pair. However, the long-range Coulomb attraction in one dimension is singular in the sense that the potential cannot hold the lowest bound state at a finite energy and at a finite size [4]. Instead the energy and the size of the lowest state becomes infinitely deep and infinitesimally small, respectively. In actual materials there is always a lower bound for the potential. For conjugated polymers it has been shown [5, 6] that the size of the lowest exciton is much smaller than the effective Bohr radius and roughly given by the correlation length (or the delocalization length) defined by $\xi = (t/\delta t)a$, where a is the bond length. As a consequence of the small exciton size, the oscillator strength of interband optical transitions is almost concentrated to the lowest exciton only [5]. This is one of the most fundamental properties of excitons in one-dimensional systems.

3.2 Exciton Polaron

In conjugated polymers the coupling between electrons and the lattice is also important. The formation of solitons and polarons has been extensively discussed in polyacetylene and other polymers [7]. It is natural to expect that a photocreated exciton is coupled to the lattice, too. In the case of the Su-Schrieffer-Heeger model for polyacetylene, an electron-hole excited state relaxes into a pair of separated solitons [8]. In the case of polymers with nondegenerate ground states, the solitons remains strongly bound and is called a neutral bipolaron or an exciton

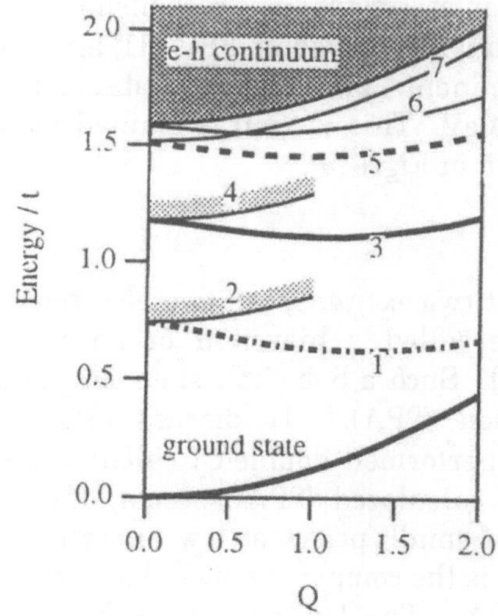

Fig. 1. Adiabatic potential curves with respect to polaron deformation Q. 1: ^{3}Bu exciton polaron. 2: ^{3}Bu free exciton. 3: ^{1}Bu exciton polaron. 4: ^{1}Bu free exciton. 5: ^{1}Ag and ^{3}Ag exciton polaron. 6: hole polaron + free electron (or electron polaron + free hole). 7: ^{1}Ag and ^{3}Ag free exciton.

polaron. The presence of the electron-hole Coulomb attraction further enhances the formation of an exciton polaron. Note that the size of a polaron in the SSH model is essentially determined by ξ [7], which is the same as the size of the lowest exciton. The polaron formation is mainly related the localization of the center-of-mass motion of the exciton, while the internal motion is not very much affected by the lattice relaxation.

Calculations for an exciton polaron in the presence of Coulomb interactions have been performed [9] by using single-CI with a variational form of the polaron deformation: $u_n = (-1)^n u_0\left[1 - Q \ \mathrm{sech}\{(n - x_c)/w\}\right]$, which contains two parameters Q and w, the depth and width of the polaron (x_c is the center position of the polaron). For an optimum configuration, w is close to ξ. An example of the calculated adiabatic potential diagram for a fixed w is shown in Fig.1 [10]. The optical transition from the ground state creates singlet free exciton (Branch 4 in Fig.1) at Q=0. This is followed by lattice relaxation to the minimum of the polaron state (Branch 3) within less than a picosecond. Luminescence from this state has been observed in poly(phenylene vinylene) and some other polymers. On the other hand, the intersystem crossing to the lower triplet exciton state (Branch 1) occurs quite efficiently. The optical absorption from the relaxed triplet state to higher triplet states is responsible for the long-lived photoinduced absorption observed in PPV [11] and polydiacetylene [12]. Especially, a prominent photoinduced absorption peak has been observed at about 1.4 eV. This is well explained as an absorption from Branch 1 to Branch 5 in Fig.1 [9].

3.3 Biexcitons

It is well known that two excitons, i.e., two electrons and two holes, can form a bound state called a biexciton or an excitonic molecule in semiconductors [13]. Such a biexciton state can be directly created by two-photon absorption (TPA). To discuss biexcitons in conjugated polymers, we have performed double-CI calculations [3]. Figure 3 is a typical example of calculated TPA spectra, Im $\chi^{(3)}(-\omega;\omega,-\omega,\omega)$. The spectrum consists of small peaks at low energies and large peaks at high energies. C_{DE} is the component of double excitations for each state contributing the TPA. The low energy peaks correspond mainly to single (one-electron one-hole) excitations, while the high energy peaks are mainly due to double (two-electron two-hole) excitations, although there is some mixing between single and double excitations in the present system. The energy of the lowest one-photon allowed exciton is about $2t$ in this calculation, so that the strong TPA peaks in Fig.2 are located well below the twice of the exciton energy. Therefore, these doubly excited states must be biexciton states.

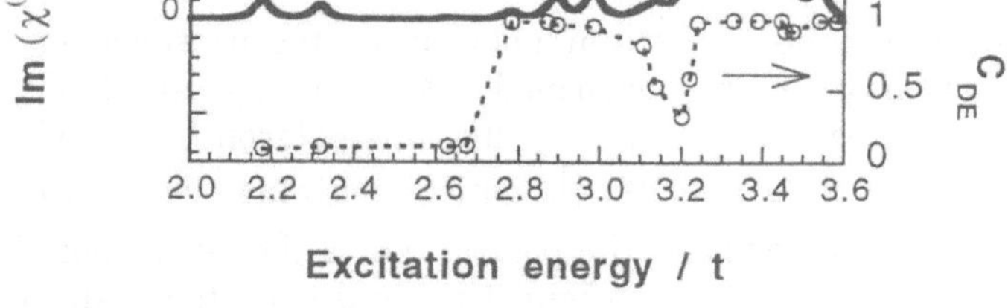

Fig. 2. Nonlinear absorption spectrum calculated with double-CI method for a ring of 40 sites with $\delta t=0.2t$, $U=2t$, and $V=t$. The component of double excitations (C_{DE}) to each two-photon allowed state is indicated by circles connected by broken line.

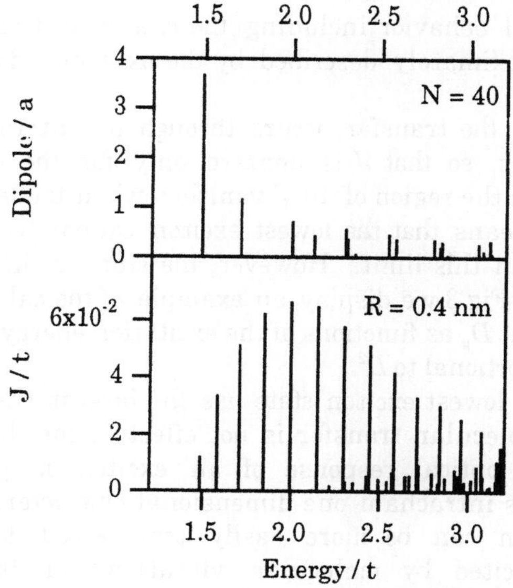

Fig. 3. Calculated transition dipoles (upper panel) in units of a (~1.4 Å) and excitation transfer J (lower panel) for 40-site chains aligned in parallel with distance R=4 Å. The unit of energy is t ~ 2 eV.

4. Interchain Delocalization of an Exciton

In this section the nature of excitations on a system of two chains is examined to address the issue of interchain delocalization of excitons. The model is a direct extension of the one used for a single chain. Additional factors are interchain transfer energies and interchain Coulomb interactions. The same form of Coulomb potentials such as the Ohno potential can be used for the latter.

4.1. Interchain Exciton Transfer

We consider the transfer of an exciton among two chains aligned in parallel, neglecting for the moment the direct overlaps of wave functions and consequently the direct transfer of electrons. The transfer from an exciton state $|\phi_{1\nu}>$ of chain 1 to an exciton state $|\phi_{2\mu}>$ of chain 2 occurs through Coulomb interactions: $J_{\mu\nu} = <\phi_{2\mu}|H_{int}|\phi_{1\nu}>$.

The transfer J can be calculated by using exciton states of each chain obtained with the single CI method. The results of such calculations [6] for the main exciton state are summarized as follows.

(a) J is proportional to L/R^3 if the distance R is much larger than the chain length L.

(b) J becomes proportional to $1/LR$ if R is much smaller than L.

(c) The overall behavior including the crossover from (a) to (b) can be approximately described by the extended dipole model [14].

In the region (a) the transfer occurs through the interaction between transition dipoles, so that J is nonzero only for the dipole-allowed exciton states. In the region of (b), J vanishes when the length L goes to infinity. This means that the lowest exciton cannot be transferred to other molecules in this limit. However, the story is different for J of higher states. In Fig.3 we display an example of the calculated $J_{\mu\mu}$ and the dipole moment D_{μ} as functions of the excitation energy E_{μ}. It is clear that J is not proportional to D^2.

Although the lowest exciton state has the largest transition dipole moment, intermolecular transfer is not effective for this state. This implies that the optical response of an exciton is predominantly determined by its intrachain one-dimensional character. On the other hand, the exciton can be more easily transferred to neighboring molecules if excited by molecular vibrations or by some other scattering mechanism to a higher state with larger J.

4.2. Interchain Charge Transfer

Next we consider the case of two chains very close to each other. For example, the nearest neighbor interchain distance in polyacetylene is about 4 Å. Then the direct interchain transfer energies among π electron orbitals must be taken into account. Consider two identical chains located such that one is translated by a distance, e.g., 4 Å, from the other in the direction (z) perpendicular to the molecular plane (xy). The pz orbital of each carbon of a chain has a small overlap with that of the corresponding carbon of the other chain. We assume small transfer energies $t_{\perp}$ between them. For Coulomb interactions we assume that the coupling $V_{\perp}$ perpendicular to the chain axis is two times stronger than that of the parallel direction, $V_{//}$, in the Ohno potential.

By taking both the interchain transfer energies and the interchain Coulomb interactions, we calculate the excited states of the two molecules with the single CI method [15]. The main results are shown in Fig.4. The absorption spectrum itself is not much different from that of a single chain as shown in Fig.4(c). That is, the states with large oscillator strength remains primarily an intrachain exciton. Interchain charge transfer exciton states are located higher than but rather close to the lowest intrachain exciton state, as displayed in Fig.4(b). They have negligible transition dipole moments. Their spatial extent along the chain direction is similar to the size of the intrachain exciton, i.e., approximately given by the correlation length ξ, as shown in Fig.4(a). This state may be related to the interchain electron-hole pairs suggested in [16].

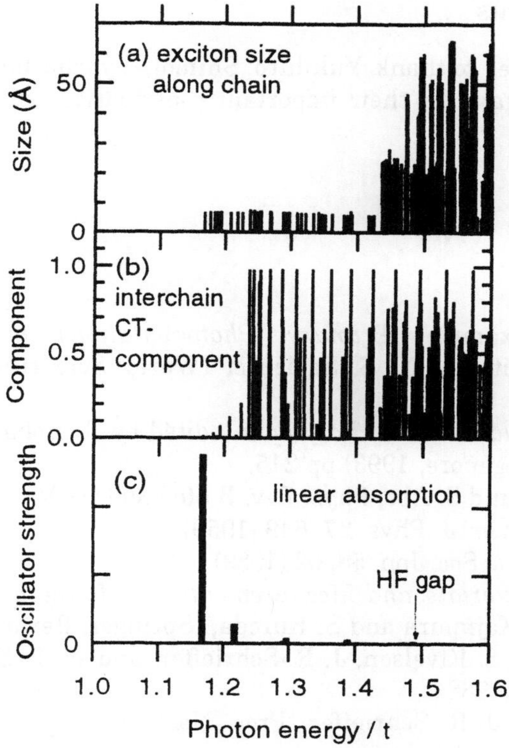

Fig. 4. Calculated excited states of a system of two chains of 100 sites with a distance 4 Å, $\delta t=0.2t$, $U=2t$, $V_{//}=t$, $V_{\perp}=2t$, $t_{\perp}=0.05t$: (a) The exciton size along the chain; (b) the interchain charge transfer component; (c) the linear absorption spectrum.

5. Conclusions

Optically excited states in conjugated polymers are primarily intrachain one-dimensional excitons, which have large binding energies. When two or more excitons are created on a single chain, they form a biexciton state, which also has a large binding energy. Intrachain lattice relaxation of an exciton is also important for photoinduced absorption experiments. Interchain excitation transfer is not very large for the lowest exciton state but is important for higher excited states, for which the simple extended dipole model is not applicable. Due to the effect of charge transfer between neighboring chains, there are interchain charge-transfer excitons located slightly above the energy of the lowest intrachain exciton, although they have almost negligible oscillator strengths in the absorption spectrum.

Acknowledgments

The author wishes to thank Yukihiro Shimoi, Vyacheslav A. Shakin, and Kikuo Harigaya for their important contributions in collaborative works.

References

[1] See, for example, *Primary Photoexcitations in Conjugated Polymers,* edited by N. S. Sariciftci (World Scientific, Singapore, 1997).
[2] S. Abe, in *Relaxation in Polymers*, edited by T. Kobayashi (World Scientific, Singapore, 1993) pp.215.
[3] V. A. Shakin and S. Abe, Phys. Rev. B **50**, 4306 (1994).
[4] R. Loudon, Amer. J. Phys. **27**, 649 (1959).
[5] S. Abe, J. Phys. Soc. Jpn. **58**, 62 (1989).
[6] S. Abe, in *Materials and Measurements in Molecular Electronics*, edited by K. Kajimura and S. Kuroda, (Springer, Berlin, 1996) pp.85.
[7] A. J. Heeger, S. Kivelson, J. R. Schrieffer, and W.-P. Su, Rev. Mod. Phys. **60**, 781 (1988).
[8] W. P. Su and J. R. Schrieffer, Proc. Natl. Acad. Sci. USA, **77**, 5626 (1980).
[9] Y. Shimoi and S. Abe, Phys. Rev. B **49**, 14113 (1994).
[10] S. Abe, Prog. Theor. Phys. Suppl. **113**, 83 (1993).
[11] N. F. Colaneri, D. D. C. Bradley, R. H. Friend, P. L. Burn, A. B. Holmes, and C. W. Spangler, Phys. Rev. B **42**, 11670 (1990).
[12] J. Orenstein, S. Etemad, and G. L. Baker, J. Phys. C **17**, L297 (1984).
[13] M. Ueta, H. Kanzaki, K. Kobayashi, Y. Toyozawa, and E. Hanamura, *Excitonic Processes in Solids* (Springer, Berlin, 1985), Chap. 2.
[14] V. Czikklely, H. D. Forsterling, and H. Kuhn, Chem. Phys. Lett. **6** (1970) 207.
[15] S. Abe, Synth. Metals, to be published.
[16] M. Yan, L. J. Rothberg, F. Papadimitrakopoulos, M. E. Galvin, and T. M. Miller, Phys. Rev. Lett. **72** (1994) 1104.

Lattice Relaxation of Photo-Excited States in Conjugated Polymers

Guo-Zhong Wen and Wu-Pei Su

Department of Physics and Texas Center for Superconductivity,
University of Houston, Houston, TX 77204-5506, USA

Abstract. Polyacetylene is a prototypical conjugated polymer which exhibits lattice dimerization and which supports solitons (kinks in the dimerization pattern). We have carried out a theoretical study of the lattice relaxation of some low-lying excited states in polyacetylene. Degenerate perturbation, strong coupling and density matrix renormalization group methods are employed for the calculations. Based on the results of those calculations, we discuss the relaxation of the major low-lying photoexcited states into solitons in their various spin and charge states.

1. Introduction

Conjugated polymers have generated a considerable amount of interest in the scientific community in the last decade or so. They are organic polymers endowed with unusual optical and electronic properties. As such, they are potentially important in many applications including polymer light-emitting diodes, which are being vigorously pursued. It turns out that the physics [1] operating behind those materials is also quite novel. Dynamical symmetry breaking, soliton formation, charge fractionalization and other profound physics ideas seem to be realized in such materials. Photoexcitation provides a particularly powerful probe of the rich relaxation dynamics in such systems.

Shortly after a noninteracting model of polyacetylene (the SSH model [2]) was introduced, it was predicted [3] that a photogenerated electron-hole pair decays within a few picoseconds into a soliton-antisoliton pair. The theoretical prediction has spurred much effort in the transient spectroscopy of polymers. While the theory is confirmed by some experiments, many observations call for a more complete explanation. It has become increasingly clear that electron-electron interactions have to be included in a quantitatively accurate theory. The inclusion of such interactions renders the problem quite challenging. Many theoretical papers have contributed to the clarification of the relaxation pathways of photoexcited states. In this article, we report on our study of this problem.

This paper is organized as follows. In Section 2, the picture in the noninteracting model is discussed. The electron-electron interactions are then treated perturbatively. As we will find out later on, such a treatment is actually quite reliable despite its simplicity. The other extreme limit of strong coupling is considered in Section 3. The actual interaction strength in polymers is probably intermediate, therefore perturbative and strong coupling approaches are likely to give complementary views. One can actually get a quite good picture of the major relaxation pathways by combining those views. In Section 4, we present

Springer Series in Solid-State Sciences, Volume 124
Relaxations of Excited States and Photo-Induced Structural Phase Transitions
Ed.: K. Nasu

accurate results based on the density matrix renormalization group method. A unified theoretical picture is attempted in Section 5, together with a brief comparison with experiment. In this article, we will focus on the prototype polymer–polyacetylene. Generalization to other polymers will be mentioned in the summary.

2. Perturbative Results

The theoretical calculations to be reported in this work are all based on the following model Hamiltonian for polyacetylene, which is the SSH model [2] supplemented by a Hubbard term (SSHH model),

$$H = \sum_{<i,j>} \{-(t-\alpha\delta r_{ij})\sum_{\sigma}[c^{\dagger}_{i,\sigma}c_{j,\sigma} + H.c.] + \frac{K}{2}(\delta r_{ij})^2\} + U\sum_{n}\rho_{n\uparrow}\rho_{n\downarrow}. \quad (1)$$

In the above equation, the first term describes π-electron hopping with a hopping integral which is linearly modulated by the change in bond length. The second term amounts to the elastic deformation energy of the σ-bonds.

Despite its simplicity, the noninteracting SSH model has led to many non-trivial predictions. Let us review two relevant results here. First, it predicts that pristine polyacetylene dimerizes into one of the two possible degenerate ground states characterized by a bond alternation wave (Fig. 1(a)) which opens up a gap (the Peierls gap) in the electronic spectrum. Second, a soliton-antisoliton pair can be generated[3] by exciting an electron-hole pair across the gap as shown schematically in Fig. 1(b).

A soliton is a domain wall between two senses of bond alternation. It is a defect in the bond alternation wave and is accompanied by a local distortion of the electronic spectrum. Specifically, a midgap state is always present in the neighborhood of a soliton. Due to topological constraints, solitons are created in pairs by photoexcitation. In an overlapping soliton-antisoliton pair, their midgap states hybridize to form a bonding and an antibonding orbital located symmetrically below and above the center of the gap.

To understand the photogeneration of solitons, it is insightful to consider the adiabatic potential surface of a pristine polyacetylene chain in its first excited state, an odd parity (B_u) state. The configurational coordinate is the separation $2x_0$ between a soliton and its antisoliton as described by the following form of the dimerization order parameter,

$$\Delta_n = \Delta\{1 + \tanh(\frac{2x_0}{\xi})[\tanh(\frac{na - x_0}{\xi}) - \tanh(\frac{na + x_0}{\xi})]\}. \quad (2)$$

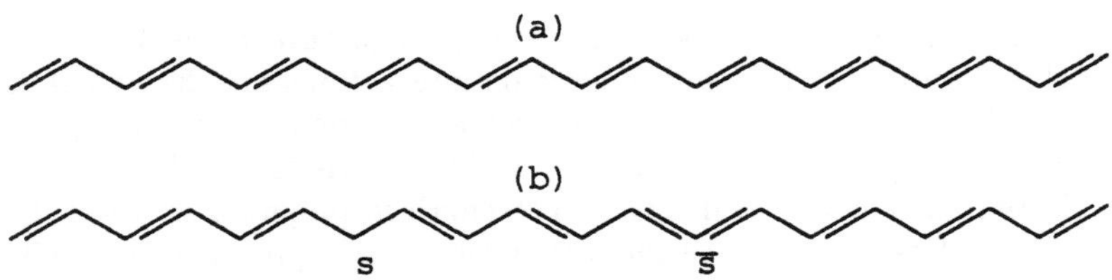

FIG. 1. (a) Bond alternation wave in pristine polyacetylene. (b) A photogenerated soliton-antisoliton pair.

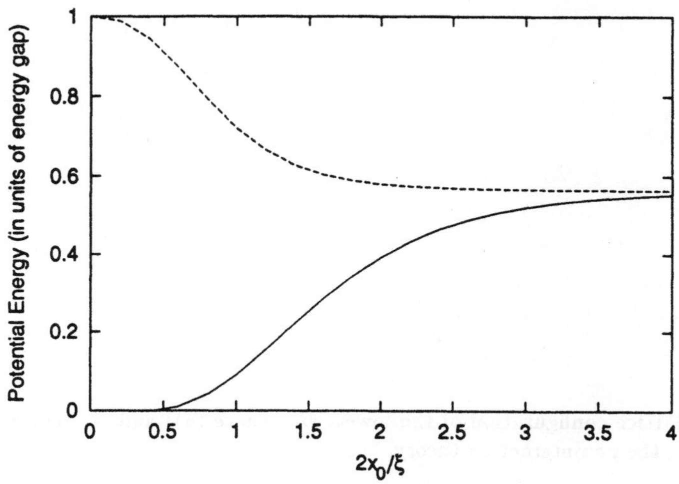

FIG. 2. Potential energy curves in the two soliton sector calculated in the noninteracting theory.

The above one-parameter family of $S\bar{S}$ configurations smoothly interpolate between a perfectly dimerized chain (for $x_0 = 0$) and the chain with two solitons (for large x_0). As depicted in Figure 2, the excited state energy monotonically decreases with the $S\bar{S}$ separation and merges with the ground state energy for large x_0 due to the appearance of almost degenerate midgap states. The photogeneration of an $S\bar{S}$ pair is therefore a downhill reaction, the maximum energy released is about 40% of the energy gap.

A careful analysis [4] of the electronic wave functions of the gap states indicates that charged solitons rather than neutral ones are produced.

Besides the first odd parity excited state, it is also of interest to look at the relaxation of the second excited state (which is an even-parity A_g state) in the noninteracting model. Curve (b) in Figure 3 describes the relaxed lattice configuration of such an excited state. It resembles two separated polarons. As a contrast curve (a) shows the relaxed B_u state.

While the photogeneration of charged solitons was subsequently verified experimentally [5], additional observations indicated that neutral solitons are also photogenerated [6]. Moreover, neutral and charged solitons possess distinct absorption features [7]. In the noninteracting model, both shall show an absorption peak at half of the energy gap, whereas experimentally it was found that charged solitons absorb at about 0.5 eV (low energy peak) and neutral ones absorb at 1.5 eV (high energy peak). It was then recognized that electron-electron interactions need to be incorporated in the model Hamiltonian. In this section, we treat the Hubbard repulsion perturbatively. Specifically, for any $S\bar{S}$ configuration, we calculate the molecular orbitals by diagonalizing the SSH Hamiltonian. Since we are interested only in low-lying excitations, we pick four molecular orbitals near the center of the bandgap and place four electrons there. Within this subspace, the Hubbard interaction is treated exactly.

In other words, we are essentially doing a full configuration interaction (CI) for the gap states. Four states are included because we want to extend the calculation to four-soliton configurations as well.

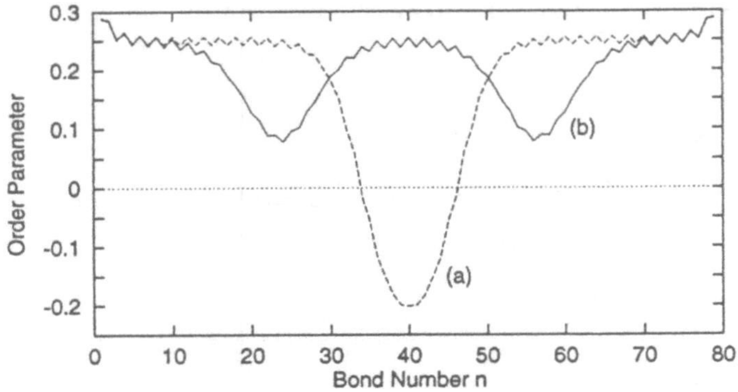

FIG. 3. Relaxed lattice configuration of the lowest B_u (curve (a)) and A_g (curve (b)) excited states in the noninteracting theory.

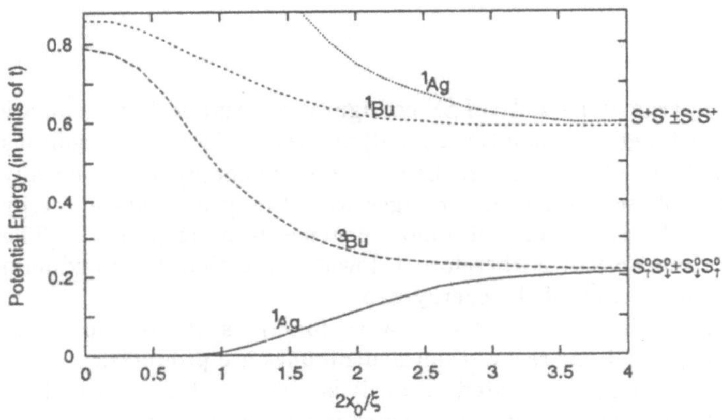

FIG. 4. Potential energy curves in the two soliton sector calculated in degenerate perturbation theory.

Figure 4 depicts the results from a calculation in which the dimensionless coupling constant is chosen to be $\lambda = \alpha^2/(Kt) = 0.37$, which implies a coherence length ξ of about 4 lattice spacings in the uniformly dimerized ground state. The repulsion strength U of the Hubbard term is taken to be 2.5 t, where t is the average hopping integral in Eq.(1).

The lowest curve in Figure 4 represents the ground state energy in the presence of a soliton-antisoliton pair. The one above it is the lowest triplet excitation curve. For large $S\bar{S}$ separation, the two curves converge to two neutral solitons in the singlet and triplet states. Our calculation is more reliable for large $S\bar{S}$ separation. In a more complete calculation (DMRG), the triplet curve should be lower than shown in Figure 4 for shorter soliton-antisoliton ($S\overline{S}$) distance. In any case, the triplet curve describes the relaxation of a triplet excited state into two neutral solitons.

The two upper curves in Figure 4 correspond to two charged solitons in the odd and even parity states. Kivelson and Wu [8] have used the degeneracy of these two states at large S$\bar{S}$ separation to argue for charge separation in the photogeneration of solitons. The gap value is about 0.8 t. For large S$\bar{S}$ separation, there is an onset of the continuum slightly below the gap value. From the energy differences between the curves, one can see that a charged soliton absorbs light at about 0.5 eV and a neutral soliton absorbs at about 1.5 eV provided we choose t=2.5 eV. The U value has been chosen to reproduce these experimentally observed absorption energies of the charged and neutral solitons. We would like to emphasize here that although more complete calculations are likely to alter the shape of those curves in Figure 4, the features at large S$\bar{S}$ separation should stay the same. We have exhausted the four possible states of two well separated solitons and their relative energies are fixed by absorption experiment. Another important feature is that the creation energy of a neutral soliton pair is only about 0.5 eV. This small value of the creation energy is actually consistent with the measurement of the triplet energy in polyenes [9]. By comparing the energy of the neutral and charged solitons, it is obvious that the energy to create four neutral solitons is less than that for creating two charged solitons. Therefore one would expect to see an even parity state relaxes into 4 neutral solitons as being more favorable than the relaxation into two charged solitons.

To confirm the above idea, we have calculated the energies of the ground state as well as the lowest even parity singlet excited state in the following two parameter set of 4-soliton configurations,

$$\Delta_n = \Delta\{1+\tanh(\frac{2x_0}{\xi})[\tanh(\frac{na-x_d-x_0}{\xi})-\tanh(\frac{na-x_d+x_0}{\xi}) \\ +\tanh(\frac{na+x_d-x_0}{\xi})-\tanh(\frac{na+x_d+x_0}{\xi})]\}. \quad (3)$$

By minimizing the energy of the lowest even-parity excited state, we find a 4-soliton bound state configuration as reported in Ref. [10]. By comparing the energy of this bound state with that of four separated solitons, the binding energy is found to be about 0.05 eV. Therefore it is quite likely that relaxation of the lowest even-parity excited state will lead to the production of four solitons.

3. Strong Coupling Results

To provide an independent check of the results in the previous section, we have considered the following spin-lattice Hamiltonian,

$$H=\sum_{<i,j>}\{J[1+\beta\delta r_{ij}]\vec{S}_i\cdot\vec{S}_j+\frac{K}{2}(\delta r_{ij})^2\}, \quad (4)$$

which can be regarded as the strong coupling limit of the SSHH Hamiltonian (1). The average exchanged coupling is of the order t^2/U. Following Hashimoto [11], the spin-lattice Hamiltonian is first transformed into a fermion-phonon model by a Jordan-Wigner transformation and the resultant model is solved by perturbation theory. Takimoto and Sasai [12] have employed this approach in comparing the creation energy for 4-soliton versus 2-soliton configurations in the excited singlet A_g manifold . They found that indeed the 4-soliton configuration has energy lower than that of the 2-soliton configuration. We have carried out a more extensive minimization of the excited state energy within the family of configurations (3). We find that a 4-soliton bound state has the lowest excited energy. The dissociation energy of the 4-soliton bound state is comparable to that reported in the previous section. We also find that the potential curve for the triplet state is a repulsive one, very similar to Fig.4. For details please refer to Ref. [10]

4. Density Matrix Renormalization Group Calculations

A recent breakthrough in the numerical solution of one-dimensional model Hamiltonians is the density matrix renormalization group (DMRG) method [13] devised by White. Previously, accurate numerical calculations of interacting models were confined to very small lattices with large finite size effects. The DMRG method allows accurate determination of low-lying excited states for large lattice size. We have employed this method to treat the SSHH Hamiltonian (1) essentially exactly to obtain the relaxed lattice configurations of the excited B_u and A_g states.

In such many-body calculations it is very useful to invoke another symmetry besides the spatial parity. It is the Pariser charge conjugation symmetry [14], the associated quantum number is called charge parity. The ground state has a negative charge parity and the electric dipole operator flips the charge parity. Therefore only a positive charge parity ${B_u}^+$ state can be directly photoexcited. The relaxed lattice configuration of the lowest electronic state of this type calculated with DMRG is shown as curve (a) in Figure 5. In contrast curve (b) shows the relaxed configuration of the lowest excited negative charge parity A_g^- state.

It is clear that curve (a) corresponds to two charged solitons, while curve (b) resembles a bound state of four neutral solitons in qualitative agreement with our results obtained by other methods. To calculate the dissociation energy of

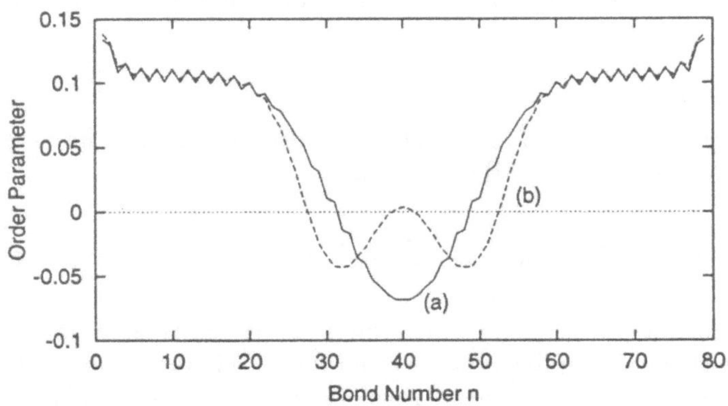

FIG. 5. Relaxed lattice configuration of the lowest B_u^+ (curve (a)) and A_g^- (curve (b)) excited states calculated with DMRG.

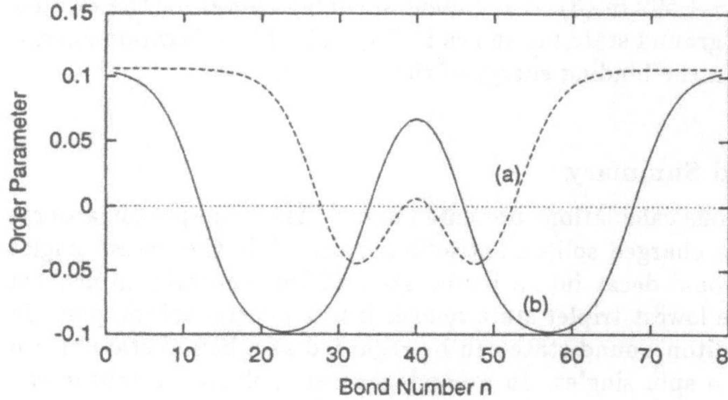

FIG. 6. Curve (a) is obtained from curve (b) of Figure 5 by fitting with formula (3); curve (b) depicts four separated solitons.

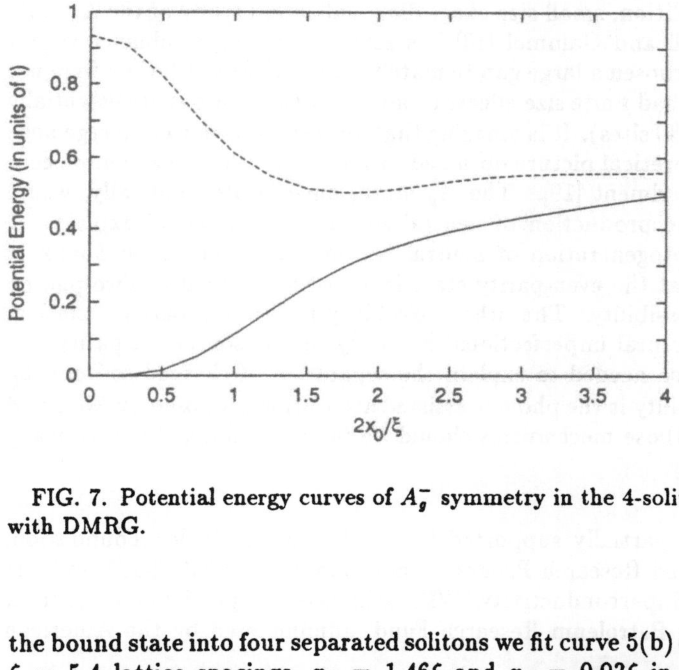

FIG. 7. Potential energy curves of A_g^- symmetry in the 4-soliton sector calculated with DMRG.

the bound state into four separated solitons we fit curve 5(b) closely by choosing $\xi = 5.4$ lattice spacings, $x_d = 1.46\xi$ and $x_0 = 0.93\xi$ in equation (3). By changing the parameters x_d and x_0 to 3.15ξ and 2.0ξ respectively, we have four separated solitons as depicted by curve (b) in Figure 6. The energy differences between configuration (b) and configuration (a) is found to be $0.057\,t$. We have chosen $U = 2.5t$ and $\lambda = 0.26$ for all DMRG calculations. The calculated optical gap for the dimerized chain is about t. To fit the experiment, we take t to be 2 eV. Therefore the 4-soliton binding energy is about 0.114 eV, twice as large as our previous estimate.

To better understand the relaxation of the $^1A_g^-$ excited state, we pick a direction in the configurational coordinate space by keeping a fixed ratio

$x_d/x_0 = 1.46/0.93 = 1.569$ in (3). The potential energy curves for the excited state as well as the ground state are shown in Figure 7. The relaxation energy is clearly larger than the binding energy of the 4-soliton.

5. Discussion and Summary

It is clear from various calculations presented above that a one-photon absorption would create a charged soliton-antisoliton pair, while the lowest singlet even parity state would decay into a bound state of four neutral solitons. On the other hand, the lowest triplet state relaxes into a neutral soliton pair. In a sense, the four soliton bound state can be regarded as a bound state of two triplets coupled to a spin singlet. In a nondegenerate polymer, a soliton can not be separated indefinitely from its antisoliton, but two triplet polarons can do so. Indeed, there is some numerical evidence [15] of that.

The relaxation of the A_g state into solitons has been extensively studied in the past by many groups including Mele and Hayden [15], Tavan and Schulten [16], Campbell [17] and others, but the results have been at odds with each other. In this paper we have obtained consistent results from three independent approaches. In addition, small size exact diagonalization study of the A_g state made by Campbell and Gammel [18] has revealed close resemblance to our result. They have chosen a large gap to match the small size, still they were not sure whether they had finite size effects or not. Our calculations are essentially bulk calculations (80 sites). It is amazing that all results seem to converge now.

Within the theoretical picture obtained so far, let us try to make more connections with experiment [19]. The A_g state, once excited optically, would lead quickly to the production of neutral solitons. That could explain the subpicosecond photogeneration of neutral solitons as observed by Shank et al. [6] provided that the even-parity state is accessible optically. Two-photon absorption is a possibility. The other possibility is through parity violation. Impurities or structural imperfections can easily be the sources of parity violation, which is also needed to explain the separation of charged solitons [8]. Yet another possibility is the phonon-assisted absorption proposed by Mele and Hayden [15]. All those mechanisms should be further examined theoretically and experimentally.

This work was partially supported by the Robert A. Welch Foundation, the Texas Advanced Research Program under grant no. 003652-183 and the Texas Center for Superconductivity. WPS acknowledges partial support from the Donors of the Petroleum Research Fund, administered by the American Chemical Society.

References

[1] A. J. Heeger, S. Kivelson, J. R. Schrieffer, and W. P. Su, Rev. Mod. Phys. **60**, 781(1988).
[2] W. P. Su, J. R. Schrieffer, and A. J. Heeger, Phys. Rev. B **22**, 2099 (1980).
[3] W. P. Su and J. R. Schrieffer, Proc. Natl. Acad. Sci. U. S. A. **77**, 5626(1980).
[4] R. Ball, W. P. Su, and J. R. Schrieffer, J. de Physique, Colloque C3 **44**, 429 (1983).
[5] L. Rothberg, T. M. Jedju, S. Etemad, and G. L. Baker, Phys. Rev. Lett. **57**, 3229(1986); L. Rothberg, T. M. Jedju, P. D. Townsend, S. Etemad, and G. L. Baker, Phys. Rev. Lett. **65**, 100(1990).
[6] C. V. Shank, Y. Ren, R. L. Fork, J. Orenstein, and G. L Baker, Phys. Rev. Lett. **49**, 1660(1982).
[7] X. Wei, B. C. Hess, Z. V. Vardeny, and F. Wudl, Phys. Rev. Lett. **68**, 666 (1992).
[8] S. Kivelson and W. K. Wu, Phys. Rev. B **34**, 5423(1986).
[9] R. V. Bensasson, E. J. Land, and T. G. Truscott, *Flash Photolysis and Pulse Radiolysis* (Pergamon, Oxford, 1983).
[10] W. P. Su, Phys. Rev. Lett. **74**, 1167(1995).
[11] K. Hashimoto, Chem. Phys. **80**, 253(1983).
[12] J. Takimoto and M. Sasai, Phys. Rev. B **39**, 8511(1989).
[13] S. R. White, Phys. Rev. Lett. **69**, 2863 (1992); Phys. Rev. B **48**, 10345 (1993).
[14] A. D. McLachlan, Mol. Phys. **2**, 271(1959).
[15] G. W. Hayden and E. J. Mele, Phys. Rev. B **34**, 5484(1986).
[16] P. Tavan and K. Schulten, Phys. Rev. B **36**, 4337(1987).
[17] D. K. Campbell, J. T. Gammel, H. Q. Lin, and E. Y. Loh, Synth. Met., **49-50**, 631(1992).
[18] J. T. Gammel and D. K. Campbell, Synth. Met. **55-57**, 4638(1993).
[19] Z. V. Vardeny, in *Relaxation in Polymers*, edited by T. Kobayashi (World Scientific Publishing Co., 1993).

Part V

Neutral-Ionic Transitions in Charge-Transfer Salts

Structural Properties, Dynamics and Pressure Effects at the Neutral-to-Ionic Transition

H. Cailleau, M. Le Cointe, and M.H. Lemée-Cailleau

Groupe Matière Condensée et Matériaux, URA au CNRS 804,
Université de Rennes 1, 35042 Rennes cedex, France

Abstract. Mixed-stack organic charge-transfer complexes which undergo a neutral-to-ionic transition are model systems to study strong cooperative effects between lattice-relaxed electronic excitations. An overview is given on structural and dynamical properties associated with this electronic-structural phase transformation which can be photo-induced. The analysis of the structural changes allows to underline the respective parts taken by the ionicity, which is a totally symmetric order parameter, and the dimerization which is the symmetry-breaking order parameter. The cooperative mechanism and the thermodynamics of the neutral-to-ionic transition are discussed on the basis of pressure effects, in particular with the observation of the tendency towards a tricritical point at high-pressure. This uncommon phase transition can be viewed as resulting from the condensation and the ordering of ionic dimers. However it needs to take into account quantum effects which are of great importance. Two types of characteristic excitation exist, on the one hand a polar vibrational soft mode of dimerization, on the other hand lattice-relaxed charge-transfer excitations which can be thermally- or photo-induced. Pretransitional dynamics is discussed in order to give a consistent view on the different results, recent and earlier.

1. Introduction

The neutral-to-ionic (N-I) transition is an electronic-structural transformation which can be not only photo-induced [1] but also temperature- [2] or pressure- induced [3]. Lattice-relaxed (LR) charge-transfer (CT) excitations are considered to play a central role in such a transformation [4-6]. Moreover it has been suggested that they are at the origin of many intriguing properties like anomalous dielectric response [7-8] or negative-resistance effects [9]. Consequently organic compounds which present a N-I transition appear to be model systems for studying strong cooperative effects between LR electronic excitations.

The N-I transition takes place in some mixed-stack CT complexes. Contrary to segregated-stack CT crystals giving rise to the well-known family of one-dimensional conductors, in mixed-stack CT crystals electron-donor (D) and electron-acceptor (A) molecules alternate along the same stack ...DADADADA... They are always non-metallic and at best small-gap semi-conductors. This uncommon phase transition is characterized by the simultaneous occurrence of an ionicity change, i.e. change of the degree of CT q, and a bond dimerization distorsion η along the stack in the I phase.

Phase N	$T > T_{N\text{-}I}$	$\ldots D^{\circ}A^{\circ}D^{\circ}A^{\circ}D^{\circ}A^{\circ}D^{\circ}A^{\circ}\ldots$
Phase I	$T < T_{N\text{-}I}$	$\ldots(D^{+}A^{-})(D^{+}A^{-})(D^{+}A^{-})(D^{+}A^{-})\ldots$

As in other low-dimensional systems this electron-induced lattice instability originates from a delicate balance between electron-transfer, electron-electron and electron-lattice interactions. The aim of this contribution is to give an overview on the structural and dynamical properties, in order to discuss the cooperative mechanism of the N-I transition and especially to underline respective parts taken by ionicity and dimerization. It is supported by recent studies [10-12] on the prototype compound tetrathiafulvalene-p-chloranil (TTF-CA).

Springer Series in Solid-State Sciences, Volume 124
Relaxations of Excited States
and Photo-Induced Structural Phase Transitions
Ed.: K. Nasu

2. Symmetry analysis and structural changes

It is well known from Landau theory of phase transitions that every density of probability describing the crystalline structure (electronic, position of nucleus,...) in the two phases can be expressed as

$$\rho = \rho_0 + \Delta\rho$$

where ρ_0 is totally symmetric (TS) with respect to the high-symmetry phase (here undimerized N phase) and $\Delta\rho$ describes the symmetry lowering arising in the low-symmetry phase (here dimerized I phase). This last term is non-zero only in this low-symmetry phase and is proportional to the symmetry-breaking (SB) order parameter. In general structural phase transitions are associated with a change of space group without drastic change of ρ_0. A typical example is given by a displacive ferroelectric phase transition where the SB order parameter is the relative static displacement of ionic sublattices. There are fewer examples of isostructural phase transition (metal-insulator Mott transition, spin transition,...) exhibiting a discontinuous change of ρ_0 but without change of space group (no SB). In this case ρ_0 plays the role of a TS order parameter, in the same sense as the density for the liquid-gas transition : it is non-zero in the two phases. The N-I transition is unusual because it involves on an equal footing the two types of order parameter, i.e. the ionicity q (TS) and the dimerization η (SB). The terms "ionicity" and "dimerization" are used as generic expressions which represent all electronic and structural changes (intra- and inter-molecular) corresponding to respectively the part preserving or not the symmetry. The first one is mainly associated with the change of electronic state and the second one with the dimerization distortion which implies the loss of inversion symmetry. Notice that, contrary to usual ferroelectrics where the ionicity does not change, the behaviour of the polarization P is here governed by these two types of order parameter : $P \sim q\eta$.

The two aspects of the N-I transition can be discussed from recent structural results on TTF-CA obtained by neutron diffraction (at LLB) and chlorine nuclear-quadrupole-resonance (NQR). Since the discovery of this transition experimental investigations have been concentrated on TTF-CA which is one of the rare cases where the phase transition occurs at atmospheric pressure. From electronic [13] and vibrational [14,15] optical spectroscopies it has been shown that at $T_{N\text{-}I} \cong 80$ K the degree of CT q increases sharply from 0.3 in the N phase to 0.7 in the I phase. The existence of a dimerization distortion in the I phase was first indicated by different spectroscopic measurements [14-16]. However, these did not allow to determine the complete nature of the SB which requires to know the change of space group. This is $P\,2_1/n$ in the N phase [17], illustrated in the diffraction pattern by two types of systematic absences, 0k0 : k = 2n + 1 due to the twofold screw axis and h0l : h + l = 2n + 1 due to the glide plane [Fig. 1]. The monoclinic unit-cell contains two TTF-CA units. Molecules are regularly stacked along **a** and are located on inversion centers. At the N-I transition there is no change of the multiplicity of the unit cell and the symmetry lowering is only characterized by the appearance of weakly intense 0k0 : k = 2n + 1 reflections [11,18]. The I space group is then Pn with two equivalent dimers related by the glide plane and the symmetry-breaking order parameter is of B_u symmetry [Fig. 1]. This implies a ferroelectric arrangement between the dimerized chains with a polarization lying in the (**a**, **c**) plane.

The complete structural determination of the I phase allowed to conclude that molecular displacements associated with the bond dimerization occurs mainly along the stacking axis **a**. The intensity of (0k0) reflections is proportional to the square of the SB order parameter η but it is exclusively sensitive to displacements along the **b** direction which are much smaller than along **a**. This explained why the symmetry lowering had been particularly difficult to establish. Instead of a regular D-A distance along the stack between molecular centers of d = 3.610 Å in the N phase (90 K), an alternation appears between an intradimer distance $d_{intra} = 3.504$ Å and an interdimer distance $d_{inter} = 3.685$ Å in the I phase (40 K). A rough estimation of the created dipole moment per unit cell, assuming a partial charge $\pm q = 0.7$ located on the molecular center positions, gives about 0.6 Debye. The resulting dipolar energy is weak in comparison with $k_B T_{N\text{-}I}$ ($W_d/k_B \cong 3$ K). This shows that long-range dipolar interactions are not sufficient to stabilize the ordering between dimers in the I phase and therefore the spin-Peierls like mechanism is probably dominant. This is a distinctive feature

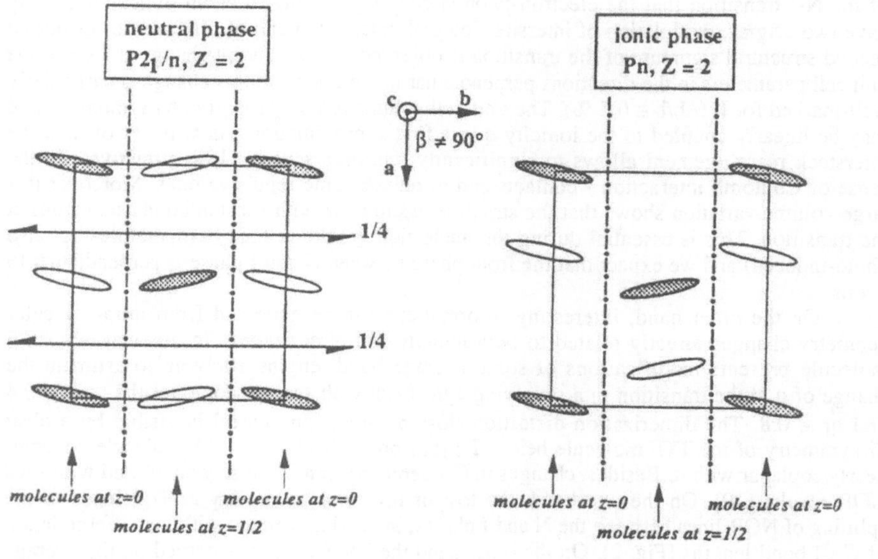

Figure 1. Schematic drawing of the TTF-CA structure in the N phase and its modification in the I phase.

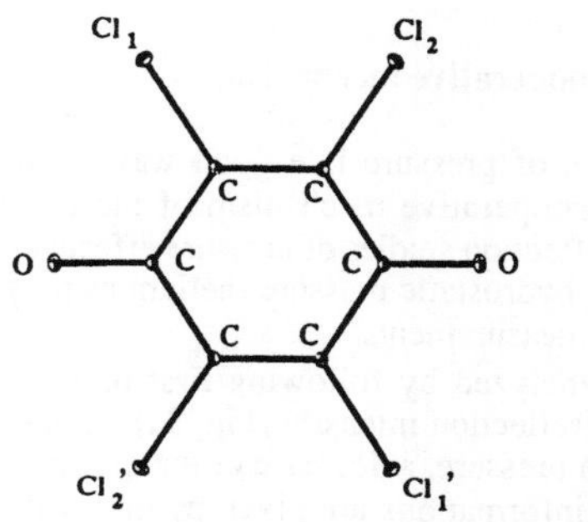

a)

b)

Bond lengths (Å)	T = 300 K	T = 90 K	T = 40 K
C-Cl_1	1.700(3)	1.708(5)	1.724(1)
C-Cl'_1			1.705(1)
C-Cl_2	1.697(3)	1.704(5)	1.711(1)
C-Cl'_2			1.714(1)

c)

	$<f_N - f_I>$ (kHz)	Δf (kHz)
high frequency (HF) line	220	120
low frequency (LF) line	245	13

Figure 2. a) Molecule of CA. b) C-Cl bond lengths at different temperatures ($T_{N\text{-}I} \equiv 80$ K) [11]. c) average of the ^{35}Cl NQR frequencies variation at the N-I transition $<f_N - f_I>$, and splitting of each line in the I phase (77 K) Δf [10].

of the N-I transition that the electron-phonon coupling leading to chain dimerization may have two origins, modulation of intersite Coulomb interaction and of CT integral. Besides a second structural signature of the transition is observed with an abrupt change at T_{N-I} of the unit cell parameters in the directions perpendicular to the stack and this change is particularly well-marked for **b** ($\Delta b/b \cong 0,5$ %). The unit-cell volume is a TS property (A_g symmetry) and may be linearly coupled to the ionicity q in a first approximation. As a matter of fact, the interstack rearrangement allows to significantly reinforce several close attractive - in the sense of Coulomb interaction - contacts and to reduce some repulsive ones. Moreover this large volume variation shows that the structure has to cope with substantial elastic strains at the transition. This is essential during the nucleation growth regime (temperature- but also photo-induced) and we expect that the front phase between N and I phase is perpendicular to **b** axis.

On the other hand, interesting informations can be extracted from intramolecular geometry changes directly related to both ionicity and dimerization. In this way, the TTF molecule presents modifications of some average bond lengths allowing to estimate the change of q at the transition in a satisfying agreement with spectroscopic results : $q_N \cong 0.4$ and $q_I \cong 0.8$. The dimerization distortion (loss of inversion center) is visible by a clear dissymmetry of the TTF molecule below T_{N-I}, as one half closer to CA molecule becomes nearly coplanar with it. Besides, changes in CA geometry can be closely connected with ^{35}Cl NQR results [10]. On the one hand, the lost of inversion symmetry at T_{N-I} leads to the splitting of NQR lines between the N and I phases, in good agreement with the differentiation of C-Cl bond lengths [Fig. 2]. On the other hand the TS change is observed on the average bond lengths evolution as well as on the frequency shift between each N line and its corresponding I doublet in average. This well illustrates that the two order parameters are on an equal footing at the level of the Cl sites.

3. Pressure effects and cooperative mechanism

The study of the influence of pressure is a good way to improve our knowledge of the thermodynamics and the cooperative mechanism of the N-I transition. An opportunity is given by recent neutron diffraction studies of pressure effects using a large high-quality single crystal and well-controlled hydrostatic pressure (helium as pressure transmitting medium), in parallel with heat capacity measurements.

These have been analyzed by following first of all the behaviour of the SB order parameter η via the (030) reflection intensity [Fig. 3a]. In addition to the increase of the N-I transition temperature with pressure, a decrease of the discontinuity at the transition is clearly observed. Other essential informations are given by the evolution under pressure of the TS order parameter via the measurement of the **b** cell parameter which provides the other most sensible structural signature of the transition [Fig. 3b]. The decrease of the discontinuity with pressure is also clearly observed. Besides, an enhancement of pretransitional phenomena is visible both in the N phase and in the I one with the rounding of this **b** evolution at the transition approach.

The (P,T) phase diagram is presented on Figure 4. By using the helium pressure cell and neutron diffraction it has been made up to 550 MPa. An additional point at room temperature giving a pressure transition of 1100 MPa have been obtained by using synchrotron radiation facilities at LURE. At high pressure the transition line rejoins the one provided by infrared (IR) [19] and conductivity [20] studies but it notably differs in the low pressure region. Such a discrepancy can be explained by the difference between the experimental methods used to pressurize the sample. Indeed contrary to helium cell, clamp and diamond-anvil cells used respectively in conductivity and IR measurements are environments known for presenting difficulties for the control of absolute pressure around the atmospheric conditions. Therefore we incline to confidence in the present results, confirmed by Raman scattering measurements performed at Rennes by using the same type of helium pressure cell [21]. Moreover as the transition is first order the observed slope dP/dT can be tested at atmospheric pressure via the Clausius-Clapeyron relation. The change of unit-cell volume at T_{N-I} can be estimated from neutron measurements to about 5-6 $Å^3$ and the entropy

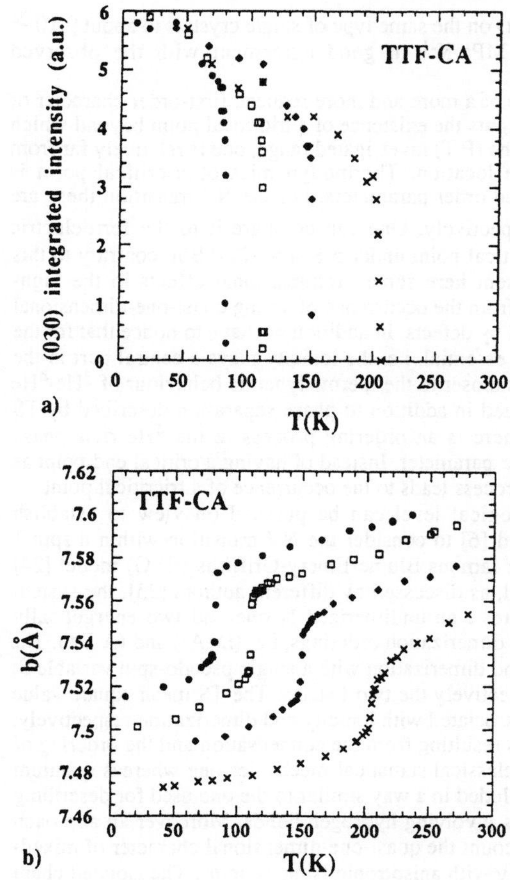

Figure 3. Evolution versus temperature at different pressures of a) the normalized intensity of (030) reflection. b) the **b** cell parameter. ● 0.1 MPa □ 100 MPa ◆ 255 MPa × 500 MPa.

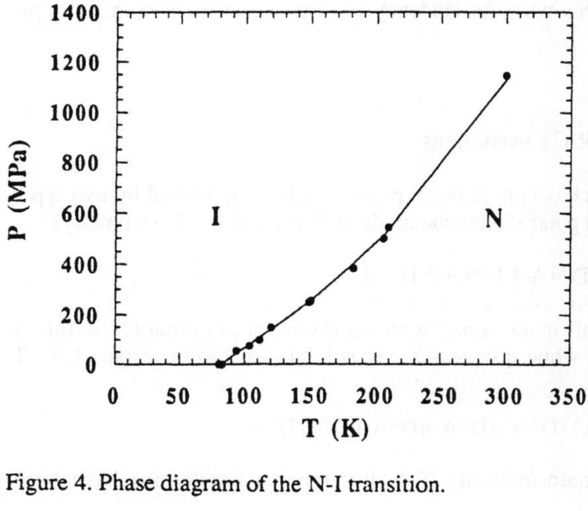

Figure 4. Phase diagram of the N-I transition.

variation from specific heat measurements on the same type of single crystals to about 9.10^{-24} $J.K^{-1}$. The deduced slope is 3.0-3.6 $MPa.K^{-1}$ in good agreement with the observed extrapolated slope 3.6 $MPa.K^{-1}$.

A major result is a clear evidence of a more and more reduced first-order character of the N-I transition under pressure. This augurs the existence of a tricritical point beyond which the transition would be second order. In the (P,T) investigated range, one is relatively far from it and it is difficult to give its precise location. Thermodynamics of tricritical point is governed by the existence of two coupled order parameters. For the N-I transition these are the SB and the TS ones, η and q respectively. One can compare it to the ferroelectric transition in KDP which exhibits a tricritical point under pressure [22]. But, contrary to this classical case, the cell parameters present here some pretransitional effects in the high-symmetry (N) phase. This can originate from the occurrence of strong quasi-one-dimensional fluctuations which can be more enhanced by defects. In addition we have to notice that for the N-I transition the TS order parameter is essential, i.e. the ionicity plays a central part in the cooperative mechanism. This situation is closer to the thermodynamic behaviour of ^3He-^4He mixtures [23] than to the KDP one. Indeed in addition to phase separation described by TS order parameter (^3He concentration) there is an ordering process in the ^4He-rich phase (superfluidity) associated with a SB order parameter. Instead of having a critical end point as for usual simple mixtures, the ordering process leads to the occurrence of a tricritical point.

This analogy on a phenomenological level can be pursued on view to establish microscopic models. Thus it can be tried [6] to consider the N-I transition within a spin-1 lattice gas model in a similar way to the famous Blume-Emery-Griffiths (BEG) model [24] introduced for ^3He-^4He mixtures. Indeed, as discussed by different authors [25], the system can be described within three local states : an undimerized N one and two energetically equivalent I ones with regards to the two dimerization orderings, i.e. (D^+A^-) and (A^-D^+). The spin-1 model describes both ionization and dimerization with a single pseudo-spin variable S_i : $S_i = 0$ for the N state, $S_i = \pm 1$ for respectively the two I states. The TS mean-square value $<S_i^2>$ and the SB mean value $<S_i>$ are associated with ionicity and dimerization respectively. Thus, the N-I transition can be viewed as resulting from the condensation and the ordering of I dimers. However the BEG model is a classical statistical-mechanics one whereas quantum effects are essential [6]. These can be included in a way similar to the one used for describing proton tunnelling in structural transitions involving hydrogen bonds. Moreover an approach of great promise [6,25] is to take into account the quasi-one-dimensional character of mixed-stack CT complexes by doing the analogy with anisotropic Ising systems. The isolated chain does not present long-range order at finite temperature but it exhibits a statistical disorder between large undimerized N and dimerized I parts. The interchain coupling enhances the ordering process and leads to a three-dimensional order between I dimers below T_{N-I}. Furthermore, pressure is expected to reinforce not only the Coulomb interactions which favours the I phase, but also the intrastack CT integral [26]. This last feature is responsible for an increase of the partial CT in the N phase and induces a smaller discontinuity of ionicity at the N-I transition [27]. This agrees with the tendency towards a tricritical point at high-pressure.

4. Dynamical properties and LR CT excitations

The pretransitional dynamics which occurs in the N phase should be governed by two types of characteristic excitation. First, a polar vibrational mode of dimerization (B_u symmetry)

$$\ldots D^{+q} A^{-q}\ D^{+q} A^{-q}\ D^{+q} A^{-q} \ldots$$
$$\rightarrow\ \leftarrow\ \ \rightarrow\ \leftarrow\ \ \rightarrow\ \leftarrow$$

may soften and condense. This soft mode comes with the evolution of partial CT in this N phase, i.e. its frequency decreases while q increases. Second, quasi-one-dimensional LR CT excitations

$$\ldots D^\circ A^\circ\ D^\circ A^\circ\ (D^+A^-)(D^+A^-)(D^+A^-)\ D^\circ A^\circ D^\circ A^\circ \ldots$$

can be thermally excited (or photo-induced). This last type of excitations have been considered by Nagaosa in terms of NI domain walls [4] or by Toyozawa in terms of self-trapped CT excitons which can self-multiply [5]. The two types of characteristic excitations may coexist. However in the hypothetical case of a continuous N-I transition at zero temperature (pure quantum regime), the softening of the dimerization mode would be alone responsible for the structural instability (displacive regime). On the other hand the LR CT excitations would become more and more efficient as the transition temperature increases (order-disorder regime).

Different recent and earlier results on dynamics of the N phase can be discussed with regards to the above considerations. Thus, for the pressure-induced transition at room temperature IR measurements [28] firstly show a regular increase of the partial CT with pressure (mainly quantum effects). Then, in approaching the transition pressure a gradual appearance of I species (up to a few ten percent) is observed. This can be attributed to the enhancement of thermally excited fluctuating dimerized I clusters. At atmospheric pressure there is no such observation. However, spin-lattice relaxation rates T_1^{-1} in NQR measurements [10] reveal the existence of pretransitional fluctuations with a predominant one-dimensional character. In the same way the high-frequency dielectric response [8] exhibits a divergent increase almost independent of frequency on approaching $T_{N\text{-}I}$. Moreover a small softening (30 to 20 cm^{-1}) of low-frequency polar lattice-mode has been recently observed by far IR spectroscopy [29]. This might have been at the origin of the above features in NQR and dielectric measurements. However, on the one hand the one-dimensional character in T_1^{-1} behaviour and, on the other hand, the very large value of dielectric constant at $T_{N\text{-}I}$ (about 500) bring out to involve the occurrence of CT excitations. We have to notice that the presence of a I dimer in a N matrix creates a very large dipole moment. In addition a small precursor effect has been observed in the behaviour of NQR frequencies [10]. This may be due to an increase of the average CT with the presence of a few percent of I dimers since for the high-frequency line the frequency change is about 5 kHz while the average frequency difference is 220 kHz between the N and I phases. These different features are consistent if the time-scale of LR CT excitations is intermediary between the IR range (quasi-static response) and the one of NQR and of dielectric response (dynamical average). Actually photo-induced effects measured at 90 K indicate that the lifetime of photo-injected domains is less than a few ns [30]. It is the typical lifetime for large-size fluctuating clusters (about 100 units). For its part recent Brillouin scattering measurements show typical relaxation in the 20 GHz range [31]. This last technique is sensitive to the whole distribution of thermally-induced CT excitations and in particular small-size clusters which live shorter. Finally, the observation of displacive to order-disorder crossover under pressure in a recent Raman scattering experiment [21] agrees with the introductory discussion at the beginning of this part.

Some additional points concerning dynamical properties can also be discussed. First, the large entropy variation at the transition, which is close to kln2 per DA pair, allows to stabilize the N phase. It is not due to statistical disorder resulting from self-trapped CT excitations owing to their weak number but rather to vibrational origin. Indeed the lattice-phonon frequencies harden in the I phase not only for the soft mode, but systematically also for all modes [29], in connection with an enhancement of the rigidity of the system due to closer interstack contacts (volume contraction). Second, the temporal annihilation of photo-generated N pairs in the I phase is much longer (ms range) [1] than the related phenomenon in the N phase. We have to remark that the I phase is ordered. The SB ordering leads to a domain structure and disorder remains only inside the domain walls. It is probable that photo-induced N species are concentrated inside these domain walls. Thus, the nature of the relaxation process in the I phase can be very different from the one in the N phase. Finally, an anomalous dielectric response with a very slow dynamics has been observed and explained in terms of NI domain walls [8]. We suggest that this phenomenon rather originates from defects or impurities (random-field effects) as in some ferroelectric compounds [32].

5. Conclusion

Some recent and earlier results on structural and dynamical properties at the N-I transition have been overviewed in order to bring out the mechanism of the transition. Thus this have given the opportunity to discuss respective parts taken by the ionicity and the dimerization, as well as those taken by the quantum effects and the thermally- or photo-induced LR CT excitations. Further experimental studies to improve pressure effects and dynamical properties are necessary. In the same way further theoretical studies will be essential both in the ground state [4,25,33], including ab-initio calculations [34], and at finite temperature with statistical-mechanics models [6,25]. Many fascinating aspects remain to be understood as the role of electric field. Finally we want to underline that photo-induced experiments give an original way to probe pretransitional cluster dynamics.

Acknowledgments. The reader will find in references the name of many colleagues with whom we have cooperated during experiments. We would like to warmly acknowledge them. We also wish to thank Tadeusz Luty for long and passionate stimulating discussions (Polish-French joint project n° 6721). Moreover we are grateful to C. Katan, C. Koenig and J. Lajzerowicz for turning our attention to some aspects. This project is for part supported by the Region Bretagne.

References

1. S. Koshihara, Y. Tokura, T. Mitani, G. Saito and T. Koda, Phys. Rev., **B42**, 6853 (1990)
2. J.B. Torrance, A. Girlando, J.J. Mayerle, J.I. Crowley, V.Y. Lee, P. Batail and S.L. La Placa, Phys. Rev. Lett., **47**, 1747 (1981)
3. J.B. Torrance, J.E. Vazquez, J.J. Mayerle and V.Y. Lee, Phys. Rev. Lett., **46**, 253 (1981)
4. N. Nagaosa, J. Phys. Soc. Jpn., **55**, 2754 (1986)
5. Y. Toyozawa, Solid State Commun., **84**, 255 (1992)
   Y. Toyozawa, Acta Phys. Polon., **A87**, 47 (1995)
6. T. Luty, this issue
7. Y. Tokura, S. Koshihara, Y. Ywasa, H. Okamoto, T. Komatsu, T. Koda, N. Iwasawa and G. Saito, Phys. Rev. Lett., **63**, 2405 (1989)
8. H. Okamoto, T. Mitani, Y. Tokura, S. Koshihara, T. Komatsu, Y. Iwasa, T. Koda and G. Saito, Phys. Rev., **B43**, 8224 (1991)
9. Y. Tokura, H. Okamoto, T. Koda, T. Mitani and G. Saito, Phys. Rev., **B38**, 2215 (1988)
10. J. Gallier, B. Toudic, Y. Délugeard, H. Cailleau, M. Gourdji, A. Péneau and L. Guibé, Phys. Rev., **B47**, 11688 (1993)
    M. Le Cointe, J. Gallier, H. Cailleau, M. Gourdji, A. Péneau, L. Guibé, Solid State Commun., **94**, 455 (1995)
11. M. Le Cointe, M.H. Lemée-Cailleau, H. Cailleau, B. Toudic, L. Toupet, G. Heger, F. Moussa, P. Schweiss, K.H. Kraft and N. Karl, Phys. Rev., **B51**, 3374 (1995)
12. M.H. Lemée-Cailleau, M. Le Cointe, H. Cailleau, B. Toudic, T. Luty, F. Moussa, C. Ayache, K.H. Kraft and N. Karl, in preparation
13. C.S. Jacobsen and J.B. Torrance, J. Chem. Phys. **78**, 112 (1983)
14. A. Girlando, F. Marzola, C. Pecile and J.B. Torrance, J. Chem. Phys., **79**, 1075 (1983)
15. Y. Tokura, Y. Kaneko, H. Okamoto, S. Tanamura, T. Koda, T. Mitani and G. Saito, Mol. Cryst. Liq. Cryst., **125**, 199 (1985)
16. T. Mitani, G. Saito, Y. Tokura and T. Koda, Phys. Rev. Lett., **53**, 842 (1984)
17. J.J. Mayerle, J.B. Torrance and J.I. Crowley, Acta Crystallogr., **B35**, 2988 (1979)

18. An independent X-ray diffraction study has given similar results : Y. Nogami, private communication
19. K. Takaoka, Y. Kaneko, H. Okamoto, Y. Tokura, T. Koda, T. Mitani and G. Saito, Phys. Rev., **B36**, 3884 (1987)
20. T. Mitani, Y. Kaneko, S. Tanuma, Y. Tokura, T. Koda and G. Saito, Phys. Rev., **B35**, 427 (1987)
21. A. Moréac, A. Girard and Y. Délugeard, J. Phys. : Condens. Matter, **8**, 3569 (1996)
22. P. Bastie, M. Vallade, C. Vettier, C.M.E. Zeyen and H. Maister, J. Physique, **42**, 445 (1981)
23. R.B. Griffiths, Phys. Rev. Lett., **24**, 715 (1970)
24 M. Blume, V.J. Emery and R.B. Griffiths, Phys. Rev., **A4**, 1071 (1971)
J. Lajzerowicz and J. Sivardière, Phys. Rev., **A11**, 2079 (1975)
25. N. Nagaosa, J. Phys. Soc. Jpn, **55**, 3488 (1986)
T. Luty, Acta Phys. Polon., **A87**, 1009 (1995)
26. A. Girlando, A. Painelli, C. Pecile, G. Calestani, C. Rizzoli and R.M. Metzger, J. Chem. Phys., **98**, 7692 (1993)
27. T. Luty and B. Kuchta, Phys. Rev., **B35**, 8542 (1987)
T. Iizuka-Sakano and Y. Toyozawa, J. Phys. Soc. Jpn, **65**, 671 (1996)
28. H. Okamoto, T. Koda, Y. Tokura, T. Mitani and G. Saito, Phys. Rev., **B39**, 10693 (1989)
29. A. Moréac, A. Girard, Y. Délugeard and Y. Marqueton, J. Phys. : Condens. Matter, **8**, 3553 (1996)
30. S. Koshihara, Y. Tokura, N. Sarukura, Y. Segawa, T. Koda and K. Takeda, Synthetic Metals, **70**, 1225 (1995)
31. D. Rouède, Y. Legrand and C. Écolivet, in preparation
32. H. Schremmer, W. Kleemann and D. Rytz, Phys. Rev. Lett., **62**, 1896 (1989)
33. A. Painelli and A. Girlando, Phys. Rev., **B37**, 5748 (1988)
34. C. Katan, P.E. Blöchl, P. Margl and C. Koenig, Phys. Rev., **B53**, 12112 (1996).

Neutral-to-Ionic Transformation as Condensation and Solidification of Charge-Transfer Excitations

T. Luty

Institute of Physical and Theoretical Chemistry,
Technical University of Wroclaw, Wroclaw, Poland

Abstract. The neutral-to-ionic transformation in organic charge-transfer crystals with mixed-stack architecture is analyzed as an electronic-structural phase transition. It is shown that, in general, the system can be described within the spin-1 lattice gas model, including quantum mixing. The thermodynamic phase diagram is calculated in two steps (approximations). A single mixed-stack, treated within the continuous approximation, is seen as a mixture of quasi-neutral, undimerized and quasi-ionic, dimerized domains. The quantum mixing determines fractional ionicities of the molecules while the observable, time and space averaged, ionicity results from a disorder of the compounds induced by thermal mixing. The competition between thermal mixing and the inter-stack interactions leads to an appearance of a tricritical point and a crossover from a displacive to an order-disorder mechanism of the structural trransition. It is concluded that the pressure-temperature phase diagram results from an interplay of quantum and thermal mixing. An evolution of the neutral-to-ionic transition towards the tricritical point might be considered as a classical to quantum crossover.

1. Introduction.

It seems appropriate to begin with the quotation, "A new age of alchemy based on the physical science has indeed started", the last sentence in Professor's Toyozawa review article on electron induced lattice instabilities [1]. In the discussion of the phenomena, a competition between electron transfer, electron-electron interaction and electron-phonon coupling has been conveniently represented on the "T-U-S triangle" as the **ground state phase diagram** [2]. In an analogy to a triangle suggested by Chakraverty [3] it can be called an **alchemical triangle**.

Most promising is the alchemy among organic, charge-transfer (CT) crystals, which undergo electronic-structural, temperature-[4] or pressure-induced [5] neutral-to-ionic transitions. For the tetrathiafulvalene (TTF)-chloranil (CA) compound the electronic transformation is manifested by an abrupt change in the degree of CT, as seen from intra- and intermolecular electronic [6] and vibrational [7,8] spectra. The structural transition is monitored by a symmetry breaking dimerization observed in the ionic phase [9].

It is an experimental fact that the ionic-to-neutral transition in the compound can be photo-induced [10]. This strongly suggests that the lattice relaxed (LR) CT excited state be treated on an equal footing with the ground state, as advocated by Toyozawa [11,12]. This view-point has proved to be very promising and useful, and the photo-induced transition has been described in terms of the condensation of LR-CT excitations [11,12]. On the other hand, following the spectroscopic studies of CT excitons in organic crystals [13] it has been postulated that the dimerization of the compounds is due to a modulation of the intersite Coulomb interaction and the ground state phase diagram has been obtained [14]. This has been proved by more detailed considerations and numerical calculations for TTF-CA, where an exact account of the long-range Coulomb interaction has been included [15].

The aim of this contribution is to demonstrate the usefulness of a global view point, in which one considers a crossover of the ground and excited CT states for the **static**,

Springer Series in Solid-State Sciences, Volume 124
Relaxations of Excited States
and Photo-Induced Structural Phase Transitions
Ed.: K. Nasu

temperature- and pressure-induced neutral-to-ionic transformation. The electronic-structural transition will be described as the **condensation** (electronic) and **crystallization** (structural) of LR-CT excitons and the system mapped onto a spin-1 lattice-gas model. The **thermodynamic phase diagram**, which reflects the main features of experiments and, in particular, the existance of a tricritical point as recently observed [16], will be constructed.

2. System and spin-1 lattice-gas model.

We consider a lattice of donor (D) and acceptor (A) molecules with stoichiometry 1:1 and organized in a typical, for CT organic compounds, sandwich-like, mixed-stack (MS) architecture. Along the stack the nearest D and A molecules are *a* distance apart. First, we treat the electronic transformation of a pair of nearest molecules, ... D^0A^0 ... ⇒ ... D^+A^- ..., which represents a CT exciton. Within the lattice-gas model,

$$H = \delta \sum_i \sigma_i + \sum_{i,j}{}' V_{ij} \sigma_i \sigma_j \ , \qquad (1)$$

where σ_i =0 or 1, denoting a neutral or ionized pair at the i-th site of the molecule D, respectively. The chemical potential of the ionized pair, δ=I-A-V, is determined by the ionization energy (I) of the D molecule, the electron affinity (A) of the A molecule and the Coulomb interaction, V ($V = e^2/a$, in the point-charge approximation). The coupling term, V_{ij}, is mostly due to long-range Coulomb interaction.

The structural change (dimerization) is described in a simplest form as a change in the intermolecular distance, $a \Rightarrow a\cdot(1\pm q)$. This modulates the chemical potential, $\delta = \delta_0 - \tau_i h$, where $h = (\partial V/\partial q)q$, and the coupling

$$V_{ij} = C_{ij} + L_{ij}(\tau_i + \tau_j) + J_{ij}\tau_i\tau_j , \qquad (2)$$

is expressed in terms of the spin variable, τ_i = +1, -1, denoting +q and -q structural changes, respectively. The parameter C_{ij} stands for the Coulomb interaction between ionized, undeformed pairs located at the i-th and j-th sites. The symmetry breaking term, $L_{ij} < 0$, prefers the configuration (..., 1, 1, ...), like (... D^+A^- ... D^+A^- ...), over (..., -1, -1, ...), like (... D^+ ... A^- D^+ ... A^- ...). The last term represents the total energy change due to the deformation, q, in the ionic phase, so that,

$$N^{-1}\sum_j J_{ij} = \frac{1}{2}\cdot M_r \omega_N^2 (1-\eta) a^2 q^2 . \qquad (3)$$

M_r is the reduced mass of the DA dimer, ω_N is the frequency of the dimerization mode in the neutral undimerized phase, and η is the electrostatic instability index, as introduced in ref. [15]. The important fact is that, as shown for TTF-CA [15], the lattice deformation energy is negative (η > 1) and stabilizes the deformed ionic pairs in the ground state.

On introducing the modulated coupling into the lattice-gas Hamiltonian and defining the spin-1 lattice-gas variable, $S_i^z = \sigma_i \tau_i$, one arrives at the general Hamiltonian,

$$H = -\Omega \sum_i S_i^x + \delta_0 \sum_i (S_i^z)^2 - h \sum_i S_i^z - \sum_{i,j}{}' C_{ij} (S_i^z)^2 (S_j^z)^2$$
$$- \sum_{i,j}{}' L_{ij} S_i^z S_j^z (S_i^z + S_j^z) - \sum_{i,j}{}' J_{ij} S_i^z S_j^z . \qquad (4)$$

The Hamiltonian contains the quantum mixing term, $\Omega = \langle DA|H|D^+A^- \rangle$, and we use convention of signs to keep the coupling constants positive. If the lattice gas Hamiltonian describes electronic transformation for a chain of molecules, ... $D^0A^0D^0A^0\,D^0A^0$ ... $\Rightarrow$ ... $D^+A^-\,D^+A^-\,D^+A^-$ ..., which represents a CT string, then the symmetry breaking fields, h and L_{ij}, are negligable and become zero for an infinite strings. The classical version of the Hamiltonian is suitable to describe photo-induced phase transitions in organic CT compounds.

The properties of the classical model are well known [17]. They have been studied by many authors, most extensively by Sivardiere and Lajzerowicz in a series of papers [18]. The appropriate order parameters are represented by $q = \langle S^z \rangle$ and $\rho = \langle (S^z)^2 \rangle$, which measure the structural dimerization and the molecular ionicity, respectively. Physical situations represented by the Hamiltonian can be interpreted as follows. At constant pressure, the model represents ternary mixture of (D^+A^-), (D^+ ... A^-) and (D^0 ... A^0) compounds, and at varied pressure it describes binary mixture of the ionic compounds. The ordering in ρ corresponds to separation of the neutral compounds from the ionic ones and this is a **condensation** of LR-CT excitons. The symmetry-breaking ordering in q corresponds to separation between the ionic compounds and it might be seen as **solidification** of the excitations. In case of photo-induced phase transition one might expect an electronic transformation (ordering in ρ) to be separated from structural phase transition (an ordering in q). The quantum mixing, which varies with pressure and determines the system's ground state, defines the compounds by their fractional ionicity.

The Hamiltonian describes a highly anisotropic system where intermolecular coupling within a MS is much stronger than between the stacks. Therefore it can be partitioned into a single MS part and the couplings between the stacks,

$$H = \sum_i H_i - \sum_{i,j}{}' C_{ij}\left(S_i^z\right)^2\left(S_j^z\right)^2 - \sum_{i,j}{}' J_{ij} S_i^z S_j^z, \tag{5}$$

where i, j indices denote stacks. A mixture of quasi-neutral and quasi-ionic compounds within the MS of organic, D and A, molecules can be imagined as **domains** separated by LR-domain walls along the one-dimensional chains, as proposed by Nagaosa [19,20]. Within such a picture it will be helpful to discuss the thermodynamics of the system in two steps. First, for the one-dimensional MS and in the next step the couplings between the stacks are included within the mean-field approximation.

3. Single mixed-stack.

Let us consider a single MS along the **a** axis (x coordinate), with highly correlated electron-transfer between A and D molecules. As suggested by Nagaosa [20], the stack is well characterized by the continuous polarization, $P(x) \sim q(x)\,\rho(x)$, the product of the dimerization displacement wave, q(x), and the charge (ionicity) density wave, $\rho(x)$. This indicates that CT crystals with MS architecture might be ferroelectric materials of an unusual type, where both the molecular ionicity and the displacement fluctuate. The polarization as a variable may be seen as continuous version of the discrete, spin-1 variable, defined by the product, $S_i = \tau_i \sigma_i$, for every dimer. The representation of molecular ionicity as a continuous distribution is well suitable for organic compounds where net charge is delocalized. It extends over the HOMO for ionized D and LUMO for ionized A molecules ($\rho(x)$ can be replaced by the so called Fukui functions). We have also the inverse relations, $q(x) \sim P(x)$ and $\rho(x) \sim P^2(x)$. It has been postulated [20], shown within the mean-field approximation [14] and finally proved by numerical calculations [15] that the ground state energy for MS with small quantum mixing, can be well described by the three-well potential,

$$V[P(x)] = a \cdot P^2(x) + b \cdot P^4(x) + c \cdot P^6(x), \tag{6}$$

with a, c > 0 and b < 0. The central well represents a (quasi-) neutral (N) stack and the side wells located at $\pm P_0$ correspond to (quasi-) ionic (I) stacks with opposite polarization (dimerization). Nagaosa [20] has shown that the ground state is characterized by lattice-relaxed domain walls (LR-DW) with energy E_{DW} (and width ζ_{DW}) of the order of room temperature, suggesting an important inhomogenity along the stack. This coincides with a picture of a MS as a mixture of N and I compounds in confined domains and moveable LR-DW allows for **thermal mixing** of the compounds.

Thermodynamical properties of a single MS have been analyzed [20] in terms of the transfer integral method which for a continuous one-dimensional model can be reduced to the Schrödinger equation [21,22]. At low temperatures one can find approximate solutions by assuming that the MS is isolated in one of the minima, located at $\varphi_{min} = P/P_0 = 0, +1, -1$. We may then approximate the potential (5) by the following harmonic forms, $V(\varphi) \approx V(\pm 1) + 1/2\, V''(\pm 1)(\varphi \pm 1)^2$ for the I wells, and $V(\varphi) \approx V(0) + 1/2\, V''(0)\, \varphi^2$ for the central well representing a N stack. The assumption that the MS is isolated in one of the wells is equivalent to separating the phase space into three phase spaces, which breaks ergodicity [23]. The situation can be realized either in low-temperatures or when an observation time is shorter than the relaxation time. For the separated phase spaces the free energies (chemical potentials) corresponding to the harmonic oscillators are calculated as $f(N) = -kT \ln(2\pi kT/\omega_N)$ and $f(I) = \delta - kT\ln(2\pi kT/\omega_I)$, where δ is the energy difference between N and I wells ($\delta < 0$ for TTF-CA compound). The free energy difference,

$$\Delta = f(I) - f(N) = \delta - kT\ln\left(\frac{2\omega_N}{\omega_I}\right), \tag{7}$$

is the chemical potential of the mesoscopic LR-CT **strings**, the domains of ionized and dimerized molecules, within a single MS. It reduces to $\Delta \approx \delta + (1/4)kT$, when only zero-point energies of the oscillators are taken into account as approximated by Nagaosa [20]. To allow for thermal mixing of the compounds along a single MS, the tunneling effect is taken into account by the matrix element, $t \approx \beta^{-1} e^{-\beta}$, where $\beta = E_{DW}/kT$.

The effective Hamiltonian for an isolated MS can be written in a spin-1 representation, $S^z = \varphi_{min} = 0, \pm 1$,

$$H_{eff,i} = -t \cdot S_i^x + \Delta\left(S_i^z\right)^2, \tag{8}$$

and its lowest eigen-value, $E_0 = 1/2[\Delta - (\Delta^2 + 8t^2)^{1/2}]$ is just the free energy of the stack. The single, isolated, one-dimensional MS is characterized by the space averaged ionicity, $\rho_0 = \langle (S^z)^2 \rangle$, as the only order parameter and is given by

$$\rho_0(T) = \frac{\partial E_0}{\partial \Delta} = \frac{1}{2} \cdot \left[1 - \Delta\left(\Delta^2 + 8t^2\right)^{-1/2}\right]. \tag{9}$$

This ionicity of the MS is due to the disorder of N and I molecules (with fractional charges determined by quantum mixing !) along the stack and induced by tunneling (**thermal mixing**). A symmetry breaking ordering in polarization (dimerization) of the stack can be induced only by a field, and this requires coupling between the stacks.

4. Inter-stack interactions.

There are two kinds of the inter-stack interactions: the Coulomb interaction and polarization interactions. Due to the quasi-one-dimensional character of CT organic compounds, the interactions are expected to be weak compared with intra-stack couplings, and can be treated within the mean-field approximation [24]. The single-stack, mean-field, effective Hamiltonian is,

$$H_{MF,i} = -tS_i^x + \Delta_{eff}\left(S_i^z\right)^2 - 2JqS_i^z + C\rho^2 + Jq^2, \tag{10}$$

where the effective chemical potential difference, $\Delta_{eff} = \Delta - 2C\rho$, includes the Coulomb crystal field. $C = \sum_j C_{ij}$ is the coupling constant for Coulomb interaction summed over stacks. $J = \sum_j J_{ij}$ and represents the deformation energy which results from the structural change, and it is mostly the polarization interaction. Due to the Coulomb crystal field, the averaged ionicity changes from ρ_0 to ρ,

$$\rho(T) = \left\langle \left(S^z\right)^2 \right\rangle_{\Delta_{eff}}, \tag{11}$$

to be determined from the self-consistent equation. The condensation of LR-CT strings results from a competition between the chemical potential difference, Δ, and the crystal field due to the Coulomb interaction, C. The first order transition terminates (the critical point) at a sufficiently large tunneling, t, which corresponds to sufficiently high temperatures. This simplest estimation of the critical density, ρ_c, gives the value 1/2, as expected. Thus, a critical point on a condensation line corresponds to a situation when the space averaged ionicity of the molecules is 1/2 and the difference in the chemical potentials of N and I compounds is compensated by the inter-stack Coulomb interaction. The **condensation phase transition** is governed by the free energy function,

$$f\{\rho(T)\} = \frac{1}{2}\left[\Delta_{eff} - \left(\Delta_{eff}^2 + 8t^2\right)^{1/2}\right]\rho + C\rho^2, \tag{12}$$

which determines the thermodynamical phase diagram for the purely electronic transformation.

The effect of inter-stack polarization interaction may be conveniently expressed as even order corrections to the free energy, $f\{\rho(T)\}$, and is calculated perturbatively. The unperturbed state is the MS with the effective chemical potential difference, Δ_{eff}. The result of the calculations gives the free energy contribution due to the dimerization of the MS,

$$f\{q,\rho(T)\} = J\left[1 - 2J\chi_{MS}\{\rho(T)\}\right]q^2 + B\{\rho(T)\}q^4 + D\{\rho(T)\}q^6 + \ldots \tag{13}$$

This function governs the structural phase transition, interpreted as a **solidification (crystallization) of charge transfer excitations**. $\chi_{MS}\{\rho(T)\}$ is the single MS polarizability,

$$\chi_{MS}\{\rho(T)\} = 2\beta\rho(1-\rho)\left[\Delta_{eff}^2 + 8t^2\right]^{-1/2}, \tag{14}$$

with a maximum value expected for $\rho_c = 1/2$. When the coupling between structural changes of the stacks is determined by polarization ineraction only, the function in square brackets of eq.(12) plays the role of a static **dielectric function**, $\varepsilon(0)$, for zero wave-vector. The static dielectric response function is thus given by,

$$X\{\rho(T)\} = \chi_{MS}\{\rho(T)\}[1 - 2J\chi_{MS}\{\rho(T)\}]^{-1}, \tag{15}$$

and can be used to understand the dielectric response measurements for a TTF-CA crystal [25] in the low-frequency limit. In particular, it may explain the significant difference between the effects of temperature (the tunneling effect, t) and pressure on the dielectric response.

5. Phase diagram.

The free energy of the system is given as a sum,

$$f_{total} = f\{\rho(T)\} + f\{q, \rho(T)\} \tag{16}$$

where the **condensation transition** is described by the first term, while the second part of the free energy describes the symmetry breaking **solidification transition**.

The **condensation** phase transition is in fact the transformation of the electronic system and the space averaged ionicity, ρ, is the order parameter. The transition is due to the competition between the inter-stack Coulomb interaction and the chemical potential difference. In low temperatures it might cause a discontinous phase transition. It terminates at some value of the matrix element, t, e.g., at the temperature when the thermal excitation of NI-domain walls overcomes the inter-stack coupling. At this temperature, the system becomes unstable against condensation of NI-domain walls and the three-dimensional crystal is seen as a weakly interacting set of MS along which there is a thermally stimulated disorder of N and I molecules (**thermal mixing**).

The **solidification** phase transition is described by the function f{q, ρ(T)}, eq.(13). The coefficients of the expansion are ρ - dependent and this is the important coupling between the electronic transformation measured by the ionicity, ρ, and the structural transition with the dimerization, q, as the order parameter. In the model, the coupling is due to **quantum mixing** as well as **thermal mixing**. In a classical approach, where quantum mixing is neglected and the thermal mixing is replaced by the entropy of mixing, the coupling between the order parameters appears as kinematical in the number of microstates of the system [18].

The analysis of the function, f{q, ρ(T)}, allows the construction of a thermodynamical phase diagram. Following an analogy of the problem to the ground state phase diagram [1,14], we shall represent it in the triangle (Δ_{eff} , t, J), Fig. 1. The coordinates of any point in the triangle are proportional to the lengths of perpendiculars there from to the Δ_{eff} J, Jt, tΔ_{eff} sides, as usually defined. The analysis of the free energy function goes as follows. The condition, $(\partial^2 f/\partial q^2) = 0$, which might be equivalent to the dielectric catastrophy ($\varepsilon(0) \Rightarrow 0$), is represented on the diagram by the curve (1-0-J) and corresponds to a set of critical points. The electronic CT instability along the curve would suggest that the van der Waals energy is no longer attractive [26]. In part (1-0) it is a continuous, symmetry breaking transition. Along the curve (0-3), conditions $(\partial f/\partial q)_{q\neq 0} = (\partial^2 f/\partial q^2)_{q\neq 0} = 0$ are fulfilled to indicate a set of critical points without symmetry breaking. The most important is the curve (0-2) obtained from the conditions, $f(q = 0) = f(q \neq 0)$ and $(\partial f/\partial q)_{q\neq 0} = 0$. This determines a discontinuous, symmetry breaking, **ferroelectric transition** between undimerized and dimerized phases. The point O is the **tricritical point** and its coordinates are found from the conditions, $(\partial^2 f/\partial q^2) = (\partial^4 f/\partial q^4) = 0$. At this point the three phases, undimerized and two ferroelectric, become identical. Tracing an evolution of the system along the equilibrium line from point 2 towards the tricritical point corresponds to increasing temperature. The ferroelectric phase transition with structural dimerization as an order parameter, can be observed when interstack polarization coupling, J, overcomes the temperature induced disorder due to tunneling between domains, t, or/and the effective crystal field, Δ_{eff} . The transition will appear as **continuous for (J-t) competition**

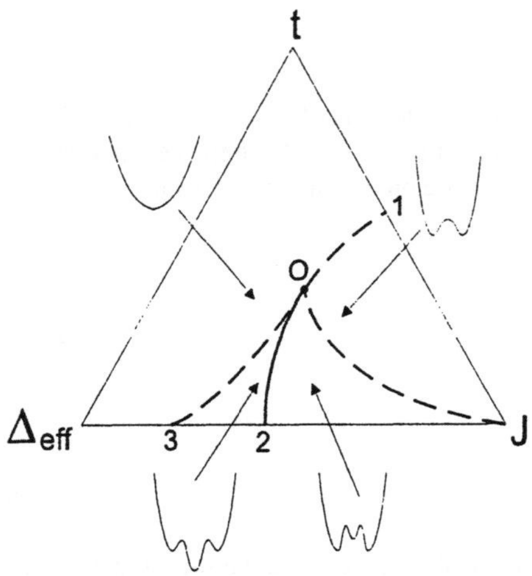

Figure 1. The thermodynamic phase diagram illustrating the analysis (details in the text) of the free energy function, eq. (13). The line (2-O) corresponds to the discontinuous, symmetry breaking, ferroelectric transition which terminates at the tricritical point (O). The icons show schematically drawn free energy functions of the dimerization order parameter, q.

(higher temperatures) **and discontinuous for the (J - Δ_{eff}) competition** (lower temperatures). The tricritical point terminates the discontinuous transitions when the thermally stimulated **disorder** (thermal mixing) along a MS overcomes the **order** due to both (!) crystal field and polarization effects. Therefore, an evolution of a system with increasing temperature towards the tricritical point will be manifested by a **displacive to order-disorder crossover** of the phase transition mechanism. This explains consistently the observations of the phase transition in TTF-CA crystal with increasing temperature [16].

6. Conclusion.

In conclusion we suggest the (p-T) phase diagram for organic CT systems, Fig. 2. The axis of pressure corresponds to **quantum mixing**, which determines molecular ionicity in the ground state ($T = 0$). The axis of temperature measures **thermal mixing**, as discussed above. In the low-pressure, low-temperature regime a discontinuous transition is expected between ordered phases, with a large discontinuity in molecular ionicity and a displacive mechanism for the structural change. The higher pressure, higher temperature regime makes the transition less discontinuous, with a smaller change in molecular ionicity and more order-disorder character for the dimerization transition. The (p - T) phase diagram is therefore seen as resulting from an interplay of **quantum and thermal mixing**. An evolution of the neutral-to-ionic transition towards the tricritical point located at higher pressures might be considered as a classical to quantum crossover. One can speculate that the equilibrium line for the continuous transition represents Peierls transition in a system where D and A molecules are ionized up to the critical value (1/2 within the simplest estimation). The slope of the equlibrium line will depend on molecular and structural details, such as sensitivity of the quantum mixing to a change of pressure, for example. The molecular ionicity in the ground state determines the $p = 0$ point on the pressure axis, as schematically indicated for TTF-CA compound.

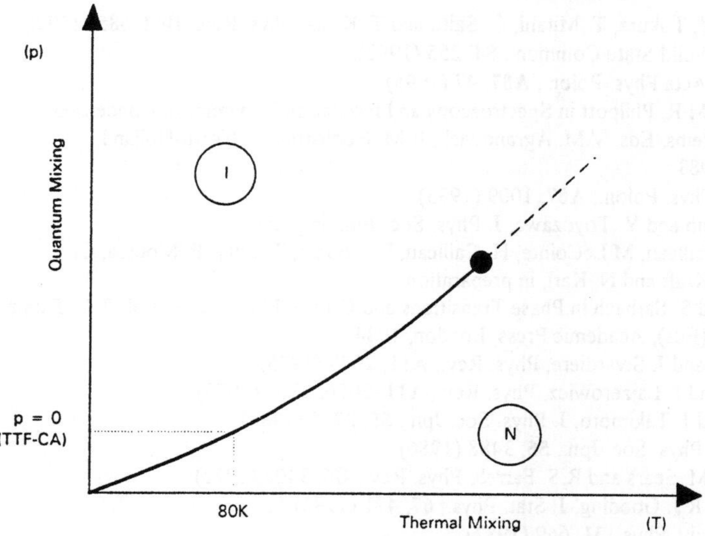

Figure 2. Suggested, pressure - temperature, phase diagram for CT systems showing the equilibrium line for neutral, undimerized - to - ionic, dimerized phase transition. The discontinuous transformation (solid line) terminates at the tricritical point and changes into continuous transition (broken line). The ground state molecular ionicity determined by the quantum mixing defines p = 0 point on the pressure axis as schematically indicated for TTF-CA compound.

Acknowledgments. The author wishes to acknowledge stimulating discussions and cooperation with Prof. Herve Cailleau (Rennes) and Prof. Naoto Nagaosa (Tokyo). The cooperation with Groupe Matiere Condense et Materiaux, University De Rennes I, is sponsored by Polish - French joint project No. 6271. The cooperation with Department of Applied Physics of the University of Tokyo has been initiated by Japan Society for Promotion of Sciences.

References

1. Y. Toyozawa, Technical Reports of Institute for Solid State Physics, University of Tokyo, No.1367, September 1983.
2. K. Nasu and Y. Toyozawa, J. Phys. Soc. Jpn., **51**, 2098, 3111 (1982).
3. B.K. Chakraverty, Nature, **287**, 393 (1980).
4. J.B. Torrance, J.E. Vazquez, J.J. Mayerle and V.Y. Lee, Phys. Rev. Lett., **46**, 253 (1981).
5. J.B. Torrance, A. Girlando, J.J. Mayerle, J.I. Crowley, V.Y. Lee, P. Batail and S.J. LaPlaca, Phys. Rev. Lett., **47**, 1747 (1981).
6. Y. Tokura, T. Koda, T. Mitani and G. Saito, Solid State Commun., **43**, 757 (1982).
7. Y. Tokura, Y. Kaneko, H. Okamoto, S. Tanuma, T. Koda, T. Mitani and G. Saito, Mol. Cryst. Liq. Cryst., **125**, 199 (1985).
8. A. Moreac, Ph.D. Thesis, University of Rennes I, Rennes, 1995.
9. M. LeCointe, M.H. Lemee-Cailleau, H. Cailleau, B. Toudic, L. Toupet, G. Heger, F. Moussa, P. Schweiss, K.H. Kraft and N. Karl, Phys. Rev., **B51**, 3374 (1995).

10. S. Koshihara, Y. Tokura, T. Mitani, G. Saito and T. Koda, Phys. Rev., **B42**, 6853 (1990).
11. Y. Toyozawa, Solid State Commun., **84**, 255 (1992).
12. Y. Toyozawa, Acta Phys. Polon., **A87**, 47 (1995).
13. D. Haarer and M.R. Philpott in Spectroscopy and Excitation Dynamics of Condensed Molecular Systems, Eds. V.M. Agranowich, R.M. Hochstrasser, North-Holland, Amsterdam, 1983.
14. T. Luty, Acta Phys. Polon., **A87**, 1009 (1995).
15. T. Iizuka-Sakano and Y. Toyozawa, J. Phys. Soc. Jpn., in press.
16. M.H. Lemee-Cailleau, M LeCointe, H. Cailleau, B. Toudic, T. Luty, F. Moussa, C. Ayache, K.H. Kraft and N. Karl, in preparation.
17. I.D. Lawrie and S. Sarbach in Phase Transitions and Critical Phenomena, Vol. 9, C. Domb, J.L. Lebowitz (Eds), Academic Press, London, 1984.
18. J. Lajzerowicz and J. Sivardiere, Phys. Rev., **A11**, 2079 (1975)
J. Sivardiere and J. Lajzerowicz, Phys. Rev., **A11**, 2090, 2101 (1975)
19. N. Nagaosa and J. Takimoto, J. Phys. Soc. Jpn., **55**, 2735 (1986).
20. N. Nagaosa, J. Phys. Soc. Jpn., **55**, 3488 (1986).
21. D.J. Scalpino, M. Sears and R.S. Farrell, Phys. Rev., **B6**, 3409 (1972).
22. J.R. Moris and R.J. Gooding, J. Stat. Phys., **67**, 471 (1992).
23. R.G. Palmer, Adv. Phys., **31**, 669 (1982).
24. D.J. Scalpino, Y. Imry and P. Pincus, Phys. Rev., **B11**, 2042 (1975)
25. H. Okamoto, T. Mitani, Y. Tokura, S. Koshihara, T. Komatsu, Y. Iwasa, T. Koda and G. Saito, Phys. Rev., **B10**, 8224 (1991)
26 G.D. Mahan, J. Chem. Phys., **43**, 1569 (1965)

Optical and Related Properties of Organic Charge-Transfer Crystals: An Overview of Experimental Results

T. Koda

Department of Mathematical and Physical Sciences, Japan Women's University, Mejirodai, Bunkyo-ku, Tokyo 112, Japan

Abstract. Some characteristic features of charge-transfer (CT) excitons and related properties are reviewed for two kinds of DA complex systems; PMDA-complexes and TMB-complexes. For PMDA complexes, discussions are made on the self-trapping of singlet CT excitons, and on the hybridization of triplet CT and triplet Frenkel excitons, on the basis of absorption, fluorescence and phosphorescence spectra. Then, anomalous dielectric and electric properties of TMB complexes are reviewed, and the results are discussed in relation with the structural instability associated with the CT excitation in the quasi-one dimensional DA complex crystals.

1. Introduction

Organic donor-acceptor (DA) crystals may be regarded as the low dimensional counterpart of inorganic binary compound crystals like, for instance, CuCl, CdS and so forth. In the latter crystals, the lowest electronic excitations are described by the Wannier exciton picture on the ground of one-electron band model. In contrast, the lowest electronic excitations in organic DA crystals are properly described as the charge-transfer (CT) excitons having a character which is intermediate between the delocalized Wannier exciton and the localized Frenkel (molecular) exciton. The CT excitons in the variety of quasi-one dimensional mixed-stack organic DA crystals are providing challenging problems for solid-state physics, but many unique features of CT excitons and the related properties are still left unclarified as yet.

In this paper, some experimental results obtaind by the author's group in the past ten years are reviewd with a hope that the results may stimulate further detailed investigations in order to get more profound knowledge on the fundamental electronic processes in orgnic DA complex crystals. We shall focus our attention here to the two kinds of 1:1 mixed-stack DA complex systems, namely the weak

Springer Series in Solid-State Sciences, Volume 124
Relaxations of Excited States and Photo-Induced Structural Phase Transitions
Ed.: K. Nasu

CT crystals composed of PMDA molecules as the common accepter, and the TMB complex system in which the degree of CT can be varied over a wide range from 0.1 to 1.0, as shown later, by the chemical modification of TCNQ-based accepter molecules.

2. Charge-transfer (CT) excitons in PMDA complexes

The 1:1 mixed stack DA crystals were prepared with pyromellitic dianhydrite (PMDA) as the common accepter and five kinds of aromatic molecules as the donor. For the latter, we used naphthalene, phenanthrene, anthracene, pyrene, and perylene. Absoption and emission spectra of these crystals are shown in **Fig.1**. The emission spectra are composed of fast fluorescence component (broken lines) and the slow phosphorescence component (dotted lines). The latter spectra were measured with the delay time of about 1 ms after the cease of chopped excitation. The fluorescence spectra are broad and show a large Stokes shift with respect to the absorption spectra (solid lines). An exception is a sharp zero-phonon line observed in the absorption and fluorescence spectra of anthracene(An)-PMDA at low temperatures (<8K), as reported before by Haarer [1]. The broad fluorescence bands are attributed to the radiative decay of the self-trapped singlet CT excitons, whereas the appearance of zero-phonon line in An-PMDA indicates that a part of CT excitons remain free at low temperatures. The band-width of the free exciton in An-PMDA was estimated before to be about 100 cm^{-1} [2].

According to Toyozawa [3], an exciton is predicted to be either free or self-trapped, depending on the value of a material parameter g, which is defined as $g = S\hbar\omega/B$, S being the exciton-lattice coupling constant, $\hbar\omega$ the energy of relevant lattice phonon mode, and B the exciton band width. When $g<g_c$ ($\sim$1), the excitons are predicted to be free, while when $g>g_c$ they are self-trapped. In an ideal one dimensional (1D) crystal, excitons must be always self-trapped, but in the actual quasi 1D DA crystals, the CT exciton may have a metastable free state if the g value is small enough.

Experimental g values were estimated for the CT excitons in PMDA complexes shown in Fig.1 assuming a 1D density of state function for the free CT exciton with the band width B as an adjustable parameter. Furthermore, the predominant lattice phonon energy $\hbar\omega$ was assumed to be 30 cm^{-1} which was estimated from the vibronic structures in An-PMDA. Experimental absorption spectra were fitted with theoretical

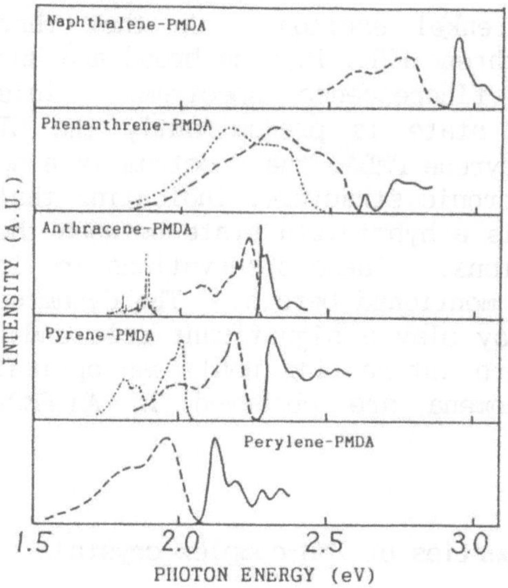

Fig.1. Absorption (solid lines), fluorescence (dashed lines) and phosphorescence spectra of PMDA-complex crystals at 2K.

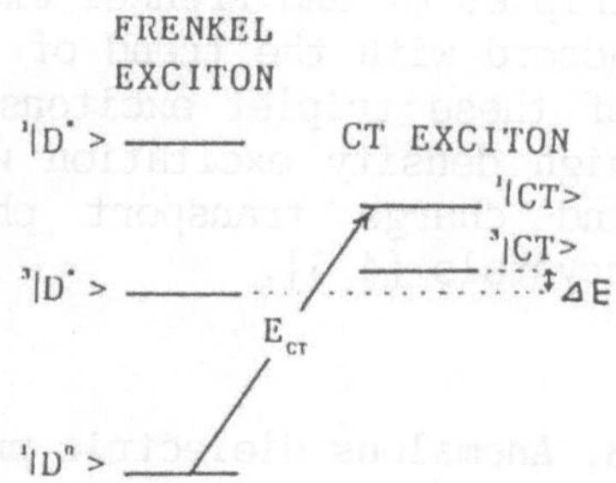

Fig.2. Schematic diagram for the singlet and triplet states of Frenkel and charge-transfer (CT) excitons.

spectra calculated by the dynamical CPA model. As a result, the g value was estimated to be 1.1 for An-PMDA, 3.2 for naphthalene-PMDA, 4.0 for perylene-PMDA, and 8.0 for pyrene-PMDA. The g-value is nearly unity in An-PMDA, which is consistent with the coexistance of free and self-trapped CT excitons. The metastable free exciton state in An-PMDA is separated from the self-trapped CT exciton state by a small energy barrier of about a few cm^{-1}.

Another interesting feature in the optical spectra of PMDA complexes is seen in the phosphorescence spectra due to the radiative decay of the lowest triplet excitons. In the DA crystals in general, the energy of triplet Frenkel exciton associated with the donor ($^3|D^*>$) is supposed to lie close to the triplet CT excioton state ($^3|CT>$), as schematically illustrated in Fig.2. As a result, the lowest triplet exciton is a hybridized state between the both types of triplet excitons. The mixing rate is determined by the energy difference $\Delta E = E(^3|CT>) - E(^3|D^*>)$. From molecular spectroscopic data, the value of ΔE is estimated to be about 0.25 eV in An-PMDA, 0.05 eV in pyrene-PMDA, and -0.2 eV in phenanthrene-PMDA. In An-PMDA, the phosphorescence spectrum is composed of sharp vibronic structures, indicating that the lowest triplet state has a predominant character of Frenkel exciton. On the other hand, the spectum in phenanthrene-PMDA is much broad and has a similar shape as the fluorescence spectrum. This indicates that the triplet state is predominantly the CT exciton in character. In pyrene-PMDA, the spectrum is also broad but shows a slight vibronic structure, indicating that the lowest triplet exciton is a hybridized state between the triplet CT and Frenkel excitons. These observations are in accord with the trend of E mentioned before. The dynamics of these triplet excitons may play a significant role under high density excitation where interesting nonlinear optical and charge transport phenomena are observed in An-PMDA crystals [4,5].

3. Anomalous dielectric properties of TMB-complex crystals

There are a number of unique features observed in DA crystals which are attributable, directly or indirectly, to the characteristic structural instability associated with the CT excitation. When crystals are located near the NI boundary, the ground state is nearly degenerate with respect to the neutral and ionic phases. Photo-generated or thermally excitated CT excitons tend to be decomposed or aggregated, wherby various kinds of solitons and NI domain walls are supposed to be formed in the quasi 1D DA stacks. In order to investigate the dynamics of such defects, measurements of dielectric and nonlinear electric responces are particularly useful [6-8]. In the following, we shall review typical experimental results on TMB complexes in oreder to demonstrate that observed features of dielectric and electric properties in DA crystals are clearly related with the structural instablity near the NI phase boundary.

The common donor in the present DA system is TMB (3,3'55' tetramethylbenzidine), having the ionization energy I_D=6.8 eV. As the accepter, series of TCNQ-based molecules (synthesized by N.Iwasawa and G.Saito) were used. The respective molecular structures are illustrated in Fig. 3, together with the values of electron affinity E_A for respective molecules. The 1:1 mixed-stack DA crystals were prepared using TMB and each of TCNQ derivatives. The crystals are referred here as TMB-F_4TCNQ, TMB-FTCNQ, TMB-TCNQ, TMB-Me_2TCNQ, and TMB-$(MeO)_2$TCNQ, in the sequence of increasing effective ionization energy $I_{eff}=I_D-E_A$. All of these TMB complexes have the same monoclinic crystal structure ($P2_1/n$). Powder samples were mostly used in the following optical and dielectric measurements. The electric conductivity measurements were made by using a

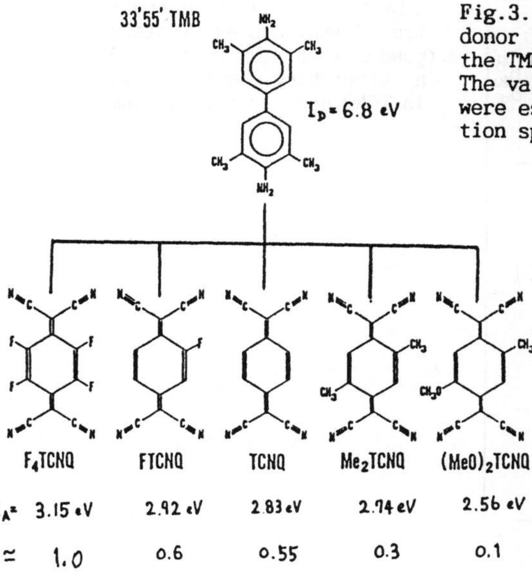

Fig.3. Molecular structures of donor (TMB) and acceptors for the TMB-complex systems. The values of degree of CT (ρ) were estimated from IR absorption spectra shown in Fig.4.

small piece of single crystals which were obtained for TMB-TCNQ, TMB-FTCNQ, TMB-F_4TCNQ, and TMB-Me_2TCNQ.

In order to estimate the degree of charge transfer ρ, infrared absorption spectra were measured in the spectral region where typical -C≡N stretching vibration modes ($b_{1,2u}$) in TCNQ derivative molecules are observed. The results are shown in Fig.4. From comparison between the peak energies of respective absorption bands, the b_{1u} (open circles) and b_{2u} (open triangles), with those in the neutral and ionic molecules shown by vertical dashed lines, experimental values for the degree of charge transfer (ρ) were estimated. The results are ρ=1.0 (TMB-F_4TCNQ), 0.6 (TMB-FTCNQ), 0.55 (TMB-TCNQ), 0.3 (TMB-Me_2TCNQ), and 0.1 (TMB-$(MeO)_2$TCNQ), as indicated at the bottom of Fig.3. Among these TMB complexes, TMB-TCNQ has been found to exhibit a temperature induced NI transition at T_c=205 K [9]. Hence, this crystal is located very close to the NI boundary for the present TMB complex system. TMB-F_4TCNQ and TMB-FTCNQ have the ρ-values larger than that of TMB-TCNQ, so that these crystals must be in the ionic phase. In fact, a strong IR absorption peak is observed in both crystals at the energy of forbidden -C≡N stretching a_g mode, as seen in Fig.4, indicating that both crystals are in the dimerized I phase. On the other hand, TMB-Me_2TCNQ and TMB-$(MeO)_2$TCNQ, having the ρ-values smaller than 0.55, must be in the neutral phase. In TMB-TCNQ, a strong absorption peak is observed at around 2220 cm^{-1}, where an -C≡N stretching a_g mode is observed.

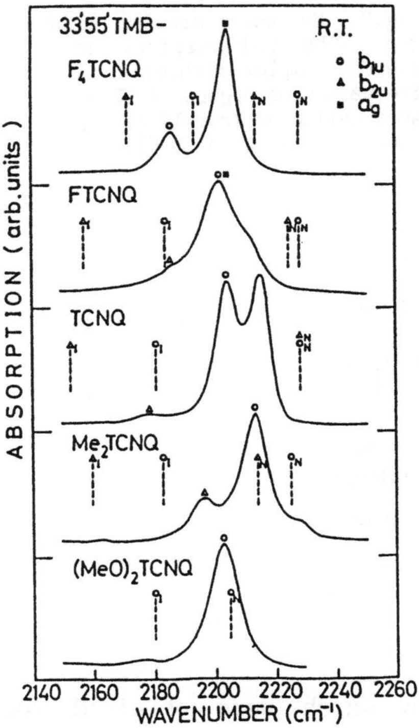

Fig.4. IR absorption spectra of TMB-complex crystals (powders) for the b_{1u}, b_{2u}, and a_g stretching modes of -C≡N in TCNQ and its derivatives.

This indicates that the lattice is dimerized at room temperature. However, according to the IR reflection spectra on single crystal, the same band becomes observable only at low temperatures below T_c for the NI transition [9]. The appearance of the same band in the absorption spectra at RT might be due to the strain effect in the powder samples used in the absorption measurement.

Low frequency dielectric constants were measued as a function of temperature for these five kinds of TMB complexes. Results are shown in Fig.5. Temperature dependence of dielectric constant is similar to that of TTF-CA reported before [7], and can be explained by a Debye-type relaxation model. As in the case of TTF-CA, the temperature dependence of relaxation constant can be expressed by a single activation energy E_a. Note that the dielectric constant at room temperatue is strongly dependent on the degree of CT, being the largest in TMB-FTCNQ and the next largest in TMB-TCNQ, while it is quite small in other three complexes.

The nonlinear electric conductivity were measured on four kinds of TMB complexes for which single crystals were available. As seen in Fig.6, remarkable negative

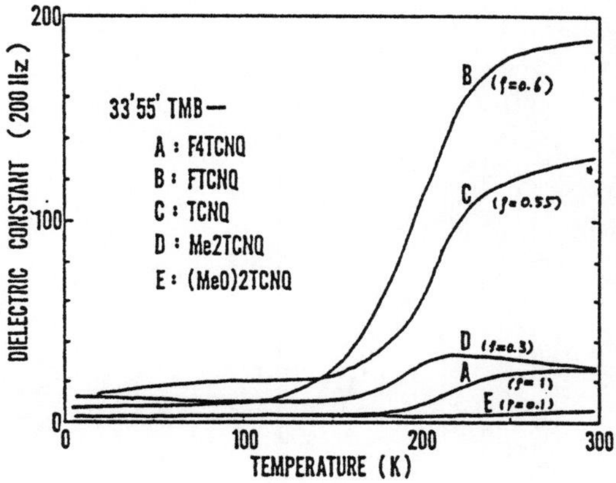

Fig.5. Dielectric constant at 200Hz vs temperature for TMB-complexe crystals.

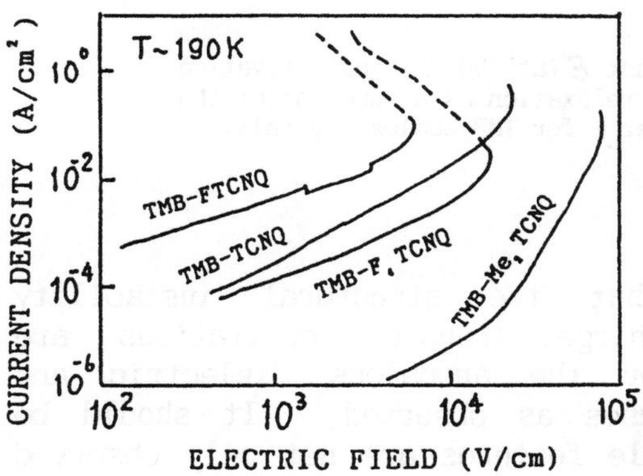

Fig.6. Nonlinear electric properties of TMB-complexe crystals.

resistance effect is seen for ionic crystals (TMB-FTCNQ and TMB-F_4TCNQ). This is attributable to the field-induced disorder effect at high electric fields within the antiferroelectrically ordered lattice in the dimerized ionic phase [8].

As a summary, the low frequency dielectric constant $\mathcal{E}$ (filled squares) and activation energy E_a for dielectric relaxation (open ellipses) are plotted in Fig.7 against for five kinds of TMB complexes investigated. The activation energy for electric conductivity is also shown by open squares. There are obvious tendencies between $\mathcal{E}$, E_a and ρ, as shown by shaded upward and downward V-shaded stripes for and E_a, respectively. The tops of stripe are located just at the NI boundary at ρ=0.55. In other words, the dielectric and electric responces of TMB complex crystals are more enhanced as the crystals are located closer to the IN boundary, irrespective of whether the crystals are in the neutral or ionic phase regions. Obviously this is an

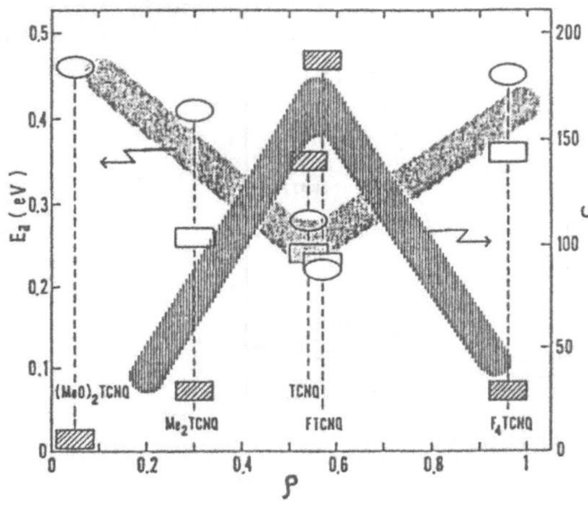

Fig.7. Dielectric constant ε (at 100 Hz) and activation energy E_a of dielectric relaxation as a function of the degree of charge-transfer ρ for TMB-comlex crystals.

experimental evidence that the structural instability associated with the charge transfer excitations are certainly responsible for the anomalous dielectric and nonlinear electric responses as observed. It should be mentioned that such notable features are not only observed in the TMB complexe system but also are seen in many other kinds of mixed-stack DA crystals in common.

Acknowledgments

This study had been done in collaboration with a number of colleagues in the author's group, in particular, T.Mitani, Y.Tokura, Y.Iwasa, S.Koshihara, H.Okamoto, T.Komatu, to whom the author is highly grateful. For preparing new chemical substances, powerful support by G.Saito and his colleagues is invaluable. As for the theoretical aspect, continuous discussions with Y.Toyozawa, K.Nasu and N.Nagaosa have been quite helpful.

References

1. D.Haarer, J. Chem. Phys. 67, 4076 (1977).
2. Y.Tokura and T.Koda, Solid State Commun. 40, 229 (1981).
3. Y.Toyozawa, in "Excitonic Processes in Solids", ed.M.Ueta et al., Springer-Verlag, 1985.
4. M.Kuwata-Gonokami et al. Nature 367, 47 (1994).
5. M.Gonokami, in this proceedings.
6. Y.Tokura, S.Koshihara, Y.Iwasa, H.Okamoto, T.Komatsu, T.Koda, N.Iwasawa, and G.Saito, Phys. Rev. Lett., 63, 2405 (1989).
7 H.Okamoto, T.Mitani, Y.Tokura, S.Koshihara, T.Komatsu, Y.Iwasa, T.Koda, and G.Saito, Phys. Rev. B44,8224 (1991).
8. Y.Iwasa, T.Koda, S.Koshihara, Y.Tokura, N.Iwasawa and G.Saito, Phys. Rev. B39, 10441 (1989).
9. Y.Iwasa, T.Koda, Y.Tokura, A.Kobayashi, N.Iwasawa and G.Saito, Phys. Rev. B42, 2374 (1990).

Theory of the Neutral-to-Ionic Transition of Quasi-one-dimensional Charge-Transfer Compounds

T. Iizuka-Sakano and Y. Toyozawa

Department of Physics, Chuo University, 1-13-27 Kasuga, Bunkyo-ku, Tokyo 112, Japan

Abstract. The neutral (N)-to-ionic (I) transition with dimerization, found in organic charge transfer compounds with mixed stacking, has been studied using a self-consistent theory which takes exact account of long-range Coulomb interactions but neglects interchain coupling. The unrestricted Hartree-Fock approximation gives a set of optimally mixed donor-acceptor bands for the N and the I phases which are characterized by the absence and the presence of sub-lattice spin, respectively. A qualitative argument is presented on the optical and magnetic excitations of this electronic system. The dimerization is ascribed to the electrostatic instability, rather than to the distance dependence of the intermolecular electronic transfer energy as is usually done. The phase diagram for the first-order N-I transition with dimerization and discontinuous change of charge transfer is presented on the parameter plane of the redox potential and the transfer energy. The theoretical predictions are consistent with the observations on those charge transfer compounds which are subject to the N-I transition.

1 Introduction

Quasi-one-dimensional organic compounds consisting of chains of mixed-stack donor (D) and acceptor (A) molecules sometimes undergo, with increasing pressure [1] or decreasing temperature [2], a neutral (N)-to-ionic (I) transition in which ρ, the degree of charge transfer from D to A, increases abruptly, as seen from intra- and intermolecular electronic [3] and vibrational [4, 5] spectra. The transition is ascribed to competition between the Madelung energy (which increases as the lattice contracts) and the redox potential Δ which is defined as the difference between the ionization energy I of D and the electron affinity A of A, as was predicted by MacConnel [6]. The success of this point charge model indicates the primary importance of long-range Coulomb interactions. The electron transfer energy t must also be playing a certain role in view of the significant value of ρ observed in the N phase, although t is small compared to the Coulomb energy $V = e^2/a$ between the nearest D and A.

According to the observation [7], the lattice is dimerized with the N-to-I transition. It has so far been ascribed to the dependence upon D-A distance of the exchange interaction $J \sim 2t^2/U$ which is the second order energy if U, the on-site Coulomb repulsion, is of the order of V. However we will point out here that the primary cause for the dimerization is the instability of the electrostatic energy which is of the first order.

In view of the importance of the electrostatic energy as mentioned above, we will present a theory [8] which (i) describes the long-range part of Coulomb interactions among ions and electrons *exactly* (the point charge model) and then (ii) explores the effect of t on the N-I transition and the associated discontinuity of charge transfer using the unrestricted Hartree-Fock (UHF) approximation.

Springer Series in Solid-State Sciences, Volume 124
Relaxations of Excited States and Photo-Induced Structural Phase Transitions
Ed.: K. Nasu

2 Exact description of long-range Coulomb interaction

We consider a chain consisting of N pairs of D^{2+} and A^0 alternately arrayed with an equal distance a along which 2N electrons are moving through the highest occupied D orbitals and the lowest unoccupied A orbitals. The electronic-ionic part H_e of our Hamiltonian is written as

$$\begin{aligned} H_e &= \sum_{m=1}^{2N} \epsilon_m n_m + \sum_{m=1}^{2N} U_m n_{m\uparrow} n_{m\downarrow} \\ &+ \sum_m \sum_{<m'} V_{mm'}(2\nu(m) - n_m)(2\nu(m') - n_{m'}) \\ &- \sum_\sigma \sum_{m=1}^{2N} t_{m-1,m}(a^\dagger_{m-1,\sigma} a_{m,\sigma} + a^\dagger_{m,\sigma} a_{m-1,\sigma}), \end{aligned} \tag{1}$$

where $a_{m,\sigma}$ and $n_{m\sigma} = a^\dagger_{m,\sigma} a_{m,\sigma}$ are respectively the annihilation and number operators of an electron with spin σ at the m-th site, and $n_m = \sum_\sigma n_{m\sigma}$. Depending on whether m is odd or even, the ionic charge $2\nu(m)$ takes the value 2 or 0, the on-site interelectron Coulomb energy U_m the value U_D or U_A and the site energy ϵ_m the value $-I_+$ or $-A$, where $I_+ = I + U_D$ represents the ionization energy of D^+. Of the possible structural changes, we consider only uniform dimerization: $a \to a \pm aq$, whereby the intersite Coulomb energy is changed to $V_{mm'} = V|(m - m') + (\nu(m) - \nu(m'))\, q|^{-1}$. The corresponding change in the transfer energy $t_{m,m-1}$ will be neglected so that $t_{m,m-1}$ will be equated to t_0 since t_0 itself is already small compared to V. The dimerization also causes a change in the short-range intermolecular interactions which will be considered up to the fourth-order term: $H_v = NM_r\omega_N^2(aq)^2/2 + Nc(aq)^4/4$.

2.1 Point charge model

Neglecting t_0 (point charge model) and taking the completely neutral phase $N^{(0)}$ as the origin of energy H_e, one finds the adiabatic energy of the completely ionized phase $I^{(0)}$ to be given by

$$\frac{E_I^{(0)}}{N} = \Delta - V\left(\alpha_M + \sum_{\nu=1}^{\infty} q^{2\nu} f_\nu(0)\right) + \frac{M_r\omega_N^2 a^2}{2} q^2 + \frac{ca^4}{4} q^4, \tag{2}$$

where α_M is the Madelung constant and

$$f_\nu(k) = \sum_{l=-\infty}^{\infty} \frac{e^{-i(2l-1)ka}}{|2l-1|^{2\nu+1}}. \tag{3}$$

Considering terms up to and including the fourth-order term of q, one finds that the lattice is destabilized when $\eta \equiv 2Vf_1(0)/(M_r\omega_N^2 a^2) > 1$, being dimerized to a finite value q_I with new vibrational frequency $\omega_I = \omega_N\sqrt{2(\eta - 1)}$ provided $\zeta \equiv 4Vf_2(0)/(ca^4) < 1$. The electrostatic instability index η readily exceeds unity in realistic situations as will be seen later. The $N^{(0)}$-$I^{(0)}$ boundary which was given by $\Delta - \alpha_M V = 0$ in the above is now slightly shifted so as to favor the $I^{(0)}$ phase.

2.2 The effect of transfer energy

When the transfer energy is considered, it is more convenient to introduce the donor and acceptor Bloch orbitals, the corresponding electron creation operators being given by:

$$\left\{ \begin{aligned} b^\dagger_{k\sigma} &= \frac{1}{\sqrt{N}} \sum_{l=1}^{N} a^\dagger_{2l-1,\sigma} e^{i(2l-1)ka}, \\ c^\dagger_{k\sigma} &= \frac{1}{\sqrt{N}} \sum_{l=1}^{N} a^\dagger_{2l,\sigma} e^{i2lka}. \end{aligned} \right. \tag{4}$$

In order to study the ground state and the one-electron excited states of this many-electron system, we use the unrestricted Hartree-Fock approximation (UHF) considering the possible existence of sublattice magnetization in the I phase where each site has a free spin at least in the zeroth approximation $I^{(0)}$. For the one-electron states, we use the charge transfer band defined by

$$d_{k\sigma}^{\dagger} = u_{k\sigma} b_{k\sigma}^{\dagger} + v_{k\sigma} c_{k\sigma}^{\dagger}, \quad |u_{k\sigma}|^2 + |v_{k\sigma}|^2 - 1 = 0. \tag{5}$$

The mixing coefficients $u_{k\sigma}$ and $v_{k\sigma}$ are determined so as to minimize the expectation value of the energy of the ground state: $E_e =< G|H_e|G >$. The ground state $|G>$ is given by $|G>= \Pi_{k\sigma}^{2N} d_{k\sigma}^{\dagger}|0>$ where the product on $(k\sigma)$ is extended over the one-electron states which are occupied according to the Pauli principle.

The extremization of E_e with respect to $u_{k\sigma}$ and $v_{k\sigma}$ with Lagrange's indeterminate coefficients $\varepsilon_{k\sigma}$ for their normalization condition leads to a two-row eigen-equation for $\varepsilon_{k\sigma}$, the Hartree Fock one electron energy. The two eigenvalues $\varepsilon_{k\sigma}^{(\mp)}$, representing the lower and the upper band energies respectively, are given as follows where the origin of energy is put at $-I$, the one electron energy of D^0.

$$\varepsilon_{k\sigma}^{(\mp)} = \frac{1}{2}\left\{\Delta - (U_D - U_A)\rho_{-\sigma} \mp \sqrt{B_{k\sigma}^2 + |C_{k\sigma}|^2}\right\}, \tag{6}$$

$$B_{k\sigma} = B_\sigma - 2VG_\sigma(k), \tag{7}$$

$$B_\sigma = \Delta + \left\{U - 2V(\alpha_M + \sum_{\nu=1}^{\infty} q^{2\nu} f_\nu(0))\right\}\rho - Us\sigma, \tag{8}$$

$$U = (U_D + U_A)/2, \tag{9}$$

$$\rho = \rho_\uparrow + \rho_\downarrow, \quad s = \rho_\uparrow - \rho_\downarrow, \quad (\rho_\sigma =< n_{2l\sigma} >), \tag{10}$$

$$G_\sigma(k) = \frac{1}{N}\sum_{k'} g(k-k')|v_{k'\sigma}|^2, \tag{11}$$

$$g(k-k') = \sum_{l=-\infty}^{\infty}{}' e^{-i2l(k-k')a}/|2l|, \tag{12}$$

$$C_{k\sigma} = 4t_0 \cos ka + 2V\sum_{\nu=0}^{\infty} q^{2\nu} F_{\nu\sigma}(k), \tag{13}$$

$$F_{\nu\sigma}(k) = (2N)^{-1}\sum_{k'} f_\nu(k-k') y_{k'\sigma}. \tag{14}$$

Here $y_{k\sigma} = 2u_{k\sigma}^* v_{k\sigma}$ and the summation on k' is over all occupied electron states. The eigen-equations for all $(k\sigma)$ must be solved self-consistently since the eigen-equation for $(k\sigma)$ depends on the occupancy of other $(k'\sigma')$'s which in turn depends on their one-electron energies $\varepsilon_{k'\sigma'}$ which are yet to be determined. The resulting nonlinear integral equation for $y_{k\sigma}$ is given by

$$|B_{k\sigma}|y_{k\sigma}(1 - |y_{k\sigma}|^2)^{-1/2} = C_{k\sigma}. \tag{15}$$

The above equations must be solved for the N phase and the I phase, separately. As already mentioned in connection with our UHF, we assume that the sublattice magnetization s vanishes in the N phase and does not vanish in the I phase, which is the only constraint imposed on our self-consistent solutions. As a result, $y_{k\sigma}$ and $\varepsilon_{k\sigma}$ depend on σ in the I phase.

3 Numerical calculation and the result

The nonlinear integral equation reduces to a set of simultaneous algebraic equations if N is taken to be finite. We adopt the damping Newton method to solve it numerically. We set $N = 80$ (for the k-summation) in order to avoid the finite size effect. However, we

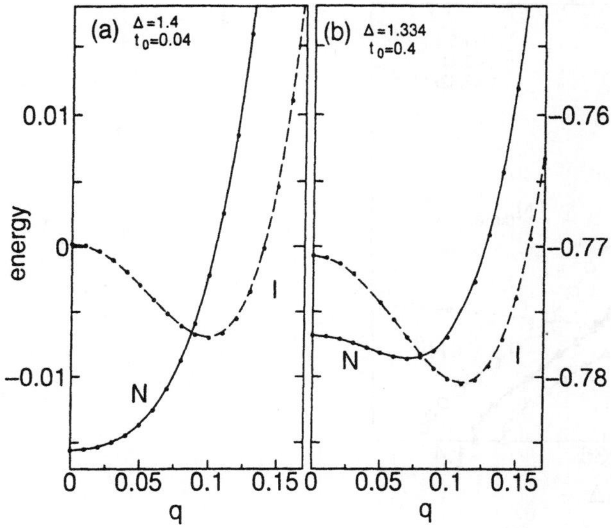

Figure 1: The calculated adiabatic potentials (a) for the N phase and (b) for the I phase.

consider the l-summation for the intersite Coulomb energies over a greater distance. For definiteness we choose the sign of s as $s \geq 0$ on the A sublattice.

As an example, we consider tetrathiafulvalene-chloranil (TTF-CA), a typical charge transfer compound which has been subject to extensive experimental study including investigation of the temperature- and pressure-induced N-I transition. Taking $a = 3.63 \times 10^{-10}$ m, the distance between the nearest TTF and CA molecules along the stacking axis at 150 K [9, 5], we obtain $V = 3.97$ eV. Using the reduced mass $M_r = 1.85 \times 10^{-25}$ kg of TTF and CA molecules and the longitudinal optical mode angular frequency $\omega_N = 5.65(\pm 0.06) \times 10^{12}$ s^{-1} obtained from neutron scattering in the N phase [5], and noting $f_1(0) = 2.1036$, we find $\eta = 3.44$, satisfying the criterion: > 1. That is, the I phase is subject to electrostatic instability towards dimerization already within the point charge model. The degree of dimerization in the I phase is $q_I \sim 0.03$ according to neutron [10] and X-ray diffraction [9] studies. However, in order to make it feasible to distinguish the adiabatic potentials of the N and I phases with the precision of our numerical calculation, we put $q_I = 0.1$. We then obtain $\zeta = 2.62 \times 10^{-2}$ (note that $f_2(0) = 2.0095$).

U_D is relatively large because it is the difference between the ionization energies of D^0 and D^+. U_A is rather small because U_A is the energy of Coulomb repulsion between the two electrons in the lowest unoccupied molecular orbital which has now a greater radius. It seems reasonable to put $U_D = 2 \cdot V$ and $U_A = 1 \cdot V$. It is only the one-electron energy $\varepsilon_{k\sigma}^{(\mp)}$ of eq. (6) which depends upon the difference $U_D - U_A$, while all the other quantities depend only on the average U of eq. (9) which now takes the value $1.5 \cdot V$. Other energies such as Δ and t_0 will also be measured in units of V.

Using the point charge model, ω_I is calculated as 1.25×10^{13} s^{-1} with these parameters. It is consistent with the experimental results [5] that ω_I is significantly larger than ω_N.

The calculated adiabatic potentials for the N and I phases are given in Fig. 1 for typical sets of (Δ, t_0)-values on both sides of the N-I boundary shown by the solid line on the phase diagram in Fig. 2. As seen from Fig. 1, the I phase is always dimerized while the N phase is not dimerized for small t_0. However, the N phase is also dimerized for large enough t_0 (Fig. 1(b)). The reason is the increase of ρ.

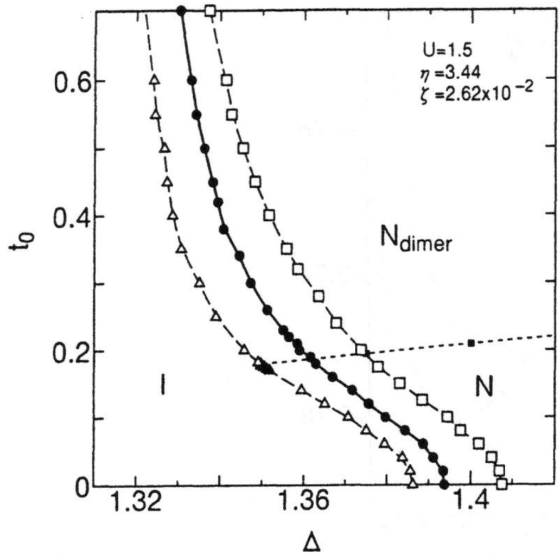

Figure 2: The phase diagram for the N-I transition and dimerization.

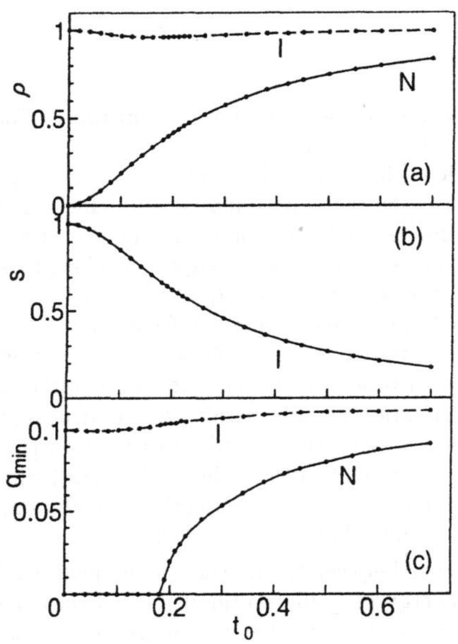

Figure 3: (a) ρ and (c) q_{min} for the N and I phases and (b) s for the I phase, taken along the N-I transition line of Fig. 2

Figure 3 shows the calculated values of ρ and q_{min} (the degree of dimerization) for the two phases and s for the I phase as functions of t_0 taken along the N-I transition line of Fig. 2. From Fig. 3(a), we can see that ρ increases with t_0 in the N phase. When t_0 takes large value such that $\rho \sim 0.4$, the second-order transition to the dimerized phase (Bond Ordered Wave (BOW) transition) takes place as shown in (c). We can understand this as follows. As ρ increases with t_0, the intersite Coulomb energy becomes large and its gain by dimerization becomes larger than the loss of the lattice energy. Thus the lattice is dimerized.

The boundary for the BOW transition in the N phase is shown by the dotted line in Fig. 2. We note in Fig. 1 that, for any fixed value of q only the lower branch of the two

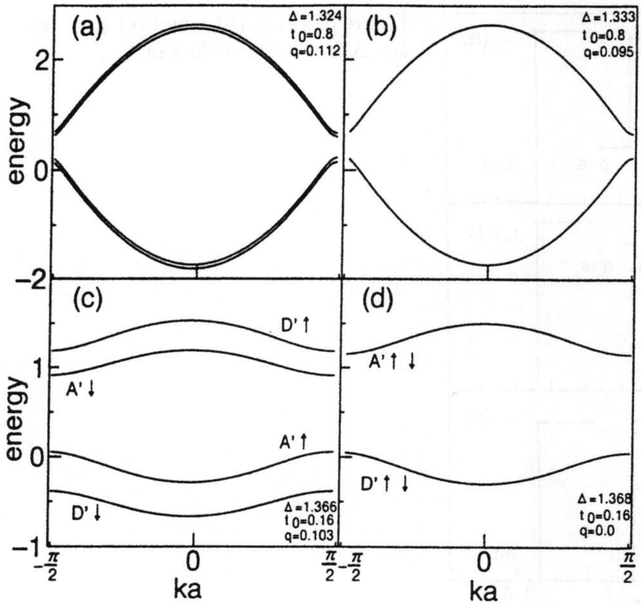

Figure 4: The sub-band energies versus k calculated with $U_D = 2$ and $U_A = 1$ for the N ((b),(d)) and I ((a),(c)) phases, for small ((c),(d)) and large ((a),(b)) values of t_0.

adiabatic curves is the stable electronic state to be realized since the UHF approximation is based on a variation principle. As we increase q, always tracing the lower branch, we have two minima, before and after the N-I crossover. Thus we have one stable and one metastable state in the case of Fig. 1. This "metastability" is realized, on the phase diagram of Fig. 2, only within the region between the two broken lines, one with open squares and the other with open triangles. Outside this region there is only the stable phase, N or I.

Figure 4 shows the sub-band energies $\varepsilon_{k\sigma}^{(\mp)}$ versus k, for the N (right) and I (left) phases at their adiabatic minima and for small (lower half) and large (upper half) values of t_0. D'(A') denotes the mixed band with D(A)-predominance. In the I phase, $\varepsilon_{k\sigma}^{(\mp)}$ depends on σ in accordance with our choice of sign: $s > 0$ on the A sublattice. In either phase, two lower bands are occupied and two upper bands unoccupied, each being spin degenerate in the N phase. The occupancy of $D'_\downarrow$ and $A'_\uparrow$ bands in the I phase corresponds in r-space to the antiferromagnetic ordering between the D and A sublattices. The energy width of each sub-band increases with t_0 as expected. However, each of the two lower and two upper bands in the I phase tends to be degenerate ((c) to (a) in Fig. 4), approaching the spin-independent bands in the N phase ((b)) asymptotically as t_0 increases.

This feature is also reflected in Fig. 3. ρ for the N phase gradually increases with t_0 and tends to ρ for the I phase which changes very little. s for the I phase tends to vanish and q_{min} for the N phase tends to the one for the I phase after the BOW transition.

We can understand these behaviors as follows. In the I phase and at $t_0 \sim 0$, one can imagine an electron with spin down on a D site and another with spin up on a neighboring A site which interact anti-ferromagnetically with each other. As t_0 increases, they are increasingly transferred to the opposite sites. Their degrees of transfer are almost the same and ρ, the average number of electrons on the A sublattice, varies very little. Thus ρ in the I phase has only a small variation with t_0, and q_{min} does, too. In contrast

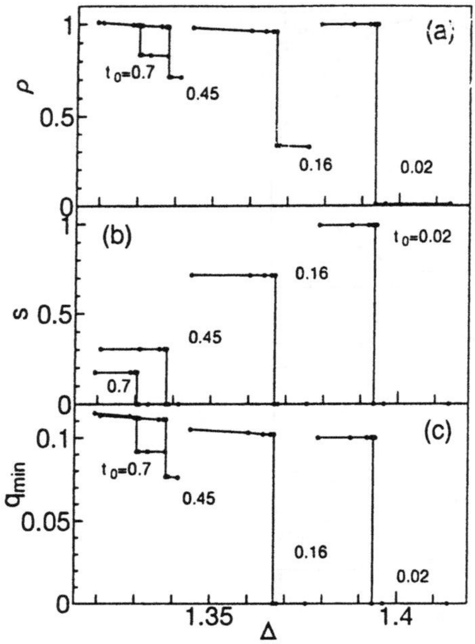

Figure 5: (a) ρ, (b) s and (c) q_{min} versus Δ for various values of t_0.

the number difference s of electrons with spin up and spin down on one sublattice is reduced with increasing t_0 due to the mutual transfer of opposite spins. However, it is not so clearly seen in Fig. 2 where one would expect the two broken lines to approach each other as t_0 increases.

These three quantities are subject to discontinuous change as Δ decreases across the N-I transition line in Fig. 2. This is shown in Fig. 5, for various values of t_0.

4 Discussion

The results of our calculation are consistent with experimental results for TTF-CA with respect to the discontinuous increase in ρ at the N-I transition and the presence of dimerization in the I phase. Comparing the observed value $\rho_N \sim 0.3$ in TTF-CA just above the transition temperature with the calculated curve in Fig. 3(a), one finds $t_0 \sim 0.15(\sim 0.6$ eV), which is larger than was estimated by other groups [11], but is still below the dotted line in Fig. 2.

It is of interest to compare our phase diagram with observation on TMB-TCNQ [13], in which dimerization transition takes place in the N phase.

Next we will make a qualitative discussion about the excited states. The occupancy of $\mathrm{D'}_\downarrow$ and $\mathrm{A'}_\uparrow$ bands in the I phase corresponds in r-space to the antiferromagnetic ordering between the D and A sublattices. As is well known, the lowest magnetic excitations from such a ground state are spin waves with energy vanishing as the wave vector. They cost energies $E^{(I)}_{MAG}$ depending on k :

$$E^{(I)}_{MAG} = \varepsilon^{A'}_{k\downarrow} - \varepsilon^{A'}_{k\uparrow} - U_A. \tag{16}$$

Here, we reduce by U_A assuming that the excited electron with inverted spin $\downarrow$ stays on the same A site as the remaining hole while they are moving together through the lattice. Then, $E^{(I)}_{MAG} = 0$ at $t_0 = 0$. When t_0 is not 0, $E^{(I)}_{MAG}$ has the minimum at $ka = \pm\pi/2$ and proportional to t_0^2 within the perturbation of t_0. Then the magnetic gap in the I phase would be very small. That is consistent with the existence of ESR signals in the I phase only [12] because they are due to the spin solitons associated with the vanishing magnetic gap.

In contrast, the energy gap for optical excitation conserving spin and wave vector is significant in both phases. In the N phase there exists a single gap corresponding to the excitation from D'_σ to A'_σ as shown in Fig. 4(d). The reduction by the on-site energy U as was considered in eq. (16) is no more operative here since we are concerned with a charge transfer exciton (D' → A'). While in the I phase, there exist two optical gaps with different energies corresponding to the excitations $A'_\uparrow \to D'_\uparrow$ and $D'_\downarrow \to A'_\downarrow$ as shown in Fig. 4(c). Their energies at $t_0 = 0$ are written as follows ;

$$h\nu_N = \Delta - V, \tag{17}$$

$$h\nu_I = 2\alpha_M V - \Delta - V, \tag{18}$$

$$h\nu_I' = -2\alpha_M V + \Delta + 2U. \tag{19}$$

Here, $h\nu_N$, $h\nu_I$, and $h\nu_I'$ are the energies for the excitation $D'_\sigma \to A'_\sigma$ in the N phase, and the excitations $A'_\uparrow \to D'_\uparrow$ and $D'_\downarrow \to A'_\downarrow$ in the I phase, respectively. Since $\Delta = \alpha_M V$ at the NI transition under neglect of dimerization, their energies become $h\nu_N = h\nu_I = \alpha_M V - V$ and $h\nu_I' = -\alpha_M V + 2U - V$. We can see from the former equation that the optical excitation energy is not changed before and after the transition. On the other hand, $h\nu_I'$ is larger than $h\nu_I$ if $U > \alpha_M V$. It shows that a new absorption peak appears on the higher energy side of the common peak ($h\nu_N = h\nu_I$) when the phase is changed to the I phase. These predictions are consistent with the observation of two exciton absorption peaks in the I phase in contrast to a single peak in the N phase [3]. We propose this explanation as an alternative to that of Jacobsen and Torrance [14] who ascribed the new peak to dimerization.

5 Conclusion

We have first studied the ground state considering the long-range Coulomb interactions exactly. In the I phase, we have found that the lattice is dimerized by its electrostatic instability. We have used the unrestricted Hartree-Fock approximation and investigated the properties of the ground state. We have also discussed the magnetic and optical excitations qualitatively.

Although the application of the mean field theory to the phase transition in one-dimensional systems is less controversial for long-range interaction than for short-range interaction, it is inclined to predict the N-I discontinuity line to persist up to too large a value of t_0. This problem is left for a future study in which intrachain fluctuations and interchain coupling will be taken into account.

Acknowledgement

We are grateful to Professor T. Koda and his collaborators for drawing our attention to this problem, and to Prof. K. Nasu, Prof. S. Abe, and Dr. Y. Shimoi for useful discussions. One of us (T. I-S,) thanks Mr. H. Sakano for helpful advice on the methods of numerical calculation.

References

[1] J. B. Torrance, J. E. Vazquez, J. J. Mayerle and V. Y. Lee: Phys. Rev. Lett. **46** (1981) 253.

[2] J. B. Torrance, A. Girlando, J. J. Mayerle, J. I. Crowley, V. Y. Lee, P. Batail and S. J. LaPlaca: Phys. Rev. Lett. **47** (1981) 1747.

[3] Y. Tokura, T. Koda, T. Mitani and G. Saito: Solid State Commun. **43** (1982) 757.

[4] Y. Tokura, Y. Kaneko, H. Okamoto, S. Tanuma, T. Koda, T. Mitani and G. Saito: Mol. Cryst. Liq. Cryst. **125** (1985) 199.

[5] A. Moréac: PhD thesis, University of Rennes I, France, (1995).

[6] H. M. MacConnel, B. M. Hoffman and R. M. Metzger: Proc. Natl. Acad. Sci. U.S.A. **53** (1965) 46.

[7] Y. Kanai, M. Tani, S. Kagoshima, T. Tokura and T. Koda: Synth. Metals **10** (1984) 157

[8] T. Iizuka-Sakano and Y. Toyozawa: J. Phys. Soc. Jpn. **65** (1996) 671

[9] Y. Nogami, M. Taoda, K. Oshima, S. Koshihara, K. Akimoto, S. Aoki, T. Nakayama and A. Miura: (unpublished).

[10] M. Le Cointe, M, H. Lemée-Cailleau, H. Cailleau, B. Toudic, L. Toupet, G. Heger, F. Moussa, P. Schweiss, K. H. Kraft and N. Karl: Phys. Rev. B **51** (1995) 3374.

[11] A. Painelli and A. Girlando : J. Chem. Phys. **87** (1987) 1705.

[12] T. Mitani, G. Saito, Y. Tokura and T. Koda: Phys. Rev. Lett. **53** (1984) 842.

[13] Y. Iwasa, T. Koda, S. Koshihara, Y. Tokura and G. Saito: Mol. Cryst. Liq. Cryst. **216** (1992) 195.

[14] C. S. Jacobsen and J. B. Torrance: J. Chem. Phys. **78** (1983) 112.

Part VI

Nonlinear Excitations in Charge-Transfer Salts

Exciton Strings and Exciton Mediated Carrier Transport in a Quasi-one-dimensional Organic Charge-Transfer Crystal

M. Kuwata-Gonokami

Department of Applied Physics, Graduate School of Engineering,
The University of Tokyo, 7-3-1, Hongo, Bunkyo-ku Tokyo, 113 Japan
email: gonokami@ap.t.u-tokyo.ac.jp

Abstract. We discuss two types of photoinduced phenomena in which Coulomb attraction between ionized donor and acceptor molecules are important in a quasi-one-dimensional weak charge transfer crystal of anthracene-PMDA. Firstly, we report the formation of multi-excitons by a strong photo-excitation. High density excitation of charge transfer excitons by sub-picosecond light pulses generate multi-exciton chains along the stack axis of the crystal. Secondly, we report a new type of photo-induced carrier transport phenomena. We apply a static electric field parallel to the stack axis of a thin single crystal at liquid helium temperature and examine the DC Stark splitting of the zero phonon line of the CT-exciton under the additional photoexcitation at the CT exciton band. With the assistance of the photo-excitation, carriers move along the stack axis and stay at the edge of the illuminated region. They eventually cancel the external field. A self trapped electron or hole attracts a photogenerated CT-exciton and forms a coupled state $A^-D^+A^-$ or $D^+A^-D^+$. Successive annihilation and creation of the CT exciton result in the charge transport along the stack axis.

1. Introduction

Collective Coulomb interaction plays an essential role in quasi-one-dimensional organic charge transfer (CT) crystals, which are composed of alternately stacked aromatic donor (D) and acceptor (A) molecules. These materials tend to show the neutral (N) phase or the ionic (I) phase at their ground states. In N-phase, we have neutral donors (D) and neutral acceptors (A). In I-phase, the electrons of donors transfer to acceptor molecules to form ionic crystals composed of D^+ and A^- molecules. In materials whose ground states are very close to the boundary between N and I phases, reversible N-to-I phase transition under pressure or temperature were observed [1,2]. Photoinduced phase transition from I-phase to N phase is reported in TTF-CA which is also close to the N-I phase boundary [3]. On the other hand, weak CT crystals which are far from the N-I boundary, have stable neutral ground state and have charge transfer excitons (CTEs) as elementary excitations of the crystals. In such neutral CT crystals, Coulomb interaction between D^+ and A^- molecules plays an important role in the photo excited states.

In this paper, we discuss two types of photoinduced phenomena in which D^+ and A^- Coulomb attraction is important in a weak charge transfer crystal of anthracene-pyromellitic dianhydride (ANT-PMDA). In section 2, we describe the various photo excited states in a neutral mixed stack CT crystal with brief summary

Springer Series in Solid-State Sciences, Volume 124
Relaxations of Excited States and Photo-Induced Structural Phase Transitions
Ed.: K. Nasu

of our results on multi-exciton states. In section 3, we present a new type of photo-induced carrier transport phenomena mediated by the exciton-charge coupled states.

2. Multi-exciton states and exciton-charge coupled states in neutral quasi-one-dimensional CT crystals

The ground state of weak charge transfer mixed stack crystal has a weak polar character. The excited state of a weak CT crystal, that is a charge transfer exciton (CTE), shows a strong polar character. CTE corresponds to a transfer of an electron from the highest occupied donor orbital to the lowest unoccupied orbital of acceptor molecule: $DADADA \rightarrow DAD^{+}A^{-}DA$ as shown in Fig.1 (a). The Coulomb attraction energy, U, between the donor ion (D^{+}) and the acceptor ion (A^{-}), which stabilizes the CTE, is the order of 0.5eV for ANT-PMDA. A strong optical excitation could create many ionized pairs in the crystal. Several CTEs can be aligned along the c-axis to form a "exciton n-string state" as shown in Fig. 1 (d, e). With the point charge model, the binding energy of the exciton n-string state is $U \cdot M(n)$ with $M(n)=\sum_{k=1}^{2n-1}(-1)^{k-1}\frac{(2n-k)}{k}$. When n becomes infinite, the value $M(n)/n$ approaches to -2ln2, the Madelung coefficient of the one dimensional ionic crystal. The energy difference between n-1 exciton string state, $E_{(n-1)}$, and n-exciton state, E_n, ; $\Delta E(n) \equiv E_n - E_{(n-1)}$ is smaller than the single CTE energy. As n increases, the $\Delta E(n)$ decreases progressively. This implies that we can observe induced absorption from (n-1)-exciton state to n-exciton state on the low energy side of the strong CTE absorption band in pump and probe spectroscopy with femtosecond pulses. We performed the experiments on a single crystal of ANT-PMDA and observed induced absorption from 1-exciton state to 2-exciton state and also 2-exciton state to 3-exciton state. The observed binding energies of 2-exciton

(a) $D^0A^0D^0A^0D^+A^-D^0A^0D^0A^0$ CT exciton
$-U$

(b) $D^0A^0D^0A^-D^+A^-D^0A^0D^0A^0$ CT exciton + electron
$-3/2U$

(c) $D^0A^0D^0A^0D^+A^-D^+A^0D^0A^0$ CT exciton + hole
$-3/2U$

(d) $D^0A^0D^0A^0D^+A^-D^+A^-D^0A^0$ 2-string

(e) $D^0A^0D^+A^- \cdots\cdots D^+A^-D^0A^0$ *n-string*

Figure 1. Coulomb interaction between donor and acceptor ions in photo-excited states in a quasi-one-dimensional neutral CT crystal state and 3-exciton state are 49meV and 59meV.[4] The larger binding energy of the 3-exciton state is in good agreement with the above prediction. Stability of the exciton *n*-string states of one-dimensional CT crystal has been theoretically examined within an extended Hubbard model with long range Coulomb interaction.[4-6] The multi-Frenkel exciton states in a one-dimensional molecular chain have also been discussed by Ezaki, Tokihiro and Hanamura, showing the existence of the bound *n*-exciton states when the Frenkel excitons have electric dipole moments along the chain direction.[7] The formation of the biexciton state in such one-dimensional Frenkel system is caused by the dipole dipole attraction between two excitons. This mechanism is almost the same as the case of one-dimensional CT crystals. However, the stabilization mechanism of *n*-exciton bound state of $n > 2$ is slightly different. This is because within the Frenkel model there is no long range interaction, and the stabilization of the multi-Frenkel exciton still originates from the nearest neighbor dipole-dipole interaction. This mechanism itself is very interesting, but has not yet been proved experimentally. We need experiments with suitable materials such as dye aggregate states.

In addition to the multi-exciton states, there exist some other interesting states which are stabilized by the Coulomb attraction in one-dimensional chain. For example, electron (A^-) or hole (D^+) attracts an exciton ($D^+ A^-$) and forms a state $A^- D^+ A^-$ or $D^+ A^- D^+$ as shown in Fig. 1 b and c. The Coulomb binding energy of these states are $-3/2U$. These complexes may play important roles in the photo assisted carrier transport phenomena which will be discussed in the next section.

3. Photo-induced carrier transport in ANT-PMDA

3-1 Charge transfer excitons of ANT-PMDA

ANT- PMDA is a typical weak charge transfer mixed-stack compound composed of alternating face-to-face donor (anthracene) and acceptor (PMDA) molecules. The optical properties of CTEs in this material have been studied extensively.[8-12] In particular, the material shows an extraordinarily narrow zero phonon line (ZPL) of CTE at low temperature and thus very precise experiments on lattice phonon coupling and external field dependence have been successfully performed [9-12]

ANT-PMDA crystals have inversion symmetry, thus the linear combination states of two polar charge transfer states, $A^-D^+A=\varphi_L$ and $AD^+A^-=\varphi_R$, are the eigen states: the even state is the dipole forbidden state and the odd state is the dipole allowed state. An external field *F* breaks the inversion symmetry so that the even state gains the oscillator strength. Thus the ZPL shows Stark splitting and shifts. By measuring the Stark spectra as a function of time and space position, we can map out the temporal behavior of electric field distribution in the crystal.

Carrier mobility measurements have also been performed on ANT-PMDA. Very large electron mobility along c-axis was observed by the time of flight measurement of photogenerated electrons at room temperature.[13] By the temperature dependence of the photoconductivity measurements, the photo carrier generation process via CTE state has activation energy of 0.2 eV, which is the CTE binding energy.[13] Because of this large activation energy, photoconductivity plummet down as the temperature decreases. Below liquid nitrogen temperature, ANT-PMDA sample is too insulating to detect photo-current directly.[14] Thus we need to find a new method to study the non-thermal carrier transport phenomena at low temperature. For this purpose, we develop a spectroscopic method with DC Stark effect as shown below.

3-2 Experimental Method

High quality ANT-PMDA single crystals were grown by vapor phase transport method from equal molar amount of zone refined ANT and PMDA polycrystals. Argon gas with 200 Torr pressure was used as a carrier. Thin single crystals with thickness of less than 50μm having (010) faces were obtained. We put gold electrodes on a crystal surface by evaporation with a slit perpendicular to the c-axis with 100 μm width. We could apply DC field parallel to the stack axis. At low temperature, the samples showed very low conductivity. For the optical measurements, samples were mounted on a sample holder being free from any strains and immersed into liquid He which was kept below lambda point. We measured the transmission spectra with probe light from a Xe lamp. In order to avoid the photo-induced effects caused by the probe beam, we eliminated the unnecessary short wavelength part by a color glass filter and kept the probe light intensity as weak as possible. With this condition, we observed very narrow ZPL of the width of less than 0.2meV. We applied the DC voltage of up to 70kV/cm. We obtained DC Stark shifts and splitting of ZPL, which were consistent with the previous report.[12] At the ZPL resonance, the transmittance is very sensitive to the electric field. Thus we used the DC Stark effects for the measurement of temporal and spatial dependent electric field distribution. Frequency doubled CW YLF laser (527nm) or Ar-ion laser of 514nm was used for the pump beam. For the pump frequency dependence measurements, we monochromatized a broad emission from a 150-W xenon lamp through a monochromator of 10 cm focal length with the band width of 2nm. When we illuminated the sample with pump light under DC field, the DC Stark spectra of ZPL changed. Figure 2 shows the typical sequence of the experiment and ZPL spectra for each stage. We start from the state without DC field. (I) Then we apply the DC field. ZPL shows Stark shift and splitting. (II) Next, we put the pump beam onto the sample and examine the change of the transmission spectrum. The spectrum gradually changes.

After several minuets, the spectrum eventually return to the zero field spectrum.(III) At this point, we stop the pump beam illumination and turn off the DC field The spectrum shows the DC Stark effect again. (IV) Then we turn on the pump beam again, the DC Stark effect gradually vanishes and return to the initial state of (I). This sequential state change is the function of optical memory. We can write and erase the data with photo irradiation under the control of bias voltage. Figure 3 shows the transmission change at the ZPL peak wave length.

From the transmission signal, we can estimate the effective internal field. Figure 4 shows the plots of the decay time versus pump intensity taken for the process of V. For both processes of III and V, decay rates are linear to the pump light power. Now we have to clarify whether the phenomena come from the space charge effect near the electrode[12], or bulk phenomena. We examined the space time dependence of the effective field by making the pump light spot size smaller

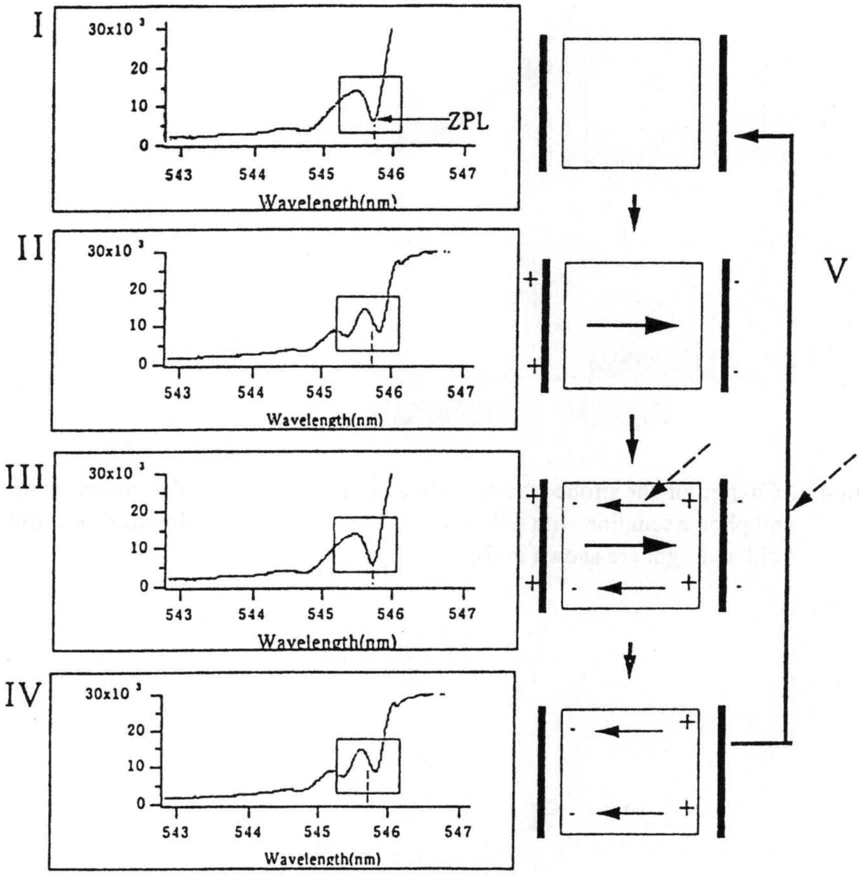

Figure 2. DC Stark spectra near the zero phonon line of ANT-PMDA at 2K under the illumination of pump light. Applied DC field is 40kV/cm.

than the slit width in order not to touch the electrode. We monochromatized the probe beam with 0.2nm band width and set its frequency to the ZPL frequency region where the transmisson was sensitive to the DC field. With this probe light, we made expanded images of the sample onto the CCD camera and recorded images for every stages of Fig. 2. We observed a clear change on the effective field only on the region of the pump irradiation. This implies that our observed phenomena relate to neither the photo carrier injection nor the space charge near the electrodes. We also measured the pump light frequency dependence. The action spectrum of the field decay rate shows an abrupt increase at the CT exciton edge and peaks at the CT exciton band.

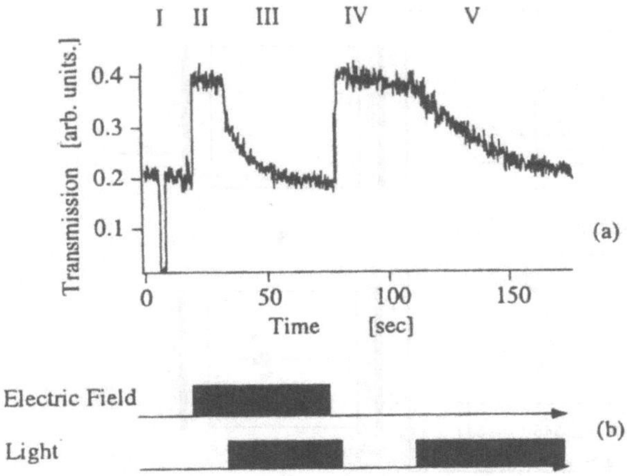

Figure 3. Change of the probe transmission at ZPL resonance under electric field and photoexcitation with CW light (527 nm) (a). The conditions of electric field and light are shown in (b).

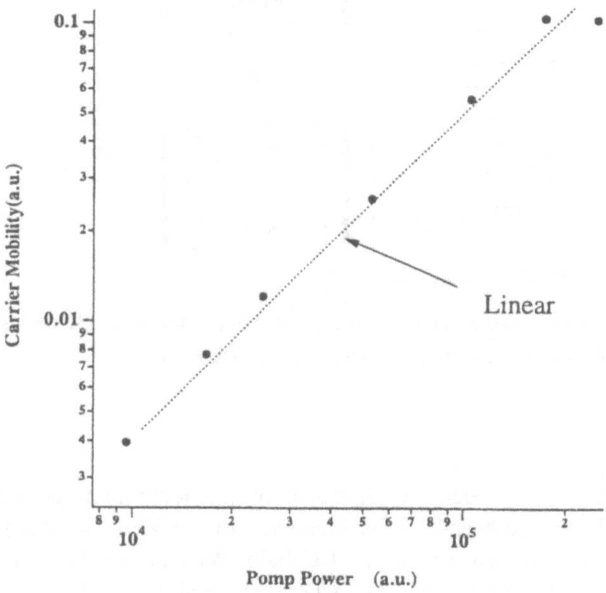

Figure 4. Photo-induced carrier mobility as a function of pump light power. Pump light is CW light with wave length of 527nm.

3-3 Results and discussion

The observed optical memory effects are caused by the photoinduced charged carrier transport. In the crystal, there are charged carriers, which we call residual charges. We measure the density of residual charges by ESR measurement and obtain the value of about 10^{-7} M/M. At low temperature, these residual charges interact strongly with lattice and remain immobile up to fields of 70kV/cm.

With the pump beam irradiation, the charged carriers start to move toward the electrodes. At the electrode, there are potential barriers which prevent the carrier injection into the crystal. Then the charges arrive near the electrodes and then they pile-up and make the reverse field which cancel the external field. In the state (IV) of Figure 2 and 3, the compensation field appear and the following photo-irradiation promotes the carrier redistribution until the compensation field vanishes. Because the decay rate shows the linear dependence to the pump beam power, the relevant optical process should be a linear process.

As we mentioned above, to generate free carrier pairs, we need to reach the band edge, which is 0.2eV above the CT exciton. The abrupt increase of the field decay rate at the CT absorption edge but not at band edge, indicates that neither the free carrier generation process by the pump beam nor photo-assisted carrier injection at the electrodes play any role. Thus the CT exciton mediated carrier transport should be the origin of the observed phenomena.

We illustrate a model to explain the mechanism of CTE mediated carrier transport in Fig. 5. When one CTE is generated close to a self-trapped electron(A^-) or hole (D^+), the self trapped charge polarizes the CTE and attracts it to form a coupled "trimer" state, $D^+A^-D^+$ or $A^-D^+A^-$, which corresponds to the state of Fig. 1 (b) or (c). With the presence of the electric field, this trimer state could split into two weakly polarized states by the lattice instability or the charge instability followed by the charge redistribution along the stack axis. In the case of $A^-D^+A^-$ state, the lower state has higher probability of CTE recombination on the left hand side than the recombination on the right hand side. This is because the wave function overlap between the electron and the hole on the left hand side is larger than that on the right hand side. If the life time of the trimer state is sufficiently long and we have enough time to establish quasi equilibrium distribution, we can get net carrier transport along the stack axis in the DC field direction. The velocity of the carrier depends on the DC field strength. From the decay rate of the effective field strength on the processes of (III) and (V), we can estimate the velocity of the carriers. The relation between carrier mobility, μ , and the field decay rate can be expressed as,

$$\frac{1}{\tau}=\frac{en\mu(P,F)}{\varepsilon} \qquad (1)$$

where P is the pump light power , F is the electric field amplitude, n is the carrier density, e is the charge and ε is the dielectric constant. At the pump power of 30[mW/cm^2] and electric field of 40[kV/cm], we obtain the mobility of 10^{-11} [cm^2/secV]. If we assume that all the photons are converted to CTEs in the crystal, the generation rate of excitons is about 10^{20}[1/cm^{3}sec]. With this value we can estimate the upper limit of the mobility under the assumption that charge can move only one site by the one CT exciton creation and successive annihilation at the adjacent site. The upper limit is the order of 10^{-13} [cm^2/secV] which is 2 order of magnitude smaller than the experimental value. There are two reasons which may cause this discrepancy. Firstly, at the formation process of trimer state, the range of the trimer formation is very large. This means that an exciton within 100 sites can

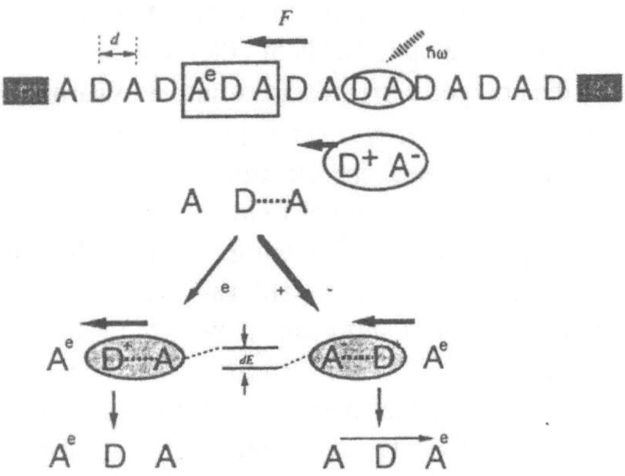

Figure 5. Schematic illustration of the CT exciton mediated carrier transport mechanism.

be dragged by a charged carrier and form a trimer state. Secondly, the polarized trimer state may move several sites before it decays. This means that the trimer state acts as a charged carrier. In the trimer state, the wave function is more delocalized than electrons or holes and thus the trimers could have reduced polaron masses and could have higher mobility.

To clarify the mechanism, we need more experiments on the electric field dependence, temperature dependence etc.

4. Conclusion

We presented two examples of photo-induced phenomena in which long range Coulomb interaction plays essential roles. One is the formation of stable multi-exciton state along the one dimensional stack axis. Although the observed multi-exciton state is highly non-equilibrium, such state is very important for the ultrafast nonlinear optical process. The other example is the exciton mediated carrier transport. In this process, the combined state of electron (or hole) and one CT exciton plays an important role. Although the detailed mechanism of the carrier transport is not yet clear, this novel process introduces a new photo-induced phenomenon in low dimensional systems. The detailed study such as the stability of such trimer state is necessary both theoretically and experimentally.

We discussed the role of the Coulomb interaction on the excited state of quasi-one dimensional system. The magnetic interaction should also be important. In particular, the multi-exciton state and trimer state should have lone pair electrons or holes, thus we need to consider the spin degree of freedom for further understanding of the nature of such states. Triplet states of charge transfer exciton in weak charge transfer crystals are observed in phosphorescence spectra. In ANT-PMDA, the triplet CT exciton is not observed, because the triplet state of donor molecule is lower than the triplet state of CT excitons. The energy level of the triplet CT exciton is estimated to be 0.1eV below singlet CT exciton level from the phosphorescence data of phenantracene-PMDA.[15] Thus phosphorescence measurements under various excitation condition will be one of the key experiments for the further understanding of the photoinduced phenomena in quasi-one dimensional organic CT crystals.

Acknowledgements

The author would like to acknowledge to Mr. T. Wada and Mr. Y. Naitoh for their effort on the experiments on photo-induced carrier transport. The author also thank to Professors N. Peyghambarian , S. Mazumdar, N. Nagaosa, Y. Toyozawa, T. Koda, E. Hanamura, Y. Tokura and K. Nasu for stimulating discussion. This work is supported by the Monbusyo International Science Research Program, Joint Research Program No.07044121, from the Ministry of Education, Science, Sports, and Culture of Japan.

References

[1]J. B. Torrance, J. E. Vazquez, J. J. Mayerle, and V. Y. Lee, Phys. Rev. Lett. 46 (1981) 253.

[2]J. B. Torrance, A. Girlando, J. J. Mayerle, J. I. Crowley, V. Y. Lee, P. Batail and S. J. Laplaca, Phys. Rev. Lett. 47 (1981) 1747.

[3]S. Koshihara, Y. Tokura, T. Mitani, G. Saito and T. Koda, Phys. Rev. B 42 (1990) 6853.

[4] M. Kuwata-Gonokami, N. Peyghambarian, K Meissner, B. Fluegel, Y. Sato, K. Ema, R. Shimano, S. Mazumdar, F. Guo, T. Tokihiro, H. Ezaki and E. Hanamura, Nature (London) **367**, 47 (1994).

[5] S. Mazumdar, F. Guo, K. Meissner, B. Fluegel, N. Peyghambarian, M. Kuwata-Gonokami, Y. Sato, K. Ema, R. Shimano, T. Tokihiro, H. Ezaki and E. Hanamura, J. Chem. Phys. **104**, 9283 (1996).

[6] S. Mazumdar, F. Guo, K. Meissner, B. Fluegel, N. Peyghambarian, M. Kuwata-Gonokami, Y. Sato, K. Ema, R. Shimano, T. Tokihiro, H. Ezaki and E. Hanamura, J. Chem. Phys. **104**, 9292 (1996).

[7] H. Ezaki, T. Tokihiro and E. Hanamura, Phys. Rev. B **50** 10506 (1994).

[8] D. Harrer and M. R. Philpott, in *Spectroscopy and Excitation Dynamics of Condensed Molecular Systems*, edited by V. M. Agranovich and R. M. Hochstrasser (North Holland, Amsterdam, 1983).

[9]D. Harrer, M. R. Philpoptt, and H. Morawitz, J. Chem. Phys. 63 (1975) 5238.

[10]A. Brillante and M. R. Philpott, J. Chem. Phys. 72 (1980) 4019.
[11]Y.Tokura and T. Koda, Solid State Commun. 40 (1981) 299.
[12]A. Elschner and G. Weiser, Chem. Phys.98 (1985) 465.
[13]N. Karl and J. Ziegler, Chem. Phys. Lett. 32 (1975) 438.
[14]D. Massa and N. Karl, Mol. Cryst. Liq. Cryst. **95**, (1989) 93.
[15]T. Koda , in this procedings.

Photophysics of Charge-Transfer Solids and π-Conjugated Polymers

M. Chandross, F. Guo, Y. Shimoi,† and S. Mazumdar*

Department of Physics, University of Arizona, Tuscon, AZ 85721, USA

* Permanent address: Department of Physics, National Chung-Hua University of Education, Chung-hua City, Taiwan 500.

† Permanent address: Electrotechnical Laboratory, Tsukuba, Ibaraki 305, Japan.

Abstract. Even though the nature of the interactions between the electrons are different in donor-acceptor organic charge-transfer solids and π-conjugated polymers, their photophysics are similar. This is a consequence of one dimensionality, which leads to strong exciton binding as well as exciton-exciton binding.

1. Introduction

Neutral mixed-stack charge-transfer (CT) solids are narrow band quasi-one-dimensional (quasi-1-d) materials in which the direct Coulomb interaction between electrons plays a very strong role[1]. One consequence of Coulomb interactions in 1-d is the occurrence of CT excitons with relatively large binding energies. It has recently been shown that the combined effects of one dimensionality and Coulomb interactions in CT solids lead to novel multiexciton strings[2]. Direct photoinduced absorption (PA) experiments demonstrated biexcitons as well as triexcitons, in a strongly neutral CT solid, anthracene-pyromellitic acid anhydride (hereafter ANT-PMDA)[2]. In addition to the high energy (HE) multiexciton states, electroabsorption (EA) measurements had earlier detected a one-photon forbidden exciton that lies about 15 cm^{-1} above the optical exciton in ANT-PMDA[3]. We present a brief theory of photoexcitations in CT solids here that explains the EA and the PA data within the same CT exciton model.

EA and PA measurements are currently also being routinely done for π-conjugated polymers. These experiments indicate that photoexcitations in conjugated polymers are quite different from that expected within the one–electron Su-Schrieffer-Heeger (SSH) model. For example, in poly(para-phenylenevinylene) (PPV), EA finds a one-photon forbidden state that is energetically removed from the optical state by at least 0.5 eV[4]. PA to this "low energy" (LE) even parity state as well as to another state at even higher energy (HE), at about 1.5 eV above the optical state, have been observed[4-6]. We point out that such excitations in conjugated polymers are the exact equivalents of the LE and HE excitations seen in ANT-PMDA. Photoexcitations in ANT-PMDA and conjugated polymers are similar, even though the former is a

Springer Series in Solid-State Sciences, Volume 124
Relaxations of Excited States and Photo-Induced Structural Phase Transitions
Ed.: K. Nasu

charge density wave (CDW) system, while the latter are bond order wave-spin density wave (BOW).

2. Mixed-stack CT solids

Mixed stack CT solids consist of planar aromatic donor (D) and acceptor (A) molecules that alternate along a 1-d stack, with the molecular planes nearly perpendicular to the stacking axis. Large intermolecular spacings (3-3.5 Å) along the stack lead to small hopping integrals (0.1-0.2 eV) between the π-molecular orbitals (MOs) of neighboring D and A molecules. Important material parameters are the ionization energy of I_D of D, the electron affinity A_A of A, and the electrostatic Madelung energy of the completely ionic state. Depending upon these parameters, it is possible to have both neutral and ionic CT solids[1], and neutral-ionic phase transitions are known[7,8]. ANT-PMDA, however, is a strongly neutral material that is far from the neutral-ionic boundary.

The standard theoretical model[1,2] for mixed stack CT solids is,

$$H_{CT} = \sum_{l,\sigma}(-1)^l \epsilon c^{\dagger}_{l\sigma} c_{l\sigma} + t \sum_{l,\sigma}(c^{\dagger}_{l\sigma} c_{l+1,\sigma} + c^{\dagger}_{l+1,\sigma} c_{l\sigma}) + \frac{1}{2} \sum_{l \neq m} V_{lm} \rho_l \rho_m \ , \quad (1)$$

where we have adopted the convention that the D(A) molecules occupy odd(even) sites. In the above, $c^{\dagger}_{l,\sigma}$ creates an electron of spin σ at the molecular site l, $2\epsilon = I_D - A_A$, and $V_{lm} = V_{|l-m|}$ the Coulomb interaction between charged sites $|l-m|$ lattice constants apart. The CT transition involves the transfer of one of the two electrons from the the highest occupied molecular orbital (HOMO) of D to the lowest unoccupied molecular orbital (LUMO) of A. Thus the quantity $\rho_l = z - \sum_\sigma c^{\dagger}_{l\sigma} c_{l\sigma}$, where $z = 2$ for a D site and 0 for an A site. Doubly charged D^{++} and A^{--} are excluded, because of the very high energies necessary to create these species.

2.1 The t = 0 limit

Qualitative insight to Eq.(1) can be obtained in the limit $t = 0$. The ground state in this limit is given by the single configuration ...DADADA..., which can be also written as the extreme CDW configuration ...202020..., where the numbers denote site occupancies by electrons. Nearest neighbor CT excitation results in the configurations $...DAD^+A^-DA...$ and $...DA^-D^+ADA...$, with site occupancies ..201120... and 211020..., respectively. Dipole-allowed transition occurs only to the odd parity "minus" combination of these two configurations, while the even parity "plus" combination is one-photon forbidden. In the $t = 0$ limit, these states are degenerate, and the excitation energy is $2\epsilon - V_1$. States in which the D^+ and A^- are far from each other have energy 2ϵ. The electron-hole pair in the lowest one- and two-photon states are therefore bound by the strong nearest neighbor electrostatic interaction V_1. We shall refer to these states as the odd and even parity 1–excitons.

The forbidden 1-exciton becomes visible in EA. For $t \neq 0$ we expect a strong dipole coupling between allowed and forbidden 1-excitons. In the presence of a static electric field, the overall Hamiltonian is the bare Hamiltonian (1) plus μF, where μ is the dipole operator and F the static field. The perturbation due to the field therefore mixes the even and odd parity excitations, and the previously forbidden higher 1-exciton becomes allowed. In CT solids this state is too close in energy to the opitcal exciton to be seen in PA.

For $t \rightarrow 0$, further consecutive CT excitations can give states with n excitations, which can consist of free excitons, as well as the bound n–exciton string. The bound 2-exciton states have site occupancies ...202011112020... There exist overall four 2-exciton states, two in each spatial parity subspace. The threshold of the continuum of two excitons is simply twice the energy of the 1-exciton, i.e., $2(2\epsilon - V_1)$. Within Eq.(1), in the limit $t = 0$, the 2-exciton occurs at energy $4\epsilon - 3V_1 + 2V_2 - V_3$, and therefore has a binding energy of $V_1 - 2V_2 + V_3$, which is positive definite for a convex potential. PA from the 1-exciton to the 2-exciton should therefore be visible in differential transmission measurements. Related conclusions for the 3-exciton may be found in the reference 2.

2.2 The t $\neq$ 0 case

Exact numerical calculations were carried out for a 14-site periodic ring within Eq.(1) for fitting the experimental data[2] on ANT-PMDA. The particular parameters we chose are: $|t| = 0.1$ eV, $\epsilon = 1.38$ eV, $V_1 = 0.5$ eV, $V_n = V_1/n$. Detailed results of the calculations of differential transmission may be found in reference 2; here we merely point out the natures of the electronic structure and the transition dipole couplings. In Fig. 1(a) we show the exact energies of the 14-site periodic ring with our parameters. The even and odd parity states are shown separately, and the double-headed arrows indicate pairs of states that are most strongly dipole coupled. Although in principle PA from the 1-exciton can occur to both the even parity 2-exciton states, only one of the dipole couplings is strong. Similarly the dipole coupling is strong to only the edge state of the two exciton continuum. The relative dipole couplings between the optical exciton and the two exciton states is shown in Fig. 1(b). As expected, the dipole coupling between the 1-exciton and the edge state of the two exciton continuum is exactly equal to that between the ground state and the exciton. The dipole coupling to the bound 2-exciton is smaller, the physical reason for which is given elsewhere[9]. The dipole coupling between the even and odd parity 1-exciton states is nearly five times stronger, indicating that even a weak electric field will be sufficient for observing the even parity 1-exciton in EA.

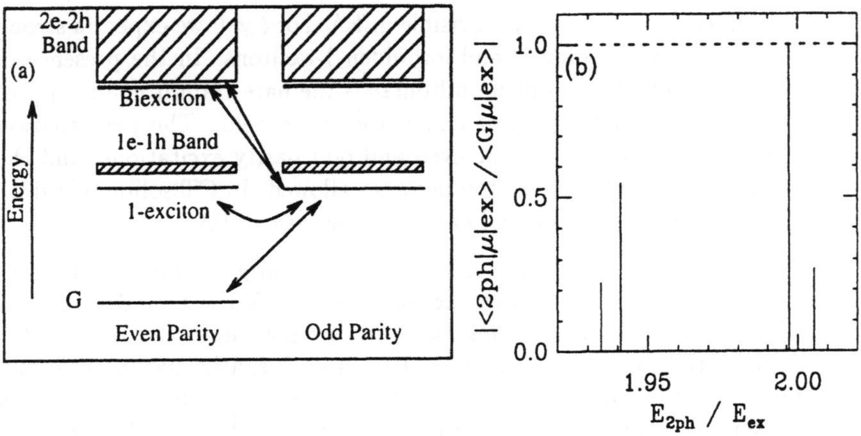

Fig. 1(a). The ground state G, exciton, $1e - 1h$ band, biexciton and two free exciton levels of a 14-site perdiodic ring within Eq.(1) for $|t| = 0.1$ eV, $\epsilon = 1.38$ eV, $V_1 = 0.5$ eV, $V_n = V_1/n$. The double-headed arrows connect states that are dipole-coupled. (b) Normalized dipole couplings between the optical exciton (ex) and the two-photon (2ph) two-exciton states of (a). The dipole coupling to the biexciton is smaller than to the state consisting of two free excitons.

3. π-conjugated polymers

We report our calculations for the simplest linear chain polymer, trans-polyacetylene, the Hamiltonian for which is written as,

$$H = U\sum_i n_{i,\uparrow}n_{i,\downarrow} + \frac{1}{2}\sum_{i,j} V_{ij}(n_i - 1)(n_j - 1) - \sum_{i,\sigma} t(1 \pm \delta)(c^\dagger_{i,\sigma}c_{i+1,\sigma} + h.c.). \quad (2)$$

Unlike Eq.(1), here all sites are equivalent. U and V_{ij} are the on-site and inter-site Coulomb interactions, $n_{i,\sigma} = c^\dagger_{i,\sigma}c_{i,\sigma}$, and $n_i = \sum_\sigma n_{i,\sigma}$. The parameter δ gives the modulation of the hopping integral due to bond alternation.

3.1 The exciton basis

Because of the many-body nature of Eq.(2) we are limited to numerical solutions. Approximate long chain calculations include configuration interactions (CI) between only certain classes of configurations[10]. Such limited CI calculations miss the well-known $2A_g$ state that occurs below the optical $1B_u$ state and are not suitable for our purpose. On the other hand, short chain spectra are necessarily discrete, and it is not possible to classify a given eigenstate as exciton or bandlike from energies alone. We have therefore developed an exciton basis approach[11] for exact short chain numerical calculations within

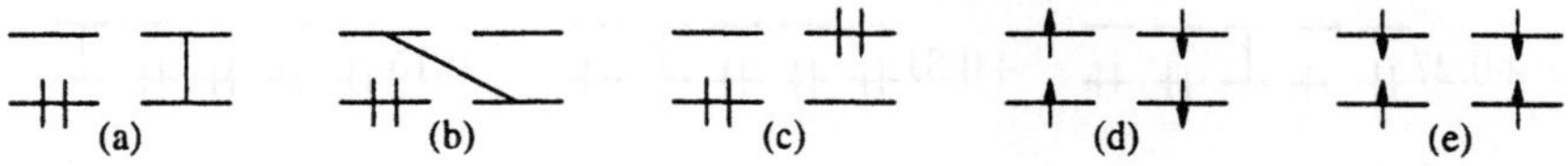

Fig. 2. Typical exciton basis excited diagrams for the 2-unit chain.

Eq.(2) that allows systematic classification of all eigenstates because of the pictorial natures of the wavefunctions.

Within the exciton basis approach we visualize polyacetylene as coupled ethylenic dimer units. Each dimer contributes one bonding and one antibonding MO, and the MOs of different units are coupled by the Coulomb interactions as well as the interunit hopping. The "noninteracting" ground state of the N-unit chain has all electrons in the N bonding MOs, and the complete basis space can be generated by repeated applications of the Hamiltonian (2) on this. We are interested only in the singlet subspace, and each pair of singlet-coupled singly occupied MOs will be shown with a bond between them.

Because of space restrictions, we describe the exact wavefunctions only in terms of the most important basis functions. These are shown in Fig. 2. Figs. 2(a) and (b) correspond to Frenkel and CT excitations from the noninteracting ground state. Fig. 2(c) describes a Frenkel biexciton. Spin is relevant, and we note in Figs. 2(d) and (e) that even within the spin singlet subspace it is possible to have two different spin arrangements for the case of two singly excited units. We will refer to configurations of the type 2(d) as triplet-triplet (TT) and those of the type 2(e) as singlet-singlet (SS). Diagram 2(c) is clearly SS. Thus all optically relevant excited states are linear combinations of TT, CT and SS diagrams. Our goal is to show that the most important eigenstates can be thought of as predominantly TT, CT or SS, in spite of considerable mixing. Optically relevant exact wavefunctions also contain small admixtures from triple, quadruple, etc. excitations, but their natures can be determined from the dominant contributions, which are of the single and double excitation variety only.

3.2 Characterization of excited states

We have calculated the exact energies and wavefunctions of the A_g and B_u subspaces within Eq.(2) for the 4- and 5-unit cases, for various extended Hubbard models and Pariser-Parr-Pople models. The most important states are discussed below for the case of $|t| = 1$, $\delta = 0.1$, $U = 3$, $V_{12} = 1$.

The $1B_u$. The dominant contributions to the N = 5 $1B_u$ state are shown in Fig. 3. The largest contributions to the wavefunctions come from the Frenkel exciton diagrams and the diagrams with nearest neighbor CT. The contributions of CT diagrams with bonds longer than nearest neighbor decrease rapidly with the length of the bond. This is a clear signature of the exciton nature of the $1B_u$ in the long chain limit.

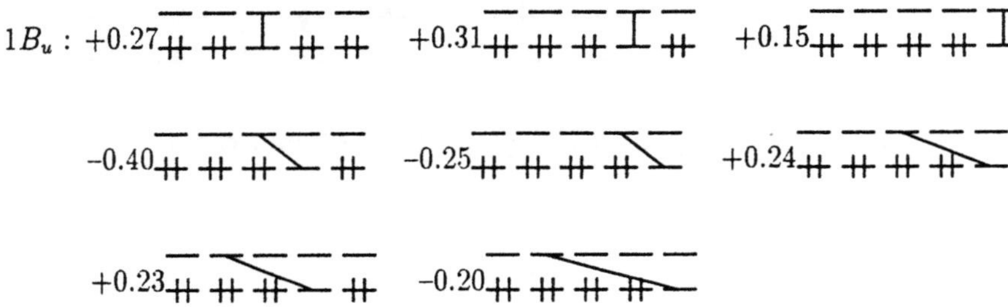

Fig. 3. The dominant contributions to the exact $1B_u$ state of the 5-unit chain within the exciton basis. Each diagram represents all equivalent diagrams obtained by applications of the mirror-plane and charge-conjugation symmetry operators. The contribution of the diagram in which the CT bond is over the entire 5 units is very small (< 0.05).

The $2A_g$. The $2A_g$ wavefunction is predominantly TT, with weak contributions from CT diagrams. We do not show the wavefunction explicitly, as its TT nature will become obvious from what follows. For the parameters we have chosen, the $3A_g$ and the $4A_g$ are also predominantly TT. These states have weak dipole couplings with the $1B_u$. This is expected, as the transition from the $1B_u$ to the TT states not only would involve a second excitation, but also spin flips within each unit, a process forbidden in optical excitation.

The mA_g. Previous calculations by several groups have shown that there exists an above gap A_g state with unusually large dipole coupling with the $1B_u$. This state is now commonly referred to as the mA_g, where m depends on the parameters of the Hamiltonian and the chain length[12]. The mA_g plays a very strong role in the photophysics of π-conjugated polymers, due to its large dipole coupling with the $1B_u$. In the present case m = 5. We find that the mA_g is substantially different from all lower A_g states, in that it is not TT but predominantly CT. The dominant CT diagrams in the mA_g have bonds longer than in the $1B_u$.

The mA_g also has substantial contributions from double excitations, and it is because of this it has also been referred to as the biexciton[13]. This is, however, incorrect. We have calculated the expectation value of the number of excitations in all eigenstates. In Fig. 4 we show a plot of this quantity for the lowest ten A_g eigenstates, relative to the number of excitations in the correlated ground state. Note that the number of excitations is rather large for the $2A_g$ - $4A_g$, but then *decreases* at the $5A_g$, indicating clearly the two excitation TT character of the $2A_g - 5A_g$ and the correlated one excitation CT character of the mA_g.

The SS states. The true biexciton should belong to the class of SS states, which occur at much higher energy. Indeed, in finite chains, they occur above $2 \times E(1B_u)$, where $E(1B_u)$ is the energy of the $1B_u$. The reason for this finite size effect is obvious. The $1B_u$ exciton has a width of about four units

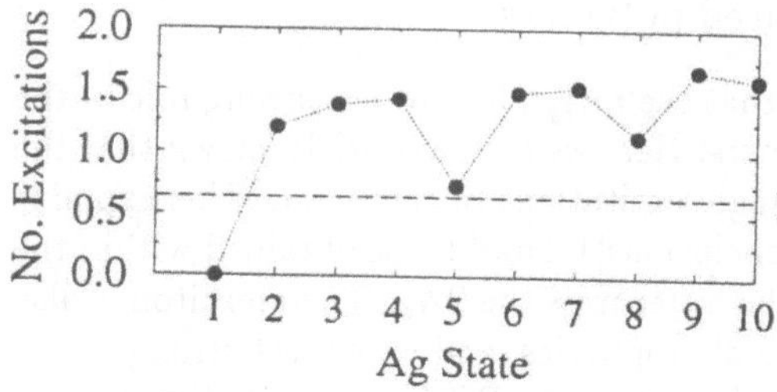

Fig. 4. The number of excitations from the exact correlated ground state for the lowest 10 excited A_g states for $N = 5$ with $U = 3t$, $V_1 = t$, and $\delta = 0.1$. The $5A_g$ is the mA_g for these parameters. The dashed line indicates the number of excitations in the $1B_u$ exciton. Like the $5A_g$, the $8A_g$ is also CT.

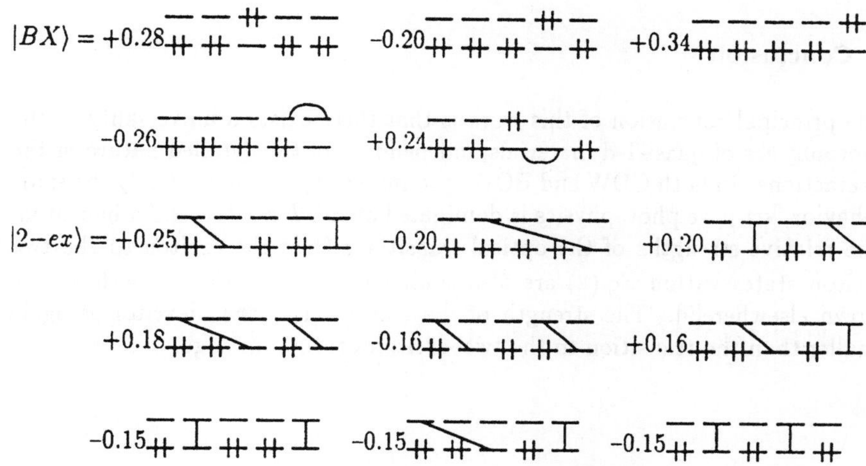

Fig. 5. The dominant contributions to the biexciton (BX) state and to the threshold state of the two-exciton continuum (2-ex). Each diagram represents all equivalent diagrams obtained by applications of the mirror-plane and charge-conjugation symmetry operators.

in Fig. 3. Individual excitons in the finite chain two exciton states have much smaller widths. This raises the energies of the two exciton states. What is more important in the present context is that we find distinct SS states which are dominated by either highly localized two exciton diagrams, or by two excitons that are separated from each other. In Fig. 5 we have shown the dominant contributions to the eigenstates that we believe to have the characters of the biexciton and two free excitons. The absence of eigenstates that have nearly equal contributions from localized as well as separated exciton pairs is a signature of biexciton formation.

3.3 Photophysics of π-conjugated polymers

Existing calculations already show that the mA_g plays a very strong role in the photophysics of π-conjugated polymers. Here we have explicitly shown that the mA_g state is predominantly CT (single excitation) in character. This explains the similarity between our earlier conclusions[12] and those obtained within the single CI studies[10], even though the latter miss the $2A_g$. The biexciton is also expected to play a strong role in the photophysics, and we expect strong PA to the lowest biexciton from the optical exciton. In PPV and related materials, PA to both the mA_g and the biexciton have been seen, while the mA_g has been reached also in EA and direct two-photon absorption. We believe that the biexciton has also been seen in PA studies of the polydiacetylenes[14].

4. Conclusion

The principal conclusion of this work is that there exists a universality in the photophysics of quasi-1-d materials, independent of the detailed nature of the interactions. In both CDW and BOW systems we expect qualitatively the same behavior, viz., the photophysics is dominated by a CT exciton and a biexciton. The relative strengths of the optical transitions from the exciton to the two exciton states within Eq.(2) are also similar to those in Eq.(1), as has been shown elsewhere[9]. The strength of the transition to the biexciton is again smaller than the transition to the two–exciton continuum edge.

5. Acknowledgements

The present work was supported by the United States NSF, AFOSR and the ONR. Collaborative interactions with Professors E. Hanamura (Tokyo), M. Kuwata-Gonokami (Tokyo), N. Peyghambarian (Arizona) and Z.V. Vardeny (Utah) are acknowledged.

References

1. Z.G. Soos, Ann, Rev. Phys. Chem. **25**, 121 (1974).

2. M. Kuwata Gonokami, N. Peyghambarian, K. Meissner, B. Fluegel, Y. Sato, K. Ema, R. Shimano, S. Mazumdar, F. Guo, T. Tokihiro, H. Ezaki, and E. Hanamura, Nature (London) **367**, 47 (1994).

3. D. Haarer, M.R. Philpott and H. Morawitz, J. Chem. Phys. **63**, 5238 (1975).

4. J.M. Leng, S. Jeglinski, X. Wei, Z.V. Vardeny, R.E. Benner, F.Y. Guo and S. Mazumdar, Phys. Rev. Lett. **72**, 156 (1994).
5. J.W.P. Hsu, M. Yan, T.M. Jedju, L.J. Rothberg, and B.R. Hsieh, Phys. Rev. B**49**, 712 (1994).
6. J.W. Blatchford, S.W. Jessen, L.B. Lin, J.J. Lih, T.L. Gustafson, A.J. Epstein, D.K. Fu, M.J. Marsella, T.W. Swager, A.G. MacDiarmid, S. Yamaguchi and H. Hamaguchi, Phys. Rev. Lett. **76**, 1513 (1996).
7. J.B. Torrance, J.E. Vasquez, J.J. Mayerle and V.Y. lee, Phys. Rev. Lett. **46**, 253 (1981).
8. S. Koshihara, Y. Tokura, T. Mitani, G. Saito and T. Koda, Phys. Rev. B**42**, 6853 (1990).
9. S. Mazumdar and F. Guo, Synth. Metals **78**, 187 (1996).
10. S. Abe, M. Schreiber, W.P. Su and J. Yu, Phys. Rev. B**45**, 9432 (1992).
11. M. Chandross and S. Mazumdar, to be published.
12. S.N. Dixit, Dandan Guo and S. Mazumdar, Phys. Rev. B**43**, 6781 (1991);
13. P.C.M. McWilliams, G.W. Hayden and Z.G. Soos, Phys. Rev. B**43**, 9777 (1991).
14. T. Kobayashi, Synth. Metals **50**, 565 (1992).

Part VII

Optical Excitations and Lattice Relaxations in Alkali Halides, Semiconductors

Femtosecond Time-Resolved Studies of Dynamics of Structural Changes Induced by Electronic Excitation in Alkali Halide Crystals

K. Tanimura

Department of Physics, Nagoya University, Furo-cho, Chikusa, Nagoya 464-01, Japan

Abstract. Dynamics of lattice-structural changes of Frenkel-pair formation upon band-gap excitation in alkali halides is discussed based on the recent femtosecond spectroscopic studies. It is shown that the relaxation process of electron-hole (e-h) pairs consists of the two distinct stages with different nature. The first stage is the relaxation of an interacting pair an electron and the relaxing hole which is under way of its relaxation towards the stable two-center type configuration. In this stage, Frenkel pairs are formed through the dynamical instability, and the self-trapped exciton (STE) with the on-center configuration is also formed. The second stage is essentially the off-center relaxation of the on-center STE into the F-H pairs and the off-center STE's. Generality of the process is discussed based on the classification of alkali halide crystals into three classes.

1. Introduction

In many solids with different structural properties and bonding nature, drastic lattice-structure modifications are induced by the excitation of electronic systems. These include Frenkel-pair production in metal halides[1-3], atomic desorption from solid surfaces[4], structural changes in amorphous solids[5], and phase transformation in low dimensional solids[6]. Based on the results of extensive studies on these photo-induced processes, it has been proposed that the interaction between or among photo-generated excited species during their relaxation plays an important role in the structural changes [1, 7-9]. In order to examine the validity of models proposed and to understand the mechanisms of these phenomena from microscopic view point, it is essential to trace the dynamics of the interaction of excited species on the time scale of lattice vibrations.

Frenkel-pair formation by the excitonic process in alkali halides has been a typical example of these photo-induced structural changes in solids. By exciting crystals to generate electron-hole (e-h) pairs, lattice structural defects of Frenkel pairs, as well as the self-trapped excitons (STE's) with different structures are formed in the process of lattice relaxation of (e-h) pairs. Based on the results of extensive experimental and theoretical studies on the mechanism [1-3], it has been proposed that the correlated relaxation of an electron and a hole, i.e., the interplay between electron-phonon and hole-phonon interaction during their relaxation plays a crucial roles in the structural changes [1]. However, the dynamics of the process of the lattice relaxation of (e-h) pairs was never resolved, and several fundamental problems remained unsolved. Recent femtosecond spectroscopic studies have just revealed

Springer Series in Solid-State Sciences, Volume 124
Relaxations of Excited States
and Photo-Induced Structural Phase Transitions
Ed.: K. Nasu

some important features of the dynamics of the relaxation [10-14]. Here we review the results of the lattice relaxation of (e-h) pairs studied by means of the femtosecond time-resolved spectroscopy.

2. Self-trapped excitons and Frenkel pairs

The lattice relaxation of holes, excitons and free (e-h) pairs have been studied extensively, and self-trapping of holes and of excitons has been well documented [2]. The self-trapped hole (STH) in all alkali halides is known to have only one stable configuration in the form of an X_2^- molecular ion (X denotes a halogen atom) occupying two neighboring halogen-lattice sites. This STH, known as the V_K center, is the small polaron in solids first observed [15]. In contrast to the hole, experimentally observed spectroscopic properties of relaxed configurations of excitons can be understood only assuming existence of several different configurations [2]. These include three different structures of the self-trapped excitons (STE's) responsible for the intrinsic luminescence bands, which are often classified as Type I, II, and III [16], and the Frenkel pair consisting of an F center, a halogen vacancy trapping an electron, and an H center, an interstitial halogen atom in the form of X_2^- molecular ion occupying a single halogen-lattice site. The Frenkel pair is called hereafter the F-H pair.

Recent studies of resonance Raman spectroscopy[17], together with spectroscopic features of optical absorption[18, 19] and luminescence[16] related to STE's, have shown clearly that the hole component of the Type-I STE maintains almost the same vibrational and electronic properties as those of the V_K center. Also, the electron at this configuration is bound with a diffused orbital, suggesting little contribution to the lattice relaxation by this trapped electron. Based on these experimental results, one can conclude that the Type-I STE is essentially a V_K center which has trapped an electron. This configuration of STE is usually called the on-center STE[2]. On the other hand, the Type-III STE is essentially a nearest-neighbor pair of an F center and an H center[17]. In this configuration, the X_2^- molecular ion is displaced significantly from the V_K -center position along a <110> direction. This displacement of the X_2^- is referred to as the off-center relaxation, and the STE which has this configuration is called the off-center STE[2]. Type-II STE is presumed to have a configuration intermediate between the two.

In Fig.1, the structures of the on- and off-center STE's and the F-H pair are shown schematically, together with the optical absorption spectra of their typical examples. The spectra show reasonably their structural features described above. Figure 1(a) shows the absorption spectrum of the STE in NaBr which is an example of Type-I STE. The band around 3 eV is due to the hole transitions, and its spectral shape is almost identical to that due to the V_K center shown by the broken curve. The electron transition causes an absorption band at around 0.2 eV, indicating its diffused property of the orbital around the hole [19]. On the other hand, the absorption spectrum of the triplet STE in KBr, which is a typical example of Type-III STE, consists of the hole-transition band similar to the H band, and the electron-transition bands which have peak energies around the F-band region; the binding energy of the electron is strongly enhanced in this configuration. The structures seen in the electron-transition bands have been ascribed reasonably to the symmetry-

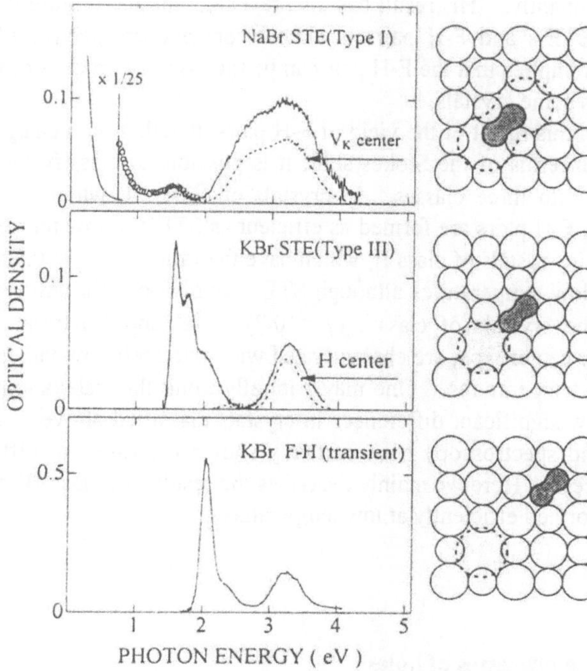

Fig.1. Optical absorption spectra of (a) the self-trapped exciton in NaBr, (b) the triplet self-trapped exciton in KBr, and (c) the transient F-H pair in KBr. The data of the spectra below 0.8 eV in NaBr was taken from Ref. 19. The schematic configurations of these centers are also shown.

split sublevels of 2p-like electron state. The configuration of the F-H pair shown in Fig.1(c) corresponds to the transient pair which has the lifetime of 100 μs at low temperatures. Evidently, the difference in these configurations is attributed primarily to the location of the X_2^- molecular ion with respect to the position of the trapped electron; the F-H pair is that in which the X_2^- molecular ion is displaced over at least a few lattice distance. In most crystals, two or more of these configurations are formed, and the presence of these different relaxed configurations of excitons, or (e-h) pairs, is nothing but a manifestation of the strong interplay between the electron-phonon and hole-phonon interactions [1].

Previous time-resolved measurements have shown that the F-H pair is formed with an efficiency which is dependent strongly on the kind of crystals; it is formed more efficiently than STE's in some crystals at low temperatures [20]. Tanimura has measured the yields of the F-H pair and of the triplet STE's per (e-h) pair for nine different crystals by means of the transient absorption spectroscopy with pulsed electron irradiation at 8 K [21]. The result has shown that in crystals with the ratio γ ($=r^+/r^-$) of ionic radius of cation (r^+) relative to that of anion (r^-) larger than 0.8, the F-H pair is formed efficiently, while in those with γ smaller than 0.8, the STE's are formed predominantly. His result has also revealed another feature that the sum of the yields of STE's and F-H pairs is almost constant irrespective of the kind of crystals. This implies that the F-H pair can be the primary product of the relaxation of (e-h) pairs in some crystals.

Based on this result of the yield of F-H pairs, together with categorization [16] of the STE's in terms of the Stokes shift, it is possible to classify the whole alkali halide crystals into three classes. In crystals of class III, which have the ratio γ larger than 0.8, F-H pairs are formed as efficient as STE's at low temperatures. On the other hand in crystals of class II, which have the ratio $0.7 < \gamma < 0.8$, F-H pairs are not formed at low temperatures although STE's involve certain degree of off-center relaxation. The crystals of class I ($\gamma < 0.7$) , NaI and NaBr, in which STE's maintain the on-center type, are characterized with extremely low yields of F-H pairs even at higher temperatures. One may naturally think that relaxation processes of (e-h) pairs show significant differences in crystals classified above. The results of the femtosecond spectroscopy have certainly shown the essential difference in the relaxation process. Here we mainly describes the results for class-III crystals where F-H pairs are formed efficiently at low temperatures.

3. Self-trapping processes of holes

Despite the fact that the holes are self-trapped into the V_K-center configuration in alkali halides, the dynamics the self-trapping processes of holes was a matter of speculation for a long time. Since this process triggers the localization and hence the lattice relaxation of (e-h) pairs, the clarification of the dynamics is of primary importance for understanding the mechanism of F-H pair formation. Iwai et al. have first studied the relaxation process of holes in KI, which belongs to class II, by means of femtosecond spectroscopy [11]. In order to observe purely the hole-relaxation process, free from the interaction with electrons during relaxation, they used KI crystals doped with NO_2^- ions which are known as effective electron-trapping impurities in alkali halide crystals. Most striking feature they have revealed is that before the holes are self-trapped in the form of the X_2^--type configuration, a transient-hole state which has two optical absorption bands at 2.1 and >3 eV is formed as the primary product within 2 ps after excitation.

Extensive theoretical studies by quantum mechanical *ab initio* calculations [11,22] have shown that the most probable candidate for this transient- hole state is a one-center type self-trapped hole. In this configuration, a hole is localized on one halogen ion, and is stabilized mainly by the polarization fields induced by the localized hole. The largest displacements are induced for the nearest-neighboring six K^+ ions with a magnitude of about 10% of the lattice constant. The theory predicts two allowed optical transitions in the energy range concerned for this one-center STH; the calculated transition energies are 2.4 and 3.3 eV. These predicted energies agrees well with the observed energies of absorption-band peaks, substantiating the assignment that the transient-hole state is the one-center type self-trapped hole. Thus, the self-trapping of holes is initiated via the one-center STH, and then it is relaxed further into the two-center type, and V_K centers are finally formed at around 20 ps after excitation at 80 K in KI.

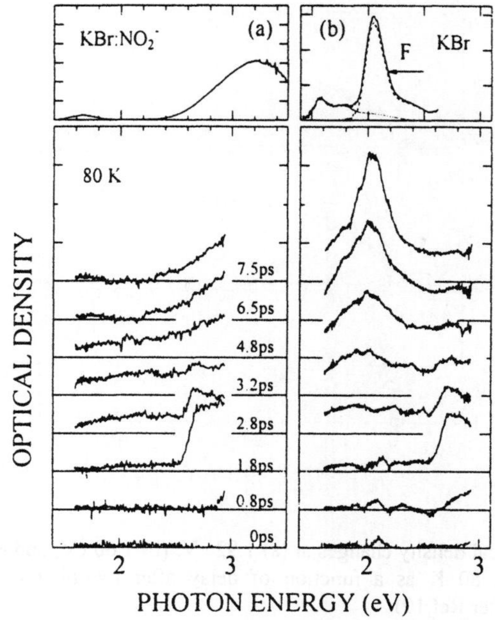

Fig.2. Time series of optical absorption spectra for (a) KBr: NO_2^- and for (b) pure KBr at 80 K; the probe pulse delay is given in each frame. The top frames of (a) and (b) show the V_K band measured at 80 K and the STE and F bands measured at 9 K. respectively.

The formation of the transient-hole state as the primary product of the self-trapping processes of holes, first observed in KI, is not specific to iodide crystals, but is more general. Fujiwara and Tanimura [14] have studied the self-trapping process of hole in NaBr (class I) and KBr (class III) in a temperature range from 5 K to 300 K, and found that the transient hole state similar to that in KI is certainly formed before forming the V_K centers in both crystals. In Fig.2(a), are shown the time-resolved spectra induced by femtosecond-laser excitation for KBr doped with NO_2^- impurities. It is clear that the transient hole state which shows the absorption band at 2.75 eV is formed at around 2 ps after excitation. It decays after 2 ps, and the induced absorption band is broadened entirely, and eventually converted into the V_K band within 10 ps. Although deeper theoretical studies have not yet been made for KBr, similarity of the results as in those in KI suggest that the transient hole state in bromides can be ascribed to the one-center type STH. These findings on the self-trapping process of holes provide a fundamental basis on which relaxation process of (e-h) pairs is studied.

4. Two distinct processes of F-H pair formation

The first extensive femtosecond time-resolved study for the Frenkel-pair formation was carried out for KBr which belongs to Class III [11,12]. In Fig.3 are shown the temporal evolution of optical absorption at 1.92eV which monitors the F

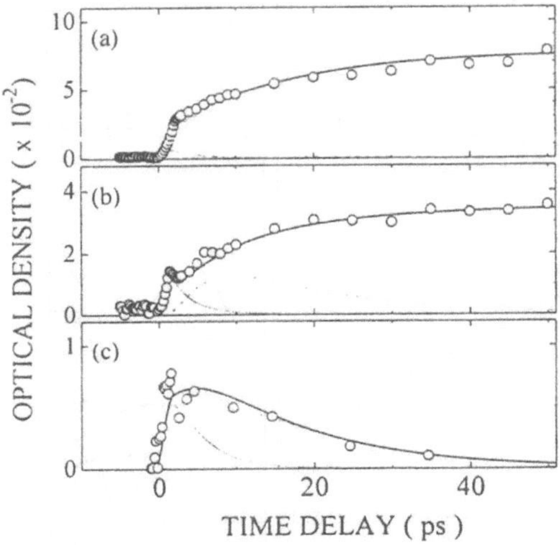

Fig.3. Optical density changes at (a) 1.92 eV, (b) 1.60 eV, and (c) 1.33 eV in pure KBr at 80 K as a function of delay after two-photon band-to-band excitation (after Ref.10)

band, and at 1.60 and 1.33 eV in KBr at 80 K. The 1.60 eV corresponds to the peak of the STE (Type III) band, and 1.33 eV monitors the long wavelength component of the absorption which is observed only in this short time experiments in KBr. Time-series of the absorption spectra until 7.5 ps are shown in Fig.2(b). One of their important findings is, as seen in Fig.3(a), that the F-H pairs are formed through two distinct processes; a fast process which takes place within a few ps after excitation, and a slow process which follows the former and lasts over 60 ps at low temperatures. On the other hand, the STE (Type III) is formed only after a few ps in parallel to the slow process of the F-center formation; the apparent rapid growth accompanied with fast decay at this photon energy is mainly due to the broad absorption band characteristic to the hole relaxation seen in Fig.2(a).

Comparing with the results of hole relaxation shown in Fig.2(a), it is evident that fast process of F-H pair formation occurs before the hole relaxation into the V_K centers is completed. It is , therefore, concluded that the instability of the fast process is originated from the interaction of an electron with a relaxing hole which is still on the way of relaxation into the V_K -center configuration. This instability may be referred to as the "dynamical instability" in the sense that the hole relaxation which is associated with violent atomic motions is still on the way.

More insight on this fast process can be obtained by comparing the temporal evolution of the absorption bands in KBr and in KBr: NO_2^- shown in Fig.2. It is evident in the figure that the 2.75-eV absorption band due to the transient-hole center is formed at 1.8 ps after excitation in pure crystals as well. The absorption band decays after 2 ps, and then the F band start to grow. An analysis of the temporal evolution of the absorption strengths of the two bands shows that the decay of the 2.75-eV band and the growth of the F band show beautiful anti correlation. If we assign that the 2.75-eV absorption is due to the one-center type STH, then this result implies that the dynamical instability resulting in the fast formation of the F-H pair is related to the configuration change of the localized hole from one-center type to two-center type, or spontaneous symmetry breaking of the localized hole in its relaxation process. It has been proposed that the electron localization which takes place in phase with the symmetry lowering of localized hole is the main origin of the dynamical instability which is responsible for the fast formation of the F-H pairs [14].

On the other hand, as seen in Fig.3, the slow process of F-H pair formation is associated with the decay of the absorption monitored at 1.33 eV; the decay time constant of the latter is identical to that governs the slow growth of the F band. Therefore, it can be preliminary concluded that the slow process of the F-H pair formation is the conversion from an intermediate state which is formed in the early stage into the F-H pair and the off-center STE's. Careful examination of the spectral changes has shown that the intermediate state shows the UV absorption band the shape of which is essentially the same as the V_K band; the absorption in UV region changes its shape from the V_K-like to the H-like bands during the process. The spectral features of the intermediate state, characterized by the V_K -like band and by the optical absorption below 1.5 eV, are essentially the same as those of Type-I STE shown in Fig.1(a). Based on these results, it has been concluded that the intermediate state is the on-center STE, and that the slow process of the F-H pair formation, which is accompanied with the Type-III STE formation, is due to the off-center relaxation of the on-center STE which is formed in parallel to the F-H pair formation through the fast process.

Thus, the relaxation process of (e-h) pairs in KBr is classified into two stages; the first stage which is characterized by the relaxation of a complex consisting of a relaxing hole and an electron, leading to the formation of F-H pair and the on-center STE, and the second stage which is essentially the off-center relaxation of the on-center STE. The relaxation pathways of (e-h) pairs are schematically shown in Fig.4.

Sugiyama et al. [12] extended the study for RbBr, which has similar features to those in KBr in many respects, and confirmed that essentially the same relaxation process is the case. They have also studied temperature dependence of the relaxation processes of (e-h) pairs extensively, and found that the most significant effect of the temperature is the enhancement of the transition (cooling) rate of a complex consisting of a relaxing hole and an electron into the on-center STE

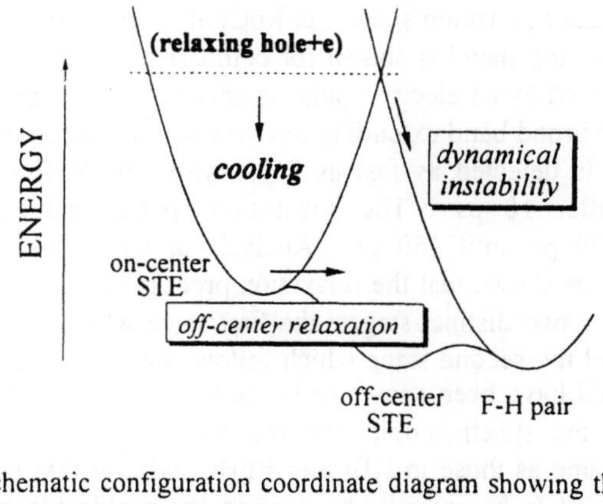

Fig.4. A schematic configuration coordinate diagram showing the relaxation pathways of (e-h) pairs (after Ref.12).

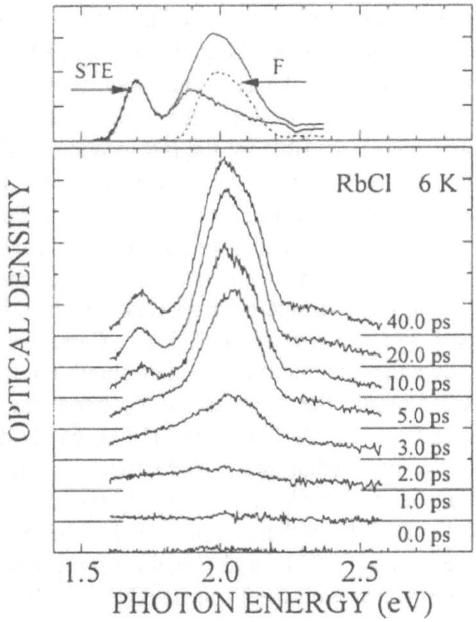

Fig.5. Time series of optical absorption spectra for RbCl at 6 K. The top frame shows the STE and F bands measured at 9 K (after Ref.13).

configuration. As a result, the yield of the F-H pair formed in the fast process is reduced, whereas that in the slow process is enhanced at higher temperatures. The enhancement of the cooling transition rate with increasing temperature has been predicted theoretically [23].

Fujiwara et al. studied the relaxation process of (e-h) pairs in KCl and RbCl both of which belong to class III [13]. One of the interesting features of these chlorides is that only Type-III STE's are formed, showing strong contrast to KBr and RbBr where both Type-I and Type-III STE's are formed [2,16]. Figure 5 shows the time-resolved optical absorption spectra in RbCl after irradiation with a femtosecond laser pulse. In the top panel is shown for comparison the STE and F band in this crystal at 9 K induced by an electron-pulse irradiation. The spectra before 2 ps are characterized by a broad band extending over the whole photon-energy range. The F band at 2.02 eV is detected as fast as 3 ps, while the STE band at 1.70 eV is generated only after 10 ps. The absorption spectra induced remains almost unchanged after 20 ps until 150 ps. Analysis of temporal evolution of optical absorption bands has shown that the relaxation process of (e-h) pairs in RbCl at 6 K can be divided into two distinct stages; the first stage which terminates within 5 ps after excitation and the second stage which follows the former and lasts over 20 ps. The results for KCl have been shown to be quite similar to those in RbCl. These characteristics of the spectroscopic features and temporal evolution, which are qualitatively the same as those in KBr and RbBr, indicate that essentially the same relaxation process of (e-h) pairs is the case in these chloride crystals. They also carried out similar measurements in a temperature range of 6 K to 300 K, and found that the yield of the fast process which is dominant at low temperatures is reduced and is compensated by that in the slow process at higher temperatures, as in bromide crystals.

One interesting difference between bromides and chlorides concerns with the formation time of the F-H pairs in the second stage; it terminates within 20 ps in KCl and RbCl, whereas it lasts over 100 ps in KBr and RbBr at low temperatures. This quantitative difference is also explained within the frame of the relaxation pathways shown in Fig.4; the height of the barrier in the off-center relaxation determines the rate of the slow process. The height of the barrier in the off-center relaxation was evaluated from the analysis of temperature dependent changes of the formation time of the F center in the slow process. These are around 30meV in bromides, whereas they are as low as about 8 meV in chlorides. The lower barrier height in chlorides, which leads to the fast rate of the off-center relaxation, is presumed to be related the lack of the luminescence from the on-center STE in these crystals.

5. Conclusion

Femtosecond time-resolved spectroscopic studies have revealed several interesting features in the dynamics of lattice relaxation of (e-h) pairs leading to the defect formation in alkali halides. In particular, presence of the dynamical instability which is induced during relaxation of the localized (e-h) pair may be one of the most important findings. It has been suggested that the instability is associated with the spontaneous symmetry breaking of the localized holes from one-center type to two-center type configurations. This instability has been proven to be common in all alkali halide crystals which belongs to class III. Our recent studies by means of the femtosecond time-resolved spectroscopy for alkali halide crystals of class II, and even for fluorite crystals like CaF_2 and BaF_2 have shown that the dynamical instability plays an important role in the relaxation process of (e-h) pairs in these crystals [24]. It may be an interesting future problem to study the dynamical processes of photo-induced structural changes in other nonmetallic solids, and to compare their features from a more general point of view.

Acknowledgments

The authors are grateful to Mr. H. Fujiwara, Dr. A. Shluger and Dr. T. Suzuki for many valuable discussions.

References

1) M. Ueta, H. Kanzaki, K. Kobayashi, Y. Toyozawa and E. Hanamura, "Excitonic Processes in Solids", (Springer, Berlin,1986).
2) A. K. S. Song and R. T. Williams, "Self-Trapped Excitons", (Springer, Berlin, 1993).
3) N. Itoh and K. Tanimura, J. Phys. Chem. Solids, **51**, 717 (1990).

4) "Desorption Induced by Electronic Transition", (Springer, Berlin, 1993), edited by A. R. Burns, E. B. Stechel, and D. R. Jennison.
5) "Amorphous Semiconductors", edited by Brodsky, (Springer, Berlin, 1985), and K. Tanaka, in this proceedings.
6) S. Koshihara, Y. Tokura, K. Takeda, and T. Koda, Phys. Rev. Lett. **68**, 1148 (1992), and S. Koshihara, in this proceedings.
8) N. Itoh and T. Nakayama, Phys. Lett. **A92**, 471 (1982).
7) P. W. Anderson, Phys. Rev. Lett. **34**, 953 (1975).
9) Y. Toyozawa, Solid State Commun. **84**, 255 (1992).
10) T. Shibata, S. Iwai, T. Tokizaki, K. Tanimura, A. Nakamura, and N. Itoh, Phys. Rev. **B49**, 13255 (1994).
11) S. Iwai, T. Tokizaki, A. Nakamura, K. Tanimura, N. Itoh, and A.L.Shluger, Phys, Rev. Lett. **76**, 1691 (1996).
12) T. Sugiyama, H. Fujiwara, T. Suzuki and K. Tanimura, Phys. Rev. B, in print.
13) H. Fujiwara, T. Suzuki, and K. Tanimura, to be published.
14) H. Fujiwara and K. Tanimura, to be published.
15) W. Kanzig, Phys. Rev. **99**, 1890 (1955).
16) K. Kan'no, K. Tanaka and T. Hayashi, Rev. Solid State Science, **4**, 383(1990).
17) T. Suzuki. K. Tanimura, and N. Itoh. Phys. Rev. **B48**, 9298 (1993), Phys. Rev. **B49**, 7233 (1994), and K. Tanimura, T. Suzuki, and N. Itoh, Phys. Rev. Lett. **68**, 635 (1992).
18) K. Tanimura and N. Itoh, Phys. Rev. **B45**, 1432 (1992).
19) S. Hirota, K. Edamatsu, and M. Hirai, Phys. Rev. Lett. **67**, 3283 (1991), and K. Edamatsu, M. Sumita, S. Hirota, and M. Hirai, Phys. Rev. **B47**, 6747 (1993).
20) R. T. Williams and M. N. Kabler, Phys. Rev. Phys. Rev. **B9**, 1897 (1974).
21) K. Tanimura, unpublished.
22) A. L. Shluger and J. D. Gale, Phys. Rev. B, in print.
23) R. H. Bartram and A. M. Stoneham, Solid State Commun. **17**, 1593 (1975).
24) H. Fujiwara and K. Tanimura, to be published.

Theoretical Studies of the Models of Self-Trapped Excitons and Holes in Ionic Crystals

A.L. Shluger

Centre for Materials Research and Department of Physics,
University College London, Gower St., London WC1E 6BT, U.K.

Abstract. The recent progress and remaining problems in the understanding of the mechanisms of self-trapping of excitons and holes and of their atomic structure in alkali halides are discussed in conjunction with recent spectroscopic experiments. The results are based on many-electron calculations using an embedded molecular cluster model and including the electron correlation. A model is proposed for the initial stages of the hole self-trapping in KI which includes its initial trapping in the one-centre state, which then transforms into the two-centre state, and finally into the V_K-centre. It is shown that the perturbations by charged and neutral hole defects induce electronic states in the valence band which are responsible for optical transitions in these centres in UV region. The electron correlation is shown to stabilise the on-centre configuration of the STE in NaCl.

1. Introduction

The self-trapping of holes and Frenkel excitons is the fundamental step in defect processes induced by electronic excitation in a wide class of insulating solids including the formation of intrinsic defects and surface decomposition. It has been observed in many insulating crystals, such as alkali-metal halides, quartz, rare-gas solids, alkaline-earth halides, perovskite-structured halides of the $KMgF_3$ family and in other systems. It drastically alters the optical, luminescent and energy transport properties of the crystal. Extensive experimental studies on self-trapped excitons (STEs) and holes (STHs) in insulators have been carried out during the last few decades using optical absorption, luminescence, ENDOR, resonant Raman scattering and other techniques, as reviewed by Song and Williams[1]. Several general models of self-trapping have been proposed starting from works by Landau, Pekar and Toyozawa. However, the initial stages of the self-trapping process and detailed atomistic and electronic structures of STEs, and in some cases of STHs, are still unclear even in alkali halides. Understanding of these issues is crucial for the development of general models of self-trapped species and of theoretical methods for their studies. In this paper we discuss the recent progress and remaining problems in this field in alkali halides, in conjunction with general mechanisms of photo-chemical processes in condensed phase.

Recently femtosecond scale pulse optical spectroscopic techniques have been applied to the study of the hole polaron and exciton self-trapping in KI, RbI[2], KBr and RbBr[3]. The results of these pump-probe spectroscopic experiments were tentatively interpreted in terms of the formation of some transient (lifetime < 5 ps)

Springer Series in Solid-State Sciences, Volume 124
Relaxations of Excited States
and Photo-Induced Structural Phase Transitions
Ed.: K. Nasu

localised species preceding the formation of self-trapped holes (V_K centres), STEs and primary defect pairs. One of the plausible candidates for such a transient hole state in alkali halides is the one-centre state which has been discussed in recent papers[3, 4]. Relaxation of this state in the lattice is similar to ionic combination reaction $X^- + X \rightarrow X_2^-$ of halide atoms in solutions[5].

Another point which has been clearly understood only recently[6, 7] is that the perturbations by charged and neutral hole defects induce electronic states in the valence band which are responsible for optical transitions in these centres in UV region.

Despite extensive studies of models of self-trapped excitons in alkali halides, the many-electron nature of the problem and the role of the electron and hole correlation in STE is not well understood. We discuss the results of recent studies[7] which demonstrate the effect of electron correlation on configuration and optical properties of STE.

Most of the results of the calculations discussed in this paper were obtained using embedded molecular cluster method which is briefly discussed in the next section.

2. Theoretical methods

The necessary condition for applicability of Embedded Molecular Cluster (EMC) method is the localisation of the defect electronic wave function within the quantum cluster region. In the case of the hole or exciton this means that it can be used only for those stages of the process when the quasiparticle is already trapped by the lattice deformation.

The detailed description of the method and corresponding software which have been developed and applied to the excitons and point defects in alkali halides can be found elsewhere[4, 8, 9]. Here we will describe qualitatively the main components of the theoretical model and the range of its applicability.

The EMC method is based on the approximation that the perfect crystal can be divided into individual ions. It allows one to combine a quantum-mechanical treatment of a part of the crystal, including a defect (quantum cluster), with the classical description of the rest of the crystal. This is made by substituting a number of classical ions by a quantum cluster, and by using a 'self-consistency' procedure based on consecutive iteration of two computational methods as discussed in ref. 4, 8, 9. To be used together, both methods must give the same lattice constants of the perfect lattice and, in the more general case, must yield the same optimised structure.

If all the ions except those in the cluster region are only slightly perturbed by the defect, the problem is reduced to the solution of the equation for the cluster wave function in the self-consistent field of the rest of the lattice as discussed in ref. 10. This external embedding potential consists of the Madelung potential, pseudopotentials of the ions surrounding the cluster and terms due to the polarisation of the crystal induced by the defect.

Point ion approximation is not valid for those ions in the rest of the lattice, which are close to the cluster boundary and have their electron densities overlapping with the cluster one. Finite ion size effects should be taken into account. The potential experienced by the cluster electrons due to these ions, besides the bare Coulomb potential, includes non-point electrostatic correction, exchange and orthogonalisation (or Pauli repulsion) terms. The former is local and depends only on the ion electron density. Two other are nonlocal, i.e. depend also on the cluster electron state, and can be expressed in terms of projection operators to the ionic wave function. The effective Pauli repulsion term keeps the cluster wave function orthogonalized to that of the remainder of the lattice and thus localised in the quantum cluster region. The normconserving semilocal pseudopotential method, which is the most efficient for the treatment of the ions in the cluster neighbourhood, is discussed in detail in ref. 9.

The lattice polarisation within the quantum cluster is treated within the accuracy of the particular method of the wave function calculation and chosen basis set of atomic orbitals. The polarisation of the rest of the lattice is most important for the charged defects, such as the V_K-centre, cation or anion vacancy. It depends on the particular charge distribution within the cluster and can be taken into account using Mott-Littleton method and the shell model[11]. In this model each ion within the finite region outside the cluster is represented by the negatively charged shell connected to its core by a spring. The outermost part of the crystal is treated as dielectric continuum. The parameters of the model including shell and core charges, elastic constant of the spring and interionic interaction potentials are fitted to the experimental data or to the *ab initio* calculations.

The response of the polarised lattice is then given in a form of the lattice polarisation energy and an electrostatic potential ϕ_i produced by the cores and shells outside the cluster at the position of each nuclei, i, inside the quantum cluster. However, this potential is calculated using not the formal ionic charges employed in the parametrisation of the shell model, but those obtained for the perfect lattice using the periodic model and the same quantum-chemical method. Since the quantum-mechanical charges are different from the formal ionic charges used in the Mott-Littleton model, this ensures a homogeneous charge distribution across the cluster border. Note that ϕ_i in this approach includes both the Madelung term and the dipole polarisation term. The diagonal matrix elements of this potential calculated on atomic orbitals μ, $\langle\mu\,|\,\phi_i\,|\,\mu\rangle$, are then added to the Fock matrix. The total energy of the whole system including the quantum cluster embedded in the infinite polarisable lattice is minimised with respect to the LCAO coefficients, positions of the nuclei inside the cluster and of the cores and shells of the rest of the crystal. This method is implemented in the ICECAP and CLUSTER95 computer codes[4, 9].

The electron correlation within the quantum cluster is treated using Møller-Plesset perturbation theory[12].

3. Results for the hole and exciton self-trapping problem

The results of extensive applications of this technique are described in ref. 7,9,13. In this paper we will summarise the results of recent studies which concern particular problems related to the initial stages of self-trapping of holes and excitons and to their structure outlined in the introduction.

3.1. One-centre Trapping of the Holes in Alkali Halide Crystals

Spectroscopic measurements were recently performed in KI on the time domain from 0.3 ps to 100 ps after the excitation pulse at both room and liquid nitrogen temperatures[3]. The two-photon excitation with an energy of about 8 eV employed in these experiments first produces an electron-hole pair in the bulk of KI. The electron is quickly trapped by the NO_2^- impurity or is delocalised in the lattice. Optical absorption spectra in the energy range from 1.5 to 3.2 eV have been observed both in pure samples of KI and those doped with the electron-trapping impurity (NO_2^-). Transformation of the optical absorption spectrum in KI and RbI is attributed to the hole and takes place via three distinct stages. i) At least two intense optical absorption bands with maxima near 2.3 and >3.2 eV were already observed at 0.3 ps after the excitation pulse in doped crystals. Their intensity rises within approximately 1 ps. With some delay (about 0.5 ps) a third absorption band at about 2.6 eV begins to rise and broadens. ii) About 3 ps after the pulse the whole spectrum transforms into a featureless broad band. iii) After that the well known optical absorption of the V_K-centre appears and rises during 10 ps with a time constant of about 3 ps. Similar transient optical absorption has been observed in RbI. The current understanding of these absorption spectra is based on the following model.

One simple argument concerns the static energetic criterion that the self-trapped state should have the energy lower than the bottom of the free hole band, i.e. E_{st} should be negative. Among possible relaxed hole states in alkali halides only one- and two-centre states are known to have negative self-trapping energies. In particular the one-centre has the largest relaxation energy, S, (about 1 eV or more) due to the polarisation of the lattice by the strongly localised hole. The kinetic energy loss due to the hole localisation in this state (localisation energy), B, is determined by the structure of the upper valence band and may be roughly estimated as a half of its width[14], which is more than 1 eV in many alkali halides. In a recent paper[4] the self-trapping energy for the one-centre hole state in KI has been estimated as $E_{st} = B - S \leq -0.3$ eV. This is close to the results obtained for other cubic crystals such as KCl (-0.2 eV), MgO (-0.4 eV), AgCl (-0.1 eV).

If the hole is equally delocalised between two or more lattice sites, the lattice polarisation is smaller and these small polaron states are less favourable than the one-centre state. However, in alkali halides and in alkali-earth fluorides, possibly in alumina and other crystals the two-centre state is stabilised by chemical bonding between the two anions sharing the hole[1,15]. This leads to formation of a quasi-molecular X_2^- state (X is the anion) where the distance between anions is much smaller than that in the perfect lattice. These are the only stable hole states observed so far in a pure lattice. This, however, does not exclude the possibility of the existence of more delocalised metastable states which are much more difficult to study.

Further argument is based on the calculation of the optical absorption energies for different hole configurations. To find the geometric and electronic structures of the hole in one-centre state, the position of the I atom carrying the hole was fixed in the lattice site whereas all other crystal ions were allowed to relax. In the energy minimum, the nearest neighbour cations are displaced outwards by about 0.1a and the next nearest-neighbour anions are displaced inwards by about 0.015a (a is the interatomic distance).

Calculations of the optical absorption energies and the matrix elements of the corresponding dipole electronic transitions for the relaxed state of the one-centre hole were made using the configuration interaction for single electron excitations technique (CIS)[16]. In the completely symmetrical configuration, the allowed transitions are from E_{1g} and A_{1g} states to the single occupied E_{1u} hole state and are accompanied by the hole delocalisation. This latter should lead to the change in the lattice polarisation which was not taken into account in our CIS calculations. To check how this can affect the calculated optical absorption energies, some of the transitions were calculated as the difference between the total energies of the ground and excited hole states with the self-consistent account of the lattice polarisation in both states. With this correction transitions calculated using CIS in the same cluster become 1.9 eV, 2.1 eV and 3.3 eV. They are in the range of the splitting and positions of the maxima of the absorption bands observed at 0.3 ps after the excitation pulse at both room and liquid nitrogen temperature in KI.

Further, the adiabatic potential for transformation of the one-centre into the two-centre state has been considered. The former state is not stable with respect to the displacement of the two anions closer to each other. However, the one-centre hole polaron state once formed does not collapse into the two-centre state immediately, but this requires the iodine ions to come closer than a critical distance of about 4.1 Å. Before that happens, the hole remains localised on one of the ions. This behaviour of the adiabatic potentials results from the competition between two main factors: i) the chemical bonding between two iodines, and ii) the lattice polarisation. The energy of the strongest $\sigma_g \rightarrow \sigma_u$ optical transition at 4.0 Å is about 1.7 eV. It is increasing up to 3.2 eV as the distance between the iodine ions is approaching the equilibrium distance for the V_K-centre.

We should note the similarity of the processes of the hole transformation from the one-centre state into the V_K-centre in KI with the $I + I^- \leftrightarrow I_2^-$ combination reaction in solutions[17]. As has been calculated in ref. [17], because of the competition between the solvent polarisation and chemical bond formation, in the latter process the transformation from (1,0) into (1/2,1/2) state also occurs at some critical distance (3.45 Å in acetonitrile). However, since the dielectric constant of the acetonitrile is 37.5, which is several times larger than in KI, the combination process has even some small barrier and the critical distance is much smaller than found in this paper for KI.

These results suggest that the one-centre polaron state in KI can exist as a transient state. However, this does not answer the question why holes produced by crystal excitation prefer to be trapped first in this state but not, for instance, in the two-centre state. To address this question one should counsider a distribution of the lattice fluctuations at a given temperature. These fluctuations are of the same nature as those leading to the exponential behaviour of the one-electron density of states[18]. For 3-dimensional crystals, a critical strength of a Gaussian fluctuation of potential necessary to produce a bound state has been evaluated in ref. 18. For a fluctuation parametrised by a depth V_o and a range α, a dimensionless quantity $2m\alpha^2V_o/h^2$ has been found to be equal 2.95, where m is the electron mass. If one assumes the range of the fluctuation, α, equal to the lattice constant, in KI this would correspond to V_o = 0.231 eV. For a simple estimate one can assume further that this fluctuation is produced by a symmetrical outward displacement of the six cations surrounding one iodine ion. For the formal ionic charge of 1.0e, the displacement necessary to produce such a fluctuation of the potential is about 0.01a, where a is the interionic distance. Although such fluctuations are readily available even at low temperatures, they correspond to rather delocalised states centred at one lattice site which are impossible to study using the present technique and localised basis set. Therefore, more detailed studies of the lattice fluctuations and corresponding holes states are needed to elaborate this point.

Another effect which has not yet been properly understood is the much lower optical density of the V_K-centres with respect to the initial absorption spectrum. One of the possibilities is that, since the self-trapping energy for the one-centre hole state is small, i.e. it is close to the bottom of the free hole band, part of the holes can be thermally delocalised. This can correspond to the intermediate case which has been discussed by Sumi[19]: while the small polaron state is energetically stable, the thermally populated large polaron states play dominant role in polaron conduction. This would explain the considerable decay of the number of holes by their fast band motion to the acceptors. However, we should note that this process competes with the two-centre state formation which immobilises the hole. Therefore, more data for different crystals, temperatures and concentrations of impurity are needed to understand this phenomenon in more details.

3.2 Optical absorption of hole centres

Although the intrinsic X_2^- hole centres in alkali halides, the V_K and H centres, are certainly amongst the best studied point defects in ionic crystals some of the fundamental optical properties of these centres remained unclear. In particular, in ref. 6 the comparison has been made of the experimentally measured and theoretically calculated absorption spectra of these centres in several alkali halides. The experimental results reveal that the $\pi_g \rightarrow \sigma_u$ transition energies are systematically larger for the H centres than for the V_K-centres in qualitative agreement with the molecular di-halogen model of both centres and with general understanding that the halogen-halogen distance for the H centre is smaller than that in the V_K-centre. However, the $\sigma_g \rightarrow \sigma_u$ transition energy for the H centres are found to be larger than those for the V_K-centre only in chlorides, nearly the same in bromides, and smaller in iodides. These results suggest that the shift of the transition energies of the UV absorption band for the H centres with respect to those for the V_K-centres is not determined just by the more shorter interatomic spacing within the X_2^- molecular ion in the H centre.

Theoretically, it has been shown[6] that correct representation of the wave-functions of the excited defect states including the interaction of the appropriate excited electronic configurations is essential for the understanding of the nature of the defect electronic transitions. It appears that this effect is particularly important for the hole centres which cause strong distortion of the surrounding lattice. The account for the interaction between the intramolecular one-electron transitions within the halogen molecular ion and these transitions from the surrounding anions to the single-occupied hole state allowed us to explain the relative transition energies for the H centre with respect to the V_K-centre. The analysis of the spin density in the excited states of the H and V_K-centres demonstrated the considerable delocalisation of these states over several anions surrounding the X_2^- molecular ion. The existence of the optical transitions of the H centre in the ultraviolet range higher than the $\sigma_g \rightarrow \sigma_u$ transition energy has been shown theoretically and confirmed experimentally.

However, some optical properties of the V_K-centres cannot be explained in the above model. In particular, that the $\sigma_g \rightarrow \sigma_u$ optical absorption band of the V_K-centres is much broader than that of the H centres. Calculations performed for NaCl with account of electron correlation have demonstrated that this band comprises two closely located Σ polarized bands. They can be attributed to the $\sigma_g \rightarrow \sigma_u$ transition in the Cl_2^- molecular ion and to the "host-to-molecule" electron transition from the surrounding anions to the empty hole state, correspondingly.

3.3 Self-trapped exciton

Exciton is an electron-hole pair which propagates in non-metallic solids. In alkali halides, if the hole is already self-trapped as a V_K centre, the electron is also localised in the vicinity of X_2^-. Since the isolated X_2^- has no affinity to the second electron, the electron is trapped not by X_2^- itself, but by the field of the distorted lattice. The repulsion between the electron and X_2 molecular ion and its attraction to one of the vacancies can lead to the separation of the hole and electron components of the exciton. In particular, the most stable atomic configuration of the self-trapped exciton in some alkali halides resembles the nearest $F - H$ pair: X_2^- is shifted from its symmetrical position towards one of the anion sites (*H* centre), and the electron is localised in the adjacent anion site (*F* centre). However, STE in alkali halides are known to have up to three equilibrium atomic configurations with different optical absorption and luminescence spectra. One of them, which is called on-centre configuration and corresponds to the electron bound to the V_K centre, has recently been proposed as a transient state in the exciton self-trapping in KBr and RbBr[3], and also as an equilibrium exciton configuration in NaBr and NaI[20].

In contrast, most of the theoretical calculations predict the existence of only one off-centre $F - H$ -like configuration as reviewed in ref. [1,7,9]. As has been demonstrated in recent calculations for the NaCl crystal[7], the electron-hole correlation is an important factor which strongly affects the structure of the STE.

In the most energetically profitable configuration of the STE in NaCl, the hole component of the STE, X_2^-, is oriented along the <110> axis and its position is intermediate between those in *H* and V_K centres, such as one of the halogens is almost in the anion lattice site. The electron is localised around the adjacent vacant anion site. The X_2^- bond length is only by 2% larger than that in the *H* centre. The calculated exciton luminescence energy 2.9 eV is in satisfactory agreement with the experimental value of 3.36 eV. The optical absorption spectrum does not correspond to that of the superposition of isolated *F* and *H* centres. The *F* centre band is splitted into three, due to the lower symmetry. The hole excitations within the X_2^- are mixed with the electron ones of the same symmetry. However, the lowest excited state of Π symmetry and mostly hole character has transition energy of 1.9 eV, close to the *H* centre excitation, and the lowest state of Σ symmetry and predominantly electron character is by 2.8 eV higher than the ground state, like for the isolated *F* centre. According to the optical absorption and luminescence spectra this configuration can be classified as type III.

Another equilibrium configuration of the STE found in this study has the D_{2h} symmetry and the V_K centre as its hole component. However, the X_2^- bond is longer, than that in the V_K centre because an excited electron occupies the bonding σ_g atomic orbital. In the Hartree-Fock approximation this configuration is unstable and results from the strong interaction of several electronic configurations corresponding to the electron and hole components of the STE. This local minimum of the adiabatic potential energy surface is higher by 0.56 eV than the off-centre one. High luminescence energy (6.2 eV) and low absorption energy of the electron component (0.6 eV) suggests that this configuration corresponds to the type I STE. Note that similar configuration has recently been proposed as a transient state for the exciton self-trapping in KBr. These results suggest that small Stokes shift of the luminescence, small absorption energy for the electron component, and the similarity of the absorption spectra of the hole component of the STE and that of the V_K centre can be characteristic for the on-centre configuration of the STE. They are indirectly supporting validity of the interpretation[20] of the transient absorption and luminescence spectra in NaBr and NaI as been due to the on-centre configuration of the STE.

4. Conclusions

Although the results of these studies provide more firm basis for our understanding of the atomic structure and spectroscopic properties of STH and STE in alkali halides, there are, however, several problems which still remain unresolved. These include the atomic configurations of singlet STEs which give rise to the σ luminescence NaCl, KBr, KI, RbBr, RbI, and the coexistence of three different configurations of the STE in such crystals as RbI. The magnetic-resonance parameters of the STE have not yet been well understood. Study of the fast channel of relaxation of the electron-hole pair recently observed[21] in KBr, which leads directly to the F - H pair creation bypassing the lowest state of the STE, requires consideration of non-relaxed configurations of the exciton.

Acknowledgements I am grateful to N. Itoh, K. Tanimura, A. M. Stoneham, K. S. Song, R. T. Williams, L. N. Kantorovich, A. I. Livshits, and E. A. Kotomin for numerous valuable discussions.

References

1 K. S. Song and R. T. Williams, *Self-Trapped Excitons* (Springer-Verlag, Berlin, 1993).
2 S. Iwai, T. Tokizaki, A. Nakamura, *et al.*, J. Luminescence **60&61**, 720 (1994).
3 S. Iwai, T. Tokizaki, A. Nakamura, *et al.*, Phys. Rev. Lett. **76**, 1691 (1996).
4 A. L. Shluger and J. D. Gale, Phys. Rev. B (at press) (1996).
5 I. Benjamin, P. F. Barbara, B. J. Gertner, *et al.*, J. Phys. Chem. **99**, 7557 (1995).

6 A. L. Shluger, V. E. Puchin, T. Suzuki, *et al.*, Phys. Rev. B **52**, 4017 (1995).
7 V. E. Puchin, A. L. Shluger, and N. Itoh, Phys. Rev. B **52**, 6254 (1995).
8 J. M. Vail, R. Pandey, and A. B. Kunz, Rev. Solid State Sci. **5**, 241 (1991).
9 A. L. Shluger, A. H. Harker, V. E. Puchin, *et al.*, Modelling Simul. Mater. Sci. Eng. **1**, 673 (1993).
10 L. N. Kantorovich, J. Phys. C: Solid State Phys. **21**, 5041 (1988).
11 These techniques have been reviewed in a special issue: C. R. A. Catlow and A. M. Stoneham (ed.) J. Chem. Soc. Faraday. Trans. II **85** (1989).
12 J. A. Pople, R. Seeger, and R. Krishnan, Int. J. Quant. Chem. Symp. **11**, 149 (1977).
13 V. E. Puchin, A. L. Shluger, K. Tanimura, *et al.*, Phys. Rev. B **47**, 6226 (1993).
14 L. Kantorovich, A. Stashans, E. Kotomin, *et al.*, Int. J. Quant. Chem. **52**, 1177 (1994).
15 A. M. Stoneham, *Theory of Defects in Solids* (Oxford University Press, Oxford, 1985).
16 J. B. Foresman, M. Head-Gordon, and J. A. Pople, J. Phys. Chem. **96**, 135 (1992).
17 R. Bianco and J. T. Hynes, J. Chem. Phys. **102**, 7885 (1995).
18 S. John, C. Soukoulis, M. H. Cohen, *et al.*, Phys. Rev. Lett. **57**, 1777 (1986).
19 H. Sumi, J. Phys. Soc. Japan **33**, 327 (1972).
20 K. Edamatsu, M. Sumita, S. Hirota, *et al.*, Phys. Rev. B **47**, 6747 (1993).
21 T. Shibata, S. Iwai, T. Tokizaki, *et al.*, Phys. Rev. B **49**, 13255 (1994).

Photo-Induced Structural Changes in Amorphous Semiconductors

K. Tanaka, A.V. Kolobov, and S. Yamasaki

Joint Research Center for Atom technology (JRCAT),
National Institute for Advanced Interdisciplinary Research (NAIR),
1-1-4, Higashi, Tsukuba-shi, Ibaraki 305, Japan
email: tanaka@jrcat.or.jp

Abstract. In this paper, firstly, an *in-situ* EXAFS study of amorphous Se (a-Se) before, under, and after light illumination is presented, where the first experimental evidence for a change of the coordination number of Se atoms under illumination is given. Secondly, our recent results of pulsed ESR measurements on hydrogenated amorphous silicon (a-Si:H) containing various isotopes are presented, through which spatial relationship between a photo-created dangling bond and near-by H atoms and other impurity atoms is described. Finally, photo-induced metastability is discussed with an emphasis towards the structural randomness, the network connectivity and the electronic structure.

1. Introduction

A variety of photo-induced phenomena have been observed in amorphous semiconductors for the past quarter century [1-10]; photo-induced changes in optical [1-5], structural [1-8], electrical [4, 6], spin-related [5,7], and chemical properties [9]. Most of them are originated from some sort of local structural changes in a disordered network. This is not surprising because amorphous semiconductors, both chalcogenide glasses and tetrahedrally-bonded amorphous films, are thermodynamically described as a solid in a non-equilibrium state, therefore, their structure and bonding configuration are not fixed but can be changed, sometimes reversibly, not only by thermal treatment but also by light illumination.

Here we pick up two main phenomena observed in two typical amorphous semiconductors; the first one is photo-induced changes in short-range order of amorphous selenium(a-Se), and the second one being photo-induced creation of metastable dangling-bond defects in hydrogenated amorphous silicon(a-Si:H). On the basis of our recent experimental results of *in-situ* EXAFS for a-Se and pulsed ESR measurements for a-Si:H, recent progress in understanding of photo-induced effects is discussed.

2. EXAFS Study of a-Se under photo-excitation

In amorphous chalcogenides such as a-Se, a-As_2S_3, a-As_2Se_3, a-GeS_2 and other multicomponent systems, independent of whether they are bulk or thin-film structure, two different sorts of photo-induced optical changes have been reported so

Springer Series in Solid-State Sciences, Volume 124
Relaxations of Excited States
and Photo-Induced Structural Phase Transitions
Ed.: K. Nasu

far [1,2,4,6]; "photodarkening" characterized by a red shift of the absorption edge [1,2,10], and "sub-bandgap absorption'"[5]. In the latter, a sub-bandgap optical absorption is induced when a material is exposed to a bandgap light at low temperatures, basically far below the glass transition temperature [5]. The induced broad-absorption band in the sub-bandgap range is originated from photo-created metastable ESR centers with their spin density being saturated at 10^{17} cm^{-3} in As_2S_3 [5]. This observation has been considered to provide a direct evidence for pair-breaking process of negatively-charged defects with a negative U_{eff}, initially predicted by Anderson [11]. Street and Mott proposed a model in terms of charged dangling bonds (D^+, D^-), which was followed by the valence-alternation-pair (VAP) model (C_1^-, C_3^+) by Kastner et al. [12] [13].

In these defect models a flexible change in coordination number of chalcogen atoms is assumed ; from the normal two-fold coordination to a singly- or three-fold coordination. This assumption is considered to be reasonable because p-like lone-pair electrons on chalcogen atoms produce an energy band at the top of the valence band and, therefore, may form a new chemical bond between closely-located chalcogen atoms in a disordered network when excited optically or thermally. Actually, according to the theoretical work on the structure of liquid selenium by Hohl and Jones, it is concluded that around 15% of total atoms are three-fold coordinated pairs [14]. However, no direct experimental evidence has been presented for the existence of three-fold coordinated chalcogen atoms so far.

In this section, we describe *in-situ* EXAFS measurements on amorphous selenium (a-Se) before, under, and after the band-gap illumination at 30K, and present the first experimental evidence for an increase in the coordination number of atoms under photo-excitation.

2.1 Experimental

Used samples were thin film structures of a-Se with a thickness range from 0.1 to 0.3 micron, which were prepared by thermal evaporation technique. The measurements were performed at BL13B station at the Photon Factory [15] using a 27-pole wiggler. Se K-absorption spectra were obtained in a fluorescence mode with a grazing angle of incidence, enabling us to use films thinner than the light penetration depth. Combination of a high brilliance photon source and a densely-packed multi-element detector array, consisting of 9-element NaI or 19-element Ge detectors, improved the efficiency of the detecting system by two orders of magnitude. *In situ* measurements under these conditions guaranteed high accuracy in a relative change in coordination number and mean-square relative displacement (MSRD) (MSRD) within±0.5%. Details will be described elsewhere [16].

The sample was mounted on an aluminum holder in an evacuated cryostat equipped with KAPTON windows for incident and fluorescent X-ray beams as well as an MYLAR window for *in situ* light irradiation. A 500-W xenon lamp with an ir-cutoff filter was used as an excitation light source, and the power density at the irradiated spot (15 mm diameter) of the sample was varied from 50 to 250 mW/cm^2. Se K-edge EXAFS spectra of a-Se were measured before, under and after the light irradiation at 30K[16].

2.2 Results and discussion

Measured Se K-EXAFS spectra were Fourier-transformed into a real space using the region extending from 4.5 to 15 A^{-1}. They clearly showed an increase in EXAFS amplitude for the sample under illumination. Detailed structural information was obtained by a least-squares curve-fitting analysis, based on a single scattering theory and FEFF amplitudes. In order to avoid the effect of correlation between the coordination and mean-square relative displacement (MSRD), a correlation analysis was made for these parameters [16]. The present photostructural change was found to be described in terms of coordination number and disorder parameter.

Figure 1 shows variations in the normalized disorder MSRD and coordination number against the sequences of sample treatment; namely, as-prepared, under illumination, after illumination and after annealing at room temperature. Every measurement was done at 30K. As shown in the figure, the MSRD increases under illumination by 25% and remains considerably large even after the cessation of light, which is compatible with the previous observations in a-As_2S_3 that the structural disorder increases after illumination [2]. This increase disappears completely by annealing at room temperature. On the other hand, the coordination number increases under illumination by about 5%, but it almost restores its original value after stopping the light, which may be the reason why previous *ex-situ* EXAFS studies were unable to detect any change in the first coordination sphere [17]. Namely, this change in the coordination number is essentially dynamical and can be observed only when the system is photo-excited. The results suggest that around 5% of total Se atoms become three-fold coordinated under illumination by the present light intensity.

Figure 2 illustrates a bond-formation model for explaining the photo-induced increase of the coordination number, where a new inter-chain bond is dynamically formed between closely-located Se atoms under photo-excitation [16]. In a simple chemical bond picture, the bandgap light excites electrons from the top of the valence band formed by lone-pair(LP) electrons of C_2^0, resulting in one electron in the anti-bonding state while the other being left unpaired in the LP orbital, i.e.,

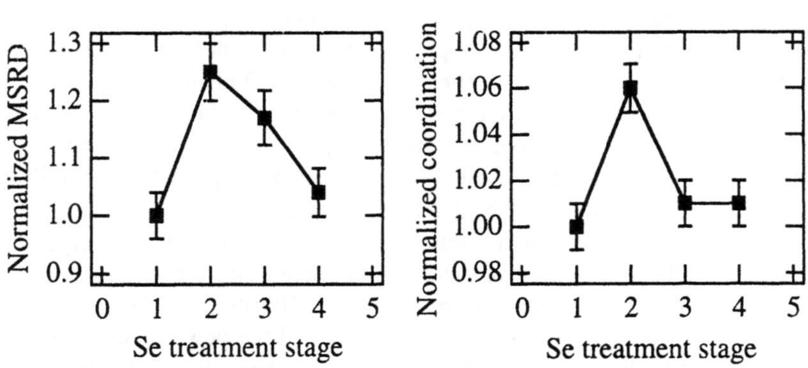

Figure 1. Results of a least-squares curve fit analysis of the Se K-EXAFS data for a-Se. The normalized disorder (MSRD) (left-hand side) and the normalized coordination number (right-hand side) for 1. as-prepared, 2. under illumination, 3. after illumination and 4. after annealing at 300K.

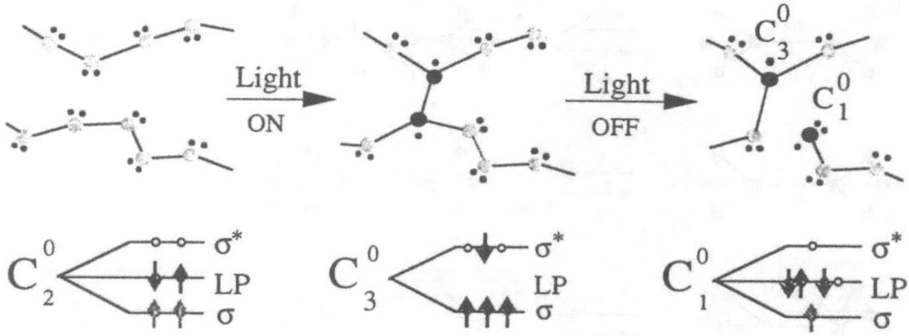

Figure 2. A schematic diagram of the local structure of a-Se network before (left), under (middle) and after (right) the light irradiation.

C_2^*. In case two excited atoms are closely located, new covalent bond will be formed between the two atoms as $C_2^*+C_2^* \rightarrow C_3^0+C_3^0$, which causes a transient increase in the average coordination number of Se atoms. It should be noted that C_3^0 has no longer the LP orbital, therefore, bond-breaking is a necessary process for the recombination of excited electrons. This argument also justifies a transient nature of C_3^0 pairs. After the cessation of the light there are several possibilities in bond rearrangements of a C_3^0 pair ; the complete restoration of two C_2^0's, the formation of a new chain topology via bond switching, and the creation of a (C_3^0,C_1^0) pair which will be followed by a subsequent conversion into a (C_3^+,C_1^-) pair at elevated temperatures. However, in any cases, the average coordination number is kept at the same value as that for as- prepared, while a structural disorder is expected to increase after the light illumination for the latter two cases. Details will be discussed elsewhere [16].

The above argument suggests that the observed photo-induced bond formation, being dynamical in nature, is quite similar to the structure of liquid Se where about 15% of atoms are three-fold coordinated pairs[14]. It is also compatible with the recent report on the photo-induced fluidity in As_2S_3 glass[18]. These data will give a clue to general understanding of photo-induced structural transformations.

3. Pulsed ESR Study of photo-created dangling bonds in a-Si:H

In contrast to a-Se, hydrogenated amorphous silicon (a-Si:H) is characterized by a rigid 3-d amorphous network mainly formed by 4-fold coordinated Si atoms. However, by the presence of hydrogen atoms amounting to 5 to 15 at.%, some sort of structural flexibility is given to the a-Si:H network, which may be the reason why photo-induced metastability occurs in the material. The photo-induced metastable ESR centers with g =2.005, increasing up to 10^{17}-10^{18} cm^{-3} with the strong bandgap light (1.8 eV), remain stable even at room temperature and are completely annealed out by thermal treatment above 150C [8]. This was originally observed as

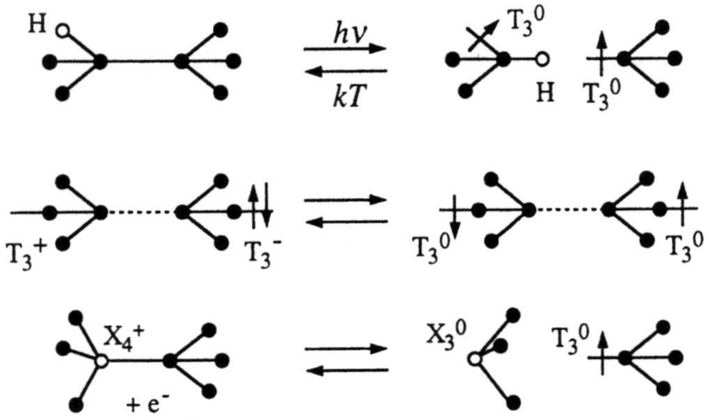

Figure 3. Schematic models for photo-creation of metastable Si dangling bonds (T_3^0) in a-Si:H proposed so far.

a photo-induced decrease of the photoconductivity by Staebler and Wronski [6]. The structure of these photo-induced defects has been known to be almost the same as that of native dangling bonds with also g=2.005 through detailed ESR and its hyperfine-structure measurements [19, 20].

Most controversial is a structural origin of the photo-created dangling bond ; what is a precursor of this defect? Several models have been proposed so far, as shown in Fig.3 ; a weak Si-Si bond neighbored by a Si-H bond [8] (top), a T_3^+-T_3^- pair with a negative U_{eff} [21] (middle), and a charged dangling bond coupled with extrinsic atoms [22] (bottom). In order to come to a unique understanding, detailed information on the spatial relationship between the metastable dangling bond and nearby atoms such as H, O and C should be clarified.

In this section, our recent results of pulsed ESR experiments are described, which were performed on undoped a-Si:H samples containing various isotopes with different concentrations [20, 23]. They suggest that a defect precursor is coupled with neither H(D) nor impurity atoms such as O and C.

3.1 Experimental

The ESR line resulting from a dilute collection of randomly oriented paramagnetic centers in an amorphous matrix is inhomogeneously broadened usually because of g-tensor anisotropy, variations of g-values at different sites and unresolved hyperfine splittings. It causes much difficulty for deconvoluting signals, separating hyperfine splittings and also for detecting small magnetic interactions. This is one of disadvantages of the conventional cw-ESR technique for amorphous materials. But these problems are cleared by introducing a variety of pulsed ESR techniques because the technique of electron-spin echo of pulsed ESR allows us to pick up a single packet from the inhomogeneously-broadened line[20].

Undoped a-Si:D as well as a-Si:H films were deposited on a metal foil at 250C by the plasma-enhanced CVD technique. D or H content was estimated to be about

9 at.%. Isotope impurities of $^{17}O(3x10^{19}\ cm^{-3}, I=5/2)$ and $^{13}C(4x10^{17} cm^{-3}, I=1/2)$ were intentionally introduced into a-Si:H. Spin density of these samples increased from $3x10^{15} cm^{-3}$(native dangling bonds) to $2x10^{17}\ cm^{-3}$ (metastable dangling bonds) after exposure to a 640-nm pulsed laser. Two-pulse and three-pulse ESEEM (electron spin echo envelope modulation) measurements were performed, by which dipole-dipole interaction between a photo-created dangling bond D^o (T_3^o) and a near-by nuclear spin as a function of a distance was detected[20].

3.2 Results and discussions

Figure 4 shows the 3-pulse ESEEM spectrum of undoped a-Si:D after light illumination, which was obtained with 3342G at the center of the spectrum at 9.38GHz. In order to simulate the spectrum, quantum mechanical calculation was carried out for the dipole-dipole interaction between the metastable dangling bond and deuterium nuclei (I=1) as a function of the spatial distribution of D atoms surrounding T_3^o. A best fit was obtained for the distribution of D atoms with respect to T_3^o, which is shown as f(r) in the inset in the figure. It clearly indicates that the photo-created metastable dangling bonds T_3^o are located in deuterium(namely, hydrogen)-depleted regions [20]. Other independent experiments on native dangling bonds have showed that f(r) of the native T_3^o is quite similar to that of the metastable one. Only one difference between two dangling bonds is in a saturated value of f(r), which corresponds to a constant D density of outer region of T_3^o; 6 at.% for the native T_3^o and 7at.% for the metastable T_3^o[20]. These values are lower than the film deuterium content of 9at.%, therefore, both native and metastable dangling bonds are thought to be formed in a randomly distributed region of deuterium, not in a deuterium clustered region.

Experimental results are summarized; (1) there is no characteristic hydrogen-T_3^o complex for metastable dangling bonds, (2) both native and metastable dangling bonds are located in hydrogen-depleted regions, and (3) metastable dangling bonds

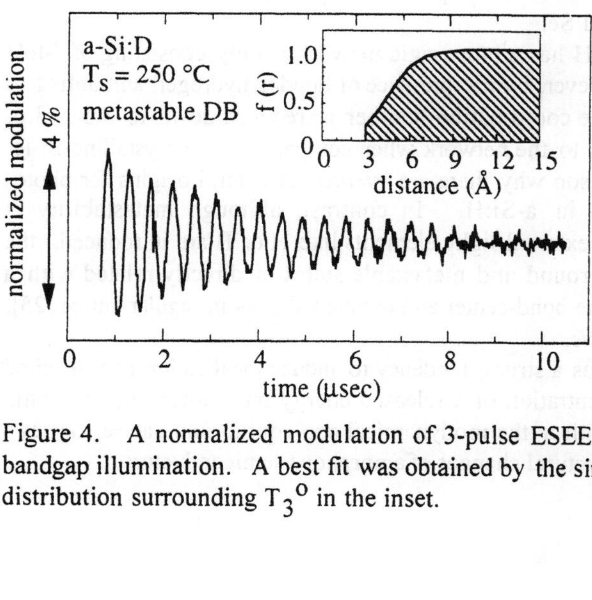

Figure 4. A normalized modulation of 3-pulse ESEEM spectrum of a-Si:D after bandgap illumination. A best fit was obtained by the simulation using a deuterium distribution surrounding T_3^o in the inset.

are located in a region with a slightly higher hydrogen content compared to native dangling bonds. The other experiment in terms of *instantaneous diffusion* provided another information; (4) metastable dangling bonds are isolated from each other (at least 100A separation on the average)[20]. Regarding the isotope(^{17}O, ^{13}C)-enriched a-Si:H sample, two-pulse ESEEM was performed to get the information on small hyperfine interactions, from which dipole-dipole interaction between $T_3^{\,o}$ and nuclear spins should be obtained if they are located in a distance from 3 to 8A. However, on a frequency-domain ESEEM spectrum obtained, no traces of signals were observed both at 1.69 MHz (^{17}O) and 3.13 MHz (^{13}C). It means that, as the item(5), neither oxygen nor carbon atoms are located in the vicinity of metastable dangling bonds[20].

Consequently, all of the experimental data indicate that photo-induced creation of metastable Si dangling bonds is not directly caused by *extrinsic* atoms such as H, O and C, but by some *intrinsic* structural reason. On the basis of this indication the models shown in Fig.3 can be examined. Any models to make two $T_3^{\,o}$'s simultaneously are unlikely, unless one can explain how to create a distant $T_3^{\,o}$ pair without forming hydrogen complexes. Any impurity-complex models are also ruled out. The $T_3^{\,+}$-$T_3^{\,-}$ model in the middle of Fig.3 does not contradict with the present experimental results if the created $T_3^{\,o}$'s are separated distantly. The recent paper by Orita et al. has proposed an intrinsic model which is compatible with the present data [24].

4. Concluding remarks

Amorphous chalcogenides are characterized by the two-fold coordination of chalcogen atoms and their lone-pair(LP) electrons forming the top of the valence band. These two factors give much structural flexibility to their random network and hence are in favour of various photo-induced structural changes observed in the materials. In particular, the presence of LP electrons causes a three-fold coordination pair in a disordered network of Se under photo-excitation, being discussed in terms of the similarity to the liquid Se.

On the other hand, a-Si:H has a more rigid network mainly consisting of 4-fold coordinated Si atoms. However, by the presence of bonded hydrogen amounting to nearly 10 at.%, the average coordination number is reduced down to 3.5 - 3.6, which gives some flexibility to the network when compared to the crystalline Si (c-Si). This might be the reason why there are *intrinsic* structural origins for photo-induced metastable defects in a-Si:H. In contrast, although metastability is produced also in c-Si when external light elements like H or B are introduced, the structural origin for both ground and metastable states is directly related with a hydrogen motion between the bond-center and tetrahedral sites in regular lattice [25]; namely, the origin is *extrinsic*.

Structural randomness has a strong tendency to induce localization of an excited electron, resulting in concentration of a released energy on a local region. This tendency, in combination with a thermodynamical non-equilibrium nature, causes a variety of photo-induced structural changes of amorphous semiconductors.

Acknowledgements

The authors would like to thank Dr. H. Oyanagi and Prof. J. Isoya for collaboration. This work, partly supported by NEDO, was performed in the Joint Research Center for Atom Technology (JRCAT) under the joint research agreement between the National Institute for Advanced Interdisciplinary Research (NAIR) and the Angstrom Technology Partnership (ATP).

References

[1] J.P. deNeufville, "Optical Properties of Solids - New Developments" ed. by B. O. Seraphin (North-Holland Pub. Co., Amsterdam 1975) p.437.
[2] K. Tanaka, Appl. Phys. Lett. **26**, 243 (1975); K. Tanaka, "Fundamental Physics of Amorphous semiconductors" ed. by F. Yonezawa (Springer-Verlag, Berlin 1981) p.104; K. Tanaka, J. Non-cryst. Solids **137&138**, 1 (1991).
[3] A.V. Kolobov and S.R. Elliot, Adv, Phys. **40**, 625 (1991).
[4] K. Shimakawa, A.V. Kolobov and S.R. Elliott, Adv. Phys. **44**, 475 (1995).
[5] S.G. Bishop, U. Strom and P.C. Taylor, Phys. Rev. Lett. **34**, 1346 (1975).
[6] D.L. Staebler and C.R. Wronski, Appl. Phys. Lett. **31**, 292 (1977).
[7] H. Dersch, J. Stuke and J. Beicher, Phys. Stat. Sol.(b) **105**, 265 (1981).
[8] M. Stutzmann, AIP Conf. Proc. No.157, ed. by B.L. Stafford and E. Sabisky (1987) p.235.
[9] A. Yoshikawa, Jpn J. Appl. Phys. **16**, Suppl. 67 (1977).
[10] Ke. Tanaka, J. Non-cryst. Solids **59&60**, 925 (1983).
[11] P.W. Anderson, Phys. Rev. Lett. **34**, 953 (1975).
[12] R.A. Street and N.F. Mott, Phys. Rev. Lett. **35**, 1293 (1975).
[13] M. Kastner, D. Adler and H. Fritzsche, Phys. Rev. Lett. **37**, 1504 (1976).
[14] D. Hohl and R.O. Jones, Phys. Rev. B **43**, 3856 (1991-I).
[15] H. Oyanagi, R. Shyoda, Y. Kuwahara and K. Haga, J. Synchrotron Radiation **2**, 99 (1995).
[16] A.V. Kolobov, H. Oyanagi, K. Tanaka and Ke Tanaka, J. Non-cryst. Solids (1996) (in press): A.V. Kolobov, H. Oyanagi, K. Tanaka, and Ke Tanaka, Phys.Rev. B (submitted).
[17] S.R. Elliott and A.V. Kolobov, Philos. Mag. B**61**, 853(1990).
[18] H. Hisakuni and Ke Tanaka, Science **270**, 974(1995).
[19] D.K. Biegelsen and M. Stutzmann, Phys. Rev. B**33**,3006(1986).
[20] S. Yamasaki, H. Okushi, A. Matsuda, K. Tanaka and S. Isoya, Phys. Rev. Lett. **65**, 756 (1990) : S. Yamasaki and J. Isoya, J. Non-cryst. Solids **164-166**,169(1993).
[21] D. Adler, J. Phys. **42**, C4-3 (1981).
[22] H. Fritzsche, J. Non-cryst. Solids **190**, 180 (1995).
[23] K. Tanaka, J. Non-cryst. Solids **137&138**, 1 (1991).
[24] N. Orita, T. Matsumura and H. Katayama-Yoshida, J. Non-cryst. Solids (in press).
[25] B. Holm, K.B. Nielsen and B.B. Nielsen, Phys. Rev. Lett. **66**, 2360 (1991).

ab initio Electronic Structure Calculation of Atomic Displacement of Defects in Semiconductors Induced by Electronic Excitation

H. Katayama-Yoshida[1] *and N. Orita*[2]

[1]Department of Condensed Matter Physics, Institute of Scientific and Industrial Research, Osaka University, Osaka 567, Japan
[2]Electrotechnical Laboratory, Tsukuba 305, Japan

Abstract. We discuss the mechanism of atomic displacements of defects induced by electronic excitation in semiconductors based upon *ab initio* quantum-molecular-dynamics simulations and a pseudo Jahn-Teller interaction. Electronic structures and dynamics of dangling bond in amorphous silicon (a-Si), hydrogenated amorphous silicon (a-Si:H), and also crystalline silicon (c-Si) are investigated based upon *ab initio* quantum-molecular-dynamics simulations. We find the bistable dangling-bonds with negative effective-U nature at normal sp^3-configuration and positive effective-U nature at flipped p_z-configuration with a large lattice-relaxation (~1.5A). It is proposed that the conversion from the negative effective-U dangling-bong (sp^3) to positive effective-U dangling-bond (p_z) induced by electronic excitation is the origin of light-induced defect-formation in a-Si:H (Staebler-Wronski effects).

1. Introduction

Chemical reaction of the defects in semiconductors induced by electronic excitation is complex because of the following three reasons; (i) presence of the multiple-charged states of the defects, (ii) the presence of the negative effective-U center, and (iii) the presence of the bistability of the defect in covalent semiconductors [1-3]. The presence of the multiple-charged states is caused by the covalent delocalization of the defect wavefunctions due to the strong hybridization between the imputy and host wavefunctions. The negative effective-U nature of the defects in semiconductors is caused by a large lattice-relaxation (Anderson's Negative-U system) or exchange correlation (Excange-correlation induced negative-U system) [3]. The bistability of the defect is caused by a psedo Jahn-Teller interaction induced by the vibronic mixture of the ground and excited states.

The purpose of the present paper is to discuss the mechanism of the atomic displacements of defects induced by electronic excitation in covalent semiconductors based upon *ab initio* quantum-molecular-dynamics simulations. In section 2, we discuss the general theory of the bistability of defects in covalent semiconductors based upon a pseudo Jahn-Teller interaction. We discuss the calculation method of *ab initio* quantum-molecuar-dynamics simulations for a-Si, a-Si:H, and c-Si in section 3. Section 4 is devoted to the calculated results and discussion. We summarize in section 5.

2. General Theory of the Bistability of Defects in Semiconductors

The general theory of the bistability of defects in covalent semiconductors is given by a pseudo Jahn-Teller interaction [4, 5]. The ground state energy $E_0(Q)$ is expanded by atomic displacement Q (see Fig. 1a),

$$E_0(Q) = E_0(0) + 1/2\,(K_s + K_v)\,Q^2\,. \tag{1}$$

Here $K_s > 0$ is the static (s) elastic force constant, and $K_v < 0$ represent the charge relaxation

Springer Series in Solid-State Sciences, Volume 124
Relaxations of Excited States and Photo-Induced Structural Phase Transitions
Ed.: K. Nasu

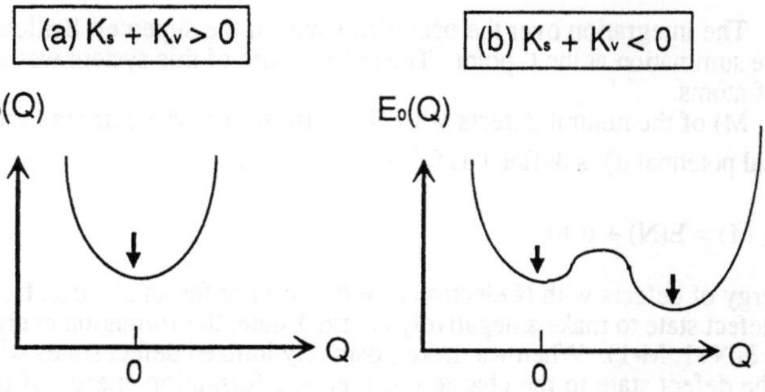

Fig. 1 Schematic explanation of the excistence of bistability due to a pseudo Jahn-Teller interaction.

due to the displacement-induced vibronic (v) coupling between the ground states |0> and excited states |i>,

$$K_v = -2 \sum_{i \neq 0} |\langle 0| \delta H/\delta Q |i\rangle|^2 / (E_i - E_0) \ . \quad (2)$$

Pseudo Jahn-Teller distortions ($K_s + K_v < 0$) are possible when the vibronic coupling competes favorably with static elastic force constant (see Fig.1b).

Equations (1) and (2) suggest that one should search for structural instabilities in (1) intrinsically "soft" semiconductors (K_s small) such as amorphous semiconductors or semiconductor surface, and (2) defects with low-lying excited states that couple strongly to the ground state through displacement ($|K_v|$ large) such as the element with p-s coupling (<0| = p, <i| = s) in second period of the periodic table or IIB-VI (IB-VII) semiconductors with d-s coupling (<0| = d, <i| = s) containing transition atoms.

Hydrogenated amorphous silicon (a-Si:H) is intrinsiccaly soft (K_s small), therfore we have a possibilty to obtain a structural instability (bistabilty) $K_s + K_v < 0$ with small p-s coupling between the states near the valence band maxmum (generally made up of atomic p-orbital) and the states near the conduction band minimum (generally s-orbitals). The impurity atoms of second period of the periodic table, such as substitutional nitrogen impurity in cystalline diamond (c-C:N), is also candidate for the structural instabilities because p-s coupling between 2p and 2s-orbitals in nitrogen atom is very large ($|K_v|$ large) with similar spatial distribution of 2p- and 2s-orbitals.

3. Calculation Method

We generate a-Si:H by rapid cooling and annealing from a liquid silicon using the dynamical simulated-annealing scheme proposed by Car and Parrinello [6] with the periodic boundary conditions of supercells containing 64 Si and 8 H atoms. We performed the dynamical simulated-annealing to determine the electronic structure and the metastable atomic coordinates of a-Si:H with rapid-cooling from the higher temperature (1800 K) to the lower temperature (300K) by the rate of 2×10^{15} K/sec after keeping it at 1800K for 0.75 psec. Then we have annealed the system for 0.25 psec at 300 K. The displacements of atoms were permitted fully in the supercell. The forces acting on the atoms were less than 0.07 eV/A in all cases.

Calculations of the total energy were performed with the norm-conserving pseudopotential method and the Ceperley-Alder-Perdew-Zunger form of the exchange and correlation energy [7]. We used a set of pseudopotentials of Si (C, N) atom by Kleinman-Bylander's [8] (Troulier and Martins [9]). Plane waves up to 163 eV (802 eV) in energy were included in the basis set for a-Si:H (C, N). The integration over the occupied states in the supercell Brillouin zone was represented by the summation at the Γ point. The temperature of this system was set by scaling of the velocity of atoms.

Formation energy F(N, M) of the neutral defects with N electrons and M electrons in the charge reserver (the chemical potential μ) is defined as follows,

$$F(N, M) = E(N) + \mu M \tag{3}$$

where E(N) is the total energy of defects with N electrons. When we tranfer an electron from the charge reserver to the defect state to make a negatively charged state, the formation energy of this system is written as F(N+1, M-1). When we make positively ionized defect states with the charge transfer from the defect state to the charge reserver, the formation energy of the system is written as F(N-1, M+1). Therefore, the first ionization energy $\Delta E(N/N-1)$ is determined by the condition of F(N,M) = F(N-1, M+1). The first ionization energy is observed when chemical potential μ satisfies following condition

$$\mu = E(N-1) - E(N) = \Delta E(N/N-1). \tag{4}$$

The first electron affinity energy of the defect $\Delta E(N+1/N)$ is determined by the condition of F(N, M) = F(N+1, M-1), and where μ satisfies following condition

$$\mu = E(N) - E(N+1) = \Delta E(N+1/N) \tag{5}$$

The effective correlation energy U (effective-U) of N electron system is defined as

$$\begin{aligned} U &= \Delta E(N+1/N) - \Delta E(N/N-1) \\ &= E(N+1) + E(N-1) - 2E(N). \end{aligned} \tag{6}$$

A positive effective-U (U>0) center means that N electron system is stable and exist in thermal equilibrium, on the other hand, negative effective-U (U<0) does that N electron system is unstable then two N electron systems are dispropotionated to (N-1) and (N+1) electron system in thermal equilibrium; 2E(N) -> E(N-1) + E(N+1) with the condition of U<0 in Eq.(6).

4. Calculated Results and Discussion

It is well known that hydrogenated amorphous silicon (a-Si:H) is degraded during the exposure to the visible light. The conductivity and photoconductivity decrease a few orders of magnitude by light soaking and recover by high-temperature annealing. This is so called Staebler-Wronski (SW) effect [10]. Accompanying with the degradation, the concentrations of the paramagnetic dangling bonds increase a few orders of magnitude. These results suggest that the dangling bonds play an important role in SW effect. Many models are proposed for the origin of SW effect; (i) the weak Si-Si back bonds adjacent to the Si-H bond [11], (ii) the negative effective-U centers of dangling bonds [12] (iii) the negative-U centers by the coupling between dangling bond and impurity [13].

The dangling bonds generated by SW effect is investigated using light-induced electron spin resonance (LESR) [14, 15]. It is well known that three kinds of dangling bonds which have different g-values, g=2.01, g=2.004 and g=2.0055, are detected by the experiment. In The light-induced signal with the g-value of g=2.01 and g=2.004 is observable only at low temperature (T<150K), suggesting negative effective-U center. In this paper, we try to explain the experiments of LESR based on the calculations of the positive- (+), neutral- (0) and negative-charged (-) dangling-bonds. Based upon the course of this study, we propose the mechanism of SW effect in a-Si and a-Si:H and give an unified physical picture to this system.

In Fig. 2, we show the total energy vs. steps during the generation of a-Si started from c-Si. We have generated liquid silicon (l-Si) with heating up to 6000K, then cooling down to

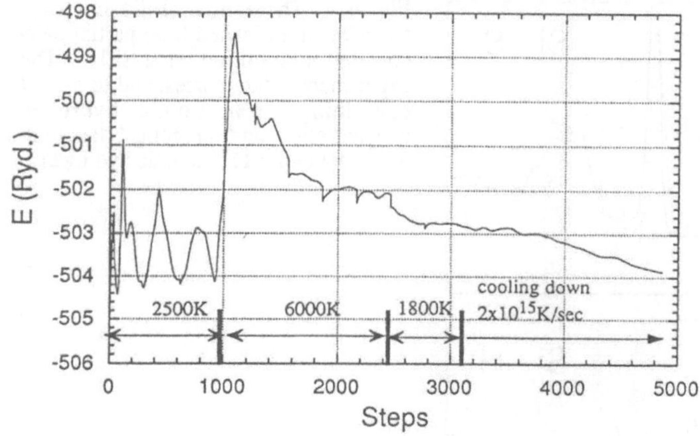

Fig. 2 Total energy vs molecular-dynamics steps during the generation of a-Si from c-Si through l-Si.

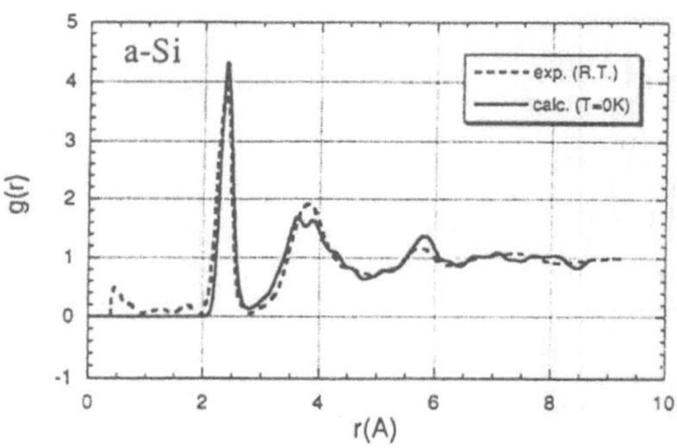

Fig. 3 Calculated pair correlation function (solid line) of amorphous silicon (a-Si) at 0K compared with the neutron experiment\cite (dashed line) at room temperature.

1800K. Averaged coordination number of the l-Si is 6 with strong covalency even in the case of the liquid metallic states. In Fig. 3, we compare the calculated pair-correlation function at 0K with the neutron-scattering experimental data [16] of a-Si (made by the sputtering method) at room temperature. The overall agreement between the calculation and experiment is satisfactory in spite of the fact that the fablication method is different between the calculation and experiment. Two peaks in the pair-correlation function in a-Si show the short-range ordering up to the second nearest neighbors, indicating the four-coordination in a-Si. Small amount of three- and five-coordinated defects are observed in the simulation. The intrinsic defects we found in the present simulations are (i) the three-coordinated dangling-bond (DB), (ii) the weak Si-Si bond, and (iii) the five-coordinated floating-bond (F_5). 91% of the Si atoms are four-coordinated and 3% of the atoms are three-coordinated dangling-bond in the simulation. The defects complexes are made by the weak Si-Si bonds and the F_5 bonds (the fifth bond of F_5 is always the Si-Si weak bond). However, a weak Si-Si bond is absent in a-Si:H. This means that the hydrogen atom can easily passivate the weak Si-Si bond in the hydrogenation process. Then the complex of the weak Si-Si bond and a hydrogen (H) is disproportionated into a four-coordinated Si-bond and a hydrogen-passivated dangling-bond (Si-H) in a-Si:H.

Figure 4 shows the pair-correlation function of the generated a-Si:H. Hydrogen-rich and -poor regions appear in this system even if we started the simulation from a homogeneous hydrogen distribution. We find that the most of the silicon dangling-bonds are terminated by

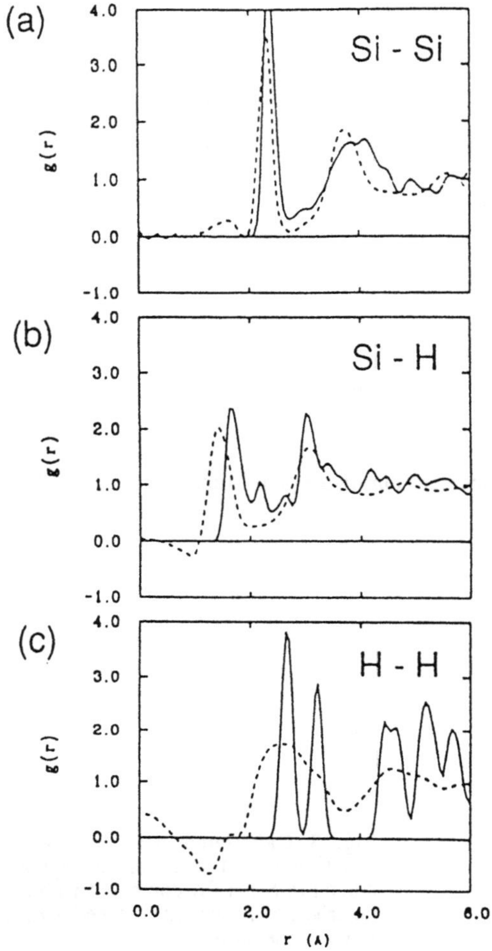

Fig. 4 Theoretical (solid line) and experimental (dashed line) partial pair-correlation function of a-Si:H. The experimental data correspond to a-Si:H containing 26% atomic hydrogen concentration, and theoretical data to a-Si:H containing 11% atomic hydrogen.

hydrogen in the hydrogen-rich region, on the other hand, the dangling bonds in the hydrogen-poor region are not terminated by hydrogen. Each atom is located at a local minimum of the total energy surface by the rapid-cooling and annealing. The intrinsic defects we found in the present a-Si:H are (i) the hydrogen-passivated dangling bond (Si-H), (ii) the positively-ionized three-centered (Si-H^+-Si) bond , (iii) the negatively-ionized three-coordinated dangling bond (D^-), suugesting negative effective-U, and (iv) the five-coordinated floating bond (F_5). The partial-pair-correlation functions of Si-Si, Si-H and H-H in a-Si:H at 300K and the experimental neutron data [16] are shown in Fig. 4. The overall agreement between the calculation and experiment is reasonable, although the hydrogen concentration is different between theory (11% atomic hydrogen) and experiment (26% atomic hydrogen). We find a broad first-peak in the Si-H partial pair correlation function which indicates the coexistence of Si-H bond and Si-H-Si three-centered bond. We also find two separated groups of peaks in the H-H partial pair-correlation-function which mainly comes from the separation of the hydrogen-rich and -poor regions in this system.

From an analysis of bond configurations, we find the existence of bistable dangling-bonds (sp^3-configuration with normal (n) dangling-bond [n-DB] and p_z-configuration with flipped (f) metastable (*)-dangling-bond [f-*DB]) as is depicted in Fig. 5a and 5c. These bistable dangling-bonds are strongly related to the mechanism of SW effect as we discuss later. The energy difference is 0.52 eV in the bistable n-DB and f-*DB with a large lattice-relaxation (~1.5 A) as is shown in Fig. 5. This bistability is similar to the case of DX-center in GaAs which is proposed by Chadi and Chang [17].

Previously, we generated a-Si:H using molecular dynamics simulation [18] to explain X-ray-irradiation-induced crystallization of amoephous silicon. In the calculation, we obtained the atomic displacement through the two types of dangling-bond, ([n-DB] and [f-*DB]) in order to explan the X-ray-irradiation-induced crystallization of amoephous silicon. Therefore, we conclude that the charged states and the different types of dangling bonds may be lelated to the origin of SW effect with high possibility.

The calculated adiabatic potential surfaces of n-DB and f-*DB with three different (+, 0, -) charge states are shown in Fig.6. All charge state (+, 0, -) shows the bistability because $K_s + K_v < 0$ in Eq.(1) with samall K_s in amorphous phase. In order to undestand this we also investigated the nature of the dangling bonds in c-Si. Only negatively charged dangling-bond shows the bistability in c-Si, and neutral and positicely charged dangling-bonds do not show the bistability because K_s is large in c-Si. We also calculated the formation energy (F) vs

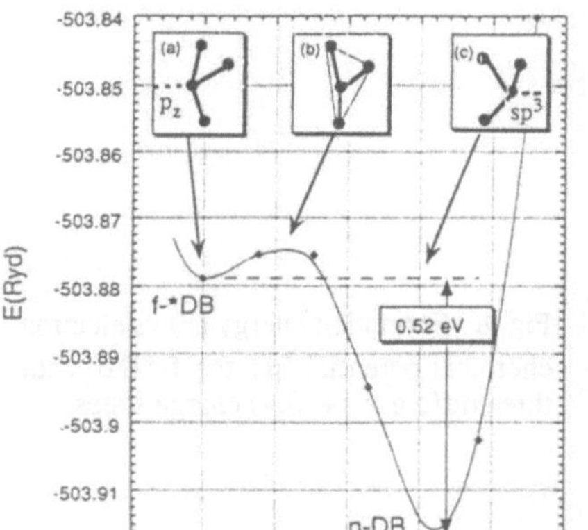

Fig. 5 Adiabatic potential surface of negatively-charged dangling-bond plotted as a function of the displacement of Si atoms. (a) Q=0A corresponds to the flipped (f) metastable (*) dangling bond (p_z-like structure, f-DB*), (b) four Si atoms located in same plane, and (c) Q=1.5A corresponds to the more stable normal (n) dangling-bond (sp^3-like structure, n-DB). Displacement (Q) of Si atom is moved toward the gravity center on the rectangular which is constructed by the three silicon-atoms at the nearest-neighbor.

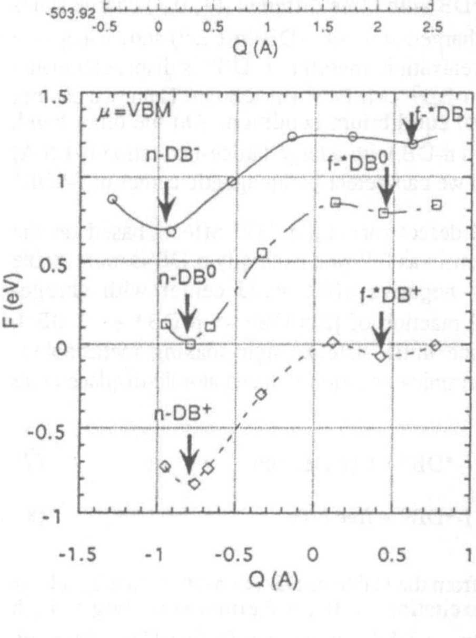

Fig. 6 Formation energy of n-DB and f-*DB with three different (+, 0, -) charge states when we chose the chemical potential (μ) is located at the valence band maxmun (VBM). Q=0A corresponds to the displacement where four Si atoms located in same plane.

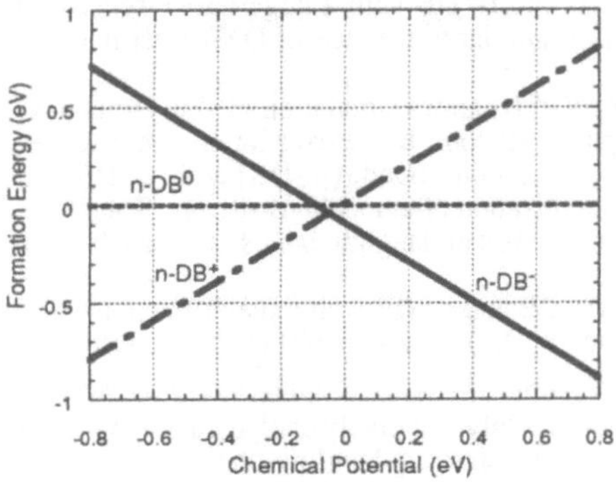

Fig. 7 Formation energy (F) vs electron chemical potential (μ) for n-DB with three different (+, 0, -) charge states.

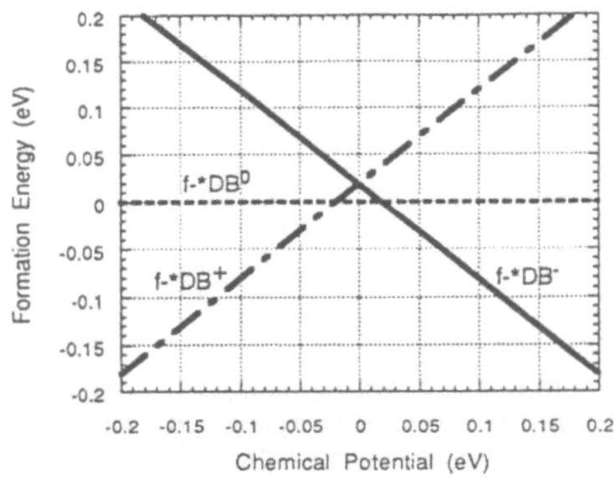

Fig. 8 Formation energy (F) vs electron chemical potential (μ) for f-*DB with three different (+, 0, -) charge states.

electron chemical potential (μ) for n-DB and f-*DB with three different (+, 0, -) charge states in a-Si as is shown in Fig. 7 and 8. Neutral charged-state of n-DB (n-DB^0) shows negative effective-U (U = -0.10 eV) with a large lattice-relaxation, therefore n-DB^0 is dispropotionated into n-DB$^+$ and n-DB$^-$ with the reaction of [2 n-DB^0 -> n-DB$^+$ + n-DB$^-$]. Then, we can not observe any paramagnetic center in the thermal equilibrium condition. On the other hand, neutral charged-state of f-*DB^0 converted from n-DB with a large lattice-relaxation (~1.5 A) shows positive effective-U (U = +0.04 eV) and we can detect paramagnetic center of f-*DB^0 (S=1/2) by ESR.

We discuss the origin of light-induced defect-formation (SW effect) based on the recombination-enhanced defect-reaction mechanism as follows; generally n-DB is more stable than the metastable f-*DB and n-DB^0 is negative effective-U center with charge-disproportionated states through the chemical reaction of [2 n-DB^0 -> n-DB$^+$ + n-DB$^-$]. Therefore, mostly n-DB$^+$ and n-DB$^-$ excists in the initial state. By light soaking, metastable f-*DB^0 is created from n-DB$^+$ and n-DB$^-$ by electronic-excitation-induced atomic-displacements (~1.5 A) as follows (depicted in Fig.9.),

n-DB$^-$ + frre hole + free electron -> f-*DB^0 + free electron (7)

n-DB$^+$ + frre hole + free electron -> f-*DB^0 + free hole (8)

At the moment metastable f-*DB^0 is generated from the stable n-DB$^-$ or n-DB$^+$ state by a large atomic displacement induced by electronic excitation. After a thermal annealing at high temperature, the metastable f-*DB^0 converte to n-DB^0 (negative effective-U center) to go

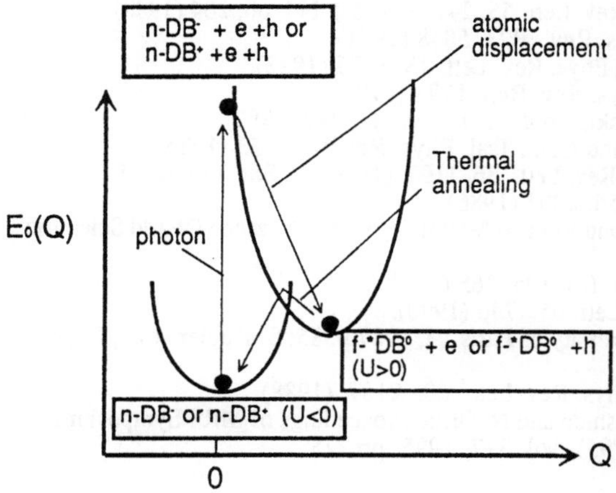

Fig. 9 Schematic explanation of Staebler-Wronski effect in a-Si:H.

beyond the potential barrier by phonons, then, 2 n-DB^0 disproportionate into n-DB$^-$ and n-DB$^+$. Metastable f-*DB^0 is observed as the defect of g=2.0055 in ESR with small s character of the dangling bond (less than 10% s character) because the dangling bond of f-*DB^0 has little s character (almost p_z-orbitals).

On one hand, at low temperature, paramagnetic centers of [n-DB$^-$ + hole (hole is trapped at n-DB$^-$ center)] or [n-DB$^+$ + electron (electron is trapped at n-DB$^+$)] can be created by shining the light and these are frozen at the metastable (positive effective-U) state since atoms can not move enough by low-temperature phonons [remember negative effective-U is caused by a large lattice-relaxatin]. However, at high temperature, [n-DB$^-$ + hole] or [n-DB$^+$ + electron] becomes n-DB^0 with negative effective-U nature by a large lattice-relaxation and the paramagnetic center is completely disappeared in ESR experiments. We identify that [n-DB$^+$ + electron] and [n-DB$^-$ + hole] at low temperature are the defects with g=2.01 and g=2.004 in LESR, respectively. This is consistent with the experiment by which the LESR signals of [n-DB$^+$ + electron] and [n-DB$^-$ + hole] at low temperature are disappered up to 150K [14, 15].

5.Summary

We have performed *ab initio* molecular-dynamics simulations of a-Si and a-Si:H, and c-Si in order to understand the mechanism of the light-induced defects formation (Staebler-Wronski effect) in a-Si:H. We propose the new mechanism of Staebler-Wronski effect for which the conversion from the negative effective-U dangling-bond to the metastable flipped positive effective-U dangling-bond induced by electronic excitation is responsible.

References

[1] Anderson and Haldane, Phys. Rev. B13, 2553 (1976).
[2] H. Katayama-Yoshida and A. Zunger, Phys. Rev. Lett. 53, 1256 (1984).
[3] H. Katayama-Yoshida and A. Zunger, Phys. Rev. Lett. 55, 1618 (1985)
[4] I.B. Bersuker, The Jahn-Teller Effect (Plenum, New York, 1984).
[5] S.H. Wei, S.B. Zhang and A. Zunger, Phys. Rev. Lett. 70, 1639 (1993).

[6] R. Car and M. Parrinello, Phys. Rev. Lett. 55, 2471 (1985); ibid 60, 204 (1988).
[7] J. P. Perdew and A. Zunger, Phys. Rev. B23, 5048 (1981).
[8] L.Kleinman and D. M. Bylander, Phys. Rev. Lett. 48, 1425 (1982).
[9] N. Troullier and J. L. Martin, Phys. Rev. B43, 1993 (1991).
[10] D. L. Staebler and C. R. Wronski, Appl. Phys. Lett. 31, 292 (1982).
[11] M. Stutzmann, W. B. Jackson, and C. C. Tsai, Phys. Rev. B32, 23 (1985).
[12] D. Alder and E. J. Yoffa, Phys. Rev. Lett. 36, 1197 (1976); Y. Bar-Yam and J. Joannopoulos, Phys. Rev. Lett. 56, 2203 (1986).
[13] H. Okushi, Amorphous Silicon and Related Materials, ed. H. Fritzsche (World Scientific Pub., 1988), p.657.
[14] H. Dersch et al., Phys. Stat. Sol. (b), 105, 265 (1981) .
[15] S. Yamasaki et al., Phys. Rev. Lett. 65, 756 (1990).
[16] A. Menelle, Ph. D. Thesis, University of Paris VI, 1987 p. 85; S. Kugler et al., Phys. Rev. B40, 8030 (1989).
[17] D. J. Chadi and K. J. Chang, Phys. Rev. Lett. 60, 2187 (1988).
[18] T. Matsumura, H. Katayama-Yoshida and N. Orita, Proccedings of MRS Symposium (Amorphous Silicon Technology '95), Vol. 377, 1995, pp.275.

Symmetry-Breaking Electronic-Atomic Processes in s-p Bonding Materials

Y. Shinozuka

Department of Applied Science, Faculty of Engineering, Yamaguchi University Tokiwadai 2557, Ube 755, Japan

Abstract. We discuss the off-center instability of a substitutional impurity in covalent semiconductors and the off-center type self-trapping of an exciton in alkali halides. Similarity and difference between their mechanisms are precisely discussed paying attention to the the Jahn-Teller effect and the driving force caused by the Coulomb interaction among electrons and ions. In the second part, two proposed mechanisms of defect reactions in semiconductors are reexamined: the recombination enhanced mechanism and the structural instability mechanism.

1. Introduction

In condensed matter, a crystal structure is realized when electrons and atoms are in the lowest ground state. In a given crystal structure, which is characterized by a space group, atoms cooperatively vibrate around each equilibrium position. The atomic elementary excitations are called phonons which are classified by the wave vector. The electronic elementary excitation in nonmetallic materials is a pair creation of a conduction electron and a valence hole: sometimes they form into an exciton. The elementary excitations both atomic and electronic are the main actors in the "stage" of the solid state physics [1]. How they behave and how they interact with each other are the interests of the solid state physics, especially in semiconductor physics. This is schematically shown in the left part of Fig. 1.

Recently, however, there have been found various phenomena which break out this "solid state physics". When we excite electrons in solids, it sometimes causes a large atomic displacement. For example, a localized electronic excitation induces large lattice distortion at a point defect, desorption of an atom from a surface, defect reaction, structural phase transition, and so on. Among them, the most attractive phenomenon might be the off-center instability found at a substitutional impurity in covalent semiconductors. When we dope IV or VI elements as a donor into III-V compound semiconductors, there appear a deep donor level in addition to a normal shallow donor level and they can be interchanged by a photoexcitation or a carrier trapping. This deep level, called DX-level is considered as the origin of the persistent photoconductivity [2]. The electronic and atomic structure of the DX level has long been unknown but it is now almost accepted that the DX level is induced by the off-center type displacement of a substitutional donor atom itself [3, 4]. This off-center instability was first regarded as unrealistic because it contradicts a common sense on the strong covalent bond. But it is now found that many impurity in semiconductors shows this off-center instability, such as EL2 in GaAs [5, 6] and N in Si [7, 8]. This on-off bistability of a substitutional donor can be understand from various point of views: the extrinsic self-trapping, the shallow-deep transition of bound electrons, the Jahn-Teller effect, Anderson's negative U and the change of the bonding between sp^3 and sp^2.

Springer Series in Solid-State Sciences, Volume 124
Relaxations of Excited States and Photo-Induced Structural Phase Transitions
Ed.: K. Nasu

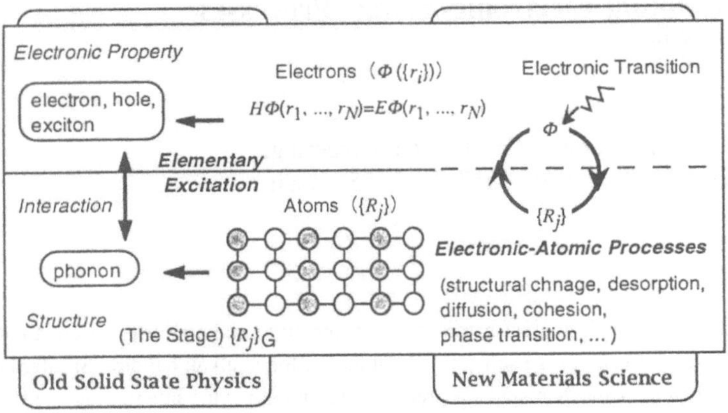

Figure 1. The physics of the elementary excitations in solids and the new stage of materials science by electronic excitation.

In the following sections, we discuss the mechanism of the off-center instability of a donor in covalent semiconductors in comparison with the off-center type self-trapping of an exciton in alkali halides: the role of the Jahn-Teller effect is discussed in section 2 and the origin of the driving force in section 3. In section 4 we briefly review the annihilation of a hydrogen-carbon complex in silicon in connection with N in Si. Two proposed mechanisms of defect reaction enhanced by electronic excitations are reexamined in terms of the configuration coordinate diagram in section 5.

2. The Jahn-Teller Effect

First we recall the Jahn-Teller (J-T) theorem [9]. Let us consider a substitutional or interstitial impurity atom in a semiconductor, whose symmetry is characterized by a point group **G**. Localized electronic states $\Psi_{\Gamma_e}(r)$ are classified with respect to the irreducible representation Γ_e of **G**. Here r denotes the set of the positions of relevant electrons $\{r_1, r_2, ...\}$. The atomic displacements $\{R_j\}$ from each equilibrium position (the suffix j=0 refers to the impurity) can be transformed into the normal modes $\{Q_\Gamma\}$ with the irreducible representation Γ of **G**. When electrons are in a state $\Psi_{\Gamma_e}(r)$, the generalized force $F_{\Gamma_e}^{\Gamma}$ with respect to Q_Γ is given by

$$F_{\Gamma_e}^{\Gamma} = -\left\langle \Psi_{\Gamma_e} \left| \frac{\partial V}{\partial Q_\Gamma} \right| \Psi_{\Gamma_e} \right\rangle, \tag{1}$$

where $V(r, Q_\Gamma)$ is the electron-atom interaction. If $\Psi_{\Gamma_e}(r)$ is degenerate, the symmetric representation $[\Gamma_e \times \Gamma_e]$ always contains at least one non-identity representation Γ. Then $F_{\Gamma_e}^{\Gamma} \neq 0$ and surrounding atoms *always* displace to lower the symmetry. In the T_d symmetry, for example, the J-T distortion occurs as E type atomic distortions for doubly degenerate (E) electronic states and E and T type atomic distortions for triply degenerate (T_1 and T_2) electronic states

If we apply the Jahn-Teller theorem to the stability of positions of an impurity atom in a crystal, the point is:

whether these J-T displacements contain R_0 of the impurity atom located at the center.

Although the number of relevant atoms is not definitely limited in solids, it is easy to find which type of mode(s) the central displacement R_0 belongs to. In the T_d symmetry, R_0 belongs to the T type distortion. Then we can conclude that the T_d position of an impurity is unstable when localized electrons are in a triply degenerate state. It would be interesting that the T_d symmetry, which is popular in four-coordination covalent semiconductors, has the inherent instability. The criterions for another point-symmetries, and the effect of spin and the number of electrons are given in Ref. [10]. It should be noted that

any position which has the inversion symmetry does not show the off-center instability caused by the J-T effect.

In systems with the inversion symmetry all the irreducible representations are classified into *gerade* and *ungerade*. The J-T distortion modes are all *gerade* while the displacement of the central atom always belongs to *ungerade* modes.

We have treated the atomic motion classically. For shallow localized states where the lattice relaxation energy E_{LR} caused by the Jahn-Teller effect is smaller than the average phonon energy $\hbar\omega$, transitions occur rapidly between the equivalent minimums of the atomic configuration (There are four C_{3v}-type minimums for the above T_d case). If the mean transition time is shorter than characteristic time of the experiment, the average lowering of the symmetry is zero (the dynamic J-T effect). On the other hand, for deep states with $E_{LR} >> \hbar\omega$, the present J-T instability really takes place to reduce the symmetry of the system.

3. Driving Force of the Off-Center Instabilities

In this section let us discuss the origin of the driving force of the off-center instability of a substitutional impurity. First we choose a particular system, the EL2 center in GaAs. The system is an antisite defect, where an As atom is located at a Ga site. The local chemical bonding of the As atom needs three valence electrons, then it behaves as a double donor. The rest two electrons feel $+2e$ positive charge of the As impurity if they are far away from it. But when two electrons come near the impurity, they feel negative charge of valence electrons around it. Thus the electron potential is attractive at a large distance but repulsive within a small distance. Then two electrons push the As impurity off and occupy a created dangling bond. Local charge neutrality of the As atom is satisfied. This on-off transition can be understand as follows: in order to satisfy the host valence, the 4-fold coordination is favorable and the on-center position (T_d, sp^3) is stabilized. On the other hand, in order to satisfy the valence of the impurity, the 3-fold coordination is favorable and the off-center position (C_{3v}, sp^2) is stabilized.

Next, let us discuss the DX-center in III-V semiconductors shown in Fig. 2(a). When a III element is replaced by a IV element such as Si, it behaves as a single donor. The local chemical bonding of Si needs three valence electrons. The rest one electron feels $+e$ positive charge of the Si impurity when it is far away from. The Si atom can attract one more electron, and if it displaces from the center, two electrons occupy a created dangling bond. The Anderson's negative U is realized. Finally four electrons which have to occupy in a Si-As bond repel each other to create a pair of dangling bonds. In this case, the As becomes neutral and satisfies the local valency, while the Si is singly charged $-e$.

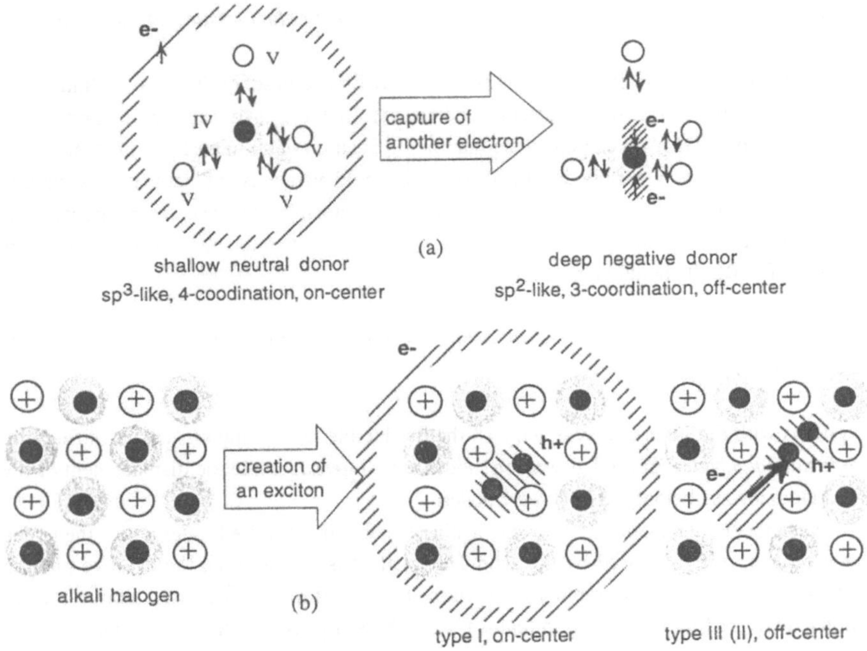

Figure 2. (a) The off-center instability of a substitutional donor (IV) in III-V semiconductors and (b) the self-trapping of an exciton in alkali halides.

In connection with the Coulomb repulsion between localized electrons discussed above, it is interesting to compare the mechanism of the self-trapping of an exciton in alkali halides [11]. It is found recently that self-trapped excitons in alkali halides have an off-center configuration (type III (II)) as well as an on-center configuration (type I) [12], as shown in Fig. 2(b). Their stabilities depend on the material. In the type I configuration, a hole is self-trapped by attracting two halogen ions to form a V_k center by which an electron is loosely bound. The effective charge of the V_k center seen by an electron away from it would be $+e$, because the charges of anions and cations are canceled. On the other hand, if an electron comes close to the V_k center, it feels the bare charge of two anions: the effective charge would be $(1\text{-}1\text{-}1)e = -e$. Thus the effective potential between an electron and a V_k center will be attractive at large distance and repulsive for short distance. Then an electron can be strongly localized pushing away the V_k center to the [110] direction to create a pair of an F center (anion vacancy) and an H center (interstitial anion pair): the type III (II) configuration. Here the symmetry changes from D_{2h}, which has the inversion symmetry, to C_{2v}, then the off-center instability of a self-trapped exciton is not due to the J-T effect but the pseudo-J-T effect. It can be also explained in terms of the different signs and modes of the electron-lattice and the hole-lattice interactions and by the imbalance in the Madelung energies for anion and cation [13].

4. Annihilation of a H-C Complex in Si

The substitutional nitrogen impurity in Si is found to show the off-center instability [7, 8]. For charge neutral states, the on-center T_d configuration is metastable and presents a shallow donor state, whereas the off-center C_{3v} configuration is stable and presents a deep-donor state. For ionized states the on-center configuration is stable and the off-center unstable. Then a carrier capture or emission by the impurity induces the interchange between the on- and off-center configurations. The off-center instability of N in Si can be explained in terms of the J-T effect as is in the DX-center, however, only one localized electron is enough to stabilize the off-center configuration.

It would be interesting to compare N in Si with H-C complex in Si. Hydrogen impurities has a key role in passivation and activation of impurities in semiconductors. A hydrogen impurity can occupy several stable sites in the silicon lattice. If it occupies a bond center site inbetween a substitutional carbon (isoelectronic) impurity and a Si atom, this H-C complex act as a shallow donor whose thermal depth is E_c-0.15eV. The positively charged H-C complex is stable above the room temperature, while the neutral state is unstable at the room temperature and the activation energy for annihilation is estimated as 0.5~0.7eV [14]. Since the "united atom" of an H-C complex is a nitrogen atom in the molecular chemistry, the effective potential for an conduction electron away from the $(H\text{-}C)^+$ complex is similar to those by the N^+ impurity. Then we can derive an analogy between the stabilities of N impurity and of H-C complex in Si. For a shallow donor state or a positively charged state (an electron is away) the central position is stable for an N impurity and the H-C complex is stable. If an electron comes close to them, on the other hand, it induces the off-center displacement of an N impurity and the annihilation of an H-C complex, that is, a hydrogen atom is pushed away.

5. Mechanisms for Defect Reaction Enhanced by Electronic Excitation

Defect reactions in semiconductors are enhanced by electronic excitations. Several models have been proposed for the mechanism and they can be classified into two categories. One is the recombination enhanced (or phonon kick) mechanism (Fig. 3(a)) where an electron-hole recombination at a defect induces a transient atomic motion (phonon) in the reaction coordinate Q_R. Examples are Weeks-Tully-Kimerling model [16] and Sumi model [17]. The other is the structural instability mechanism (Fig. 3(b)) where a special electronic state of a defect induces the instability. An electronic excitation to ex_1 in Fig 3(b) promptly induces the reaction and those to ex_2 decreases the potential barrier of the reaction. Examples are Bourgoin-Corbett model [18] and Sheinkman model [19]. In the previous section, the annihilation of an H-C complex has been explained in terms of the structural instability mechanism. Though two mechanisms (a) and (b) look different, they can be treated in a unified scheme as follows.

Carrier capture processes as well as localized electronic excitations take place via a localized electronic state whose level is located in the band gap. Therefore it is convenient for both cases to introduce the configuration coordinate diagram (CCD) [20]. A typical example is illustrated in Fig. 4: the defect has D^0 and D^- charged states in a p-type semiconductor. The defect is stationary occupied by a majority hole (D^0). A capture of a minority electron occurs by a thermally activated process which is followed by a transient lattice vibration. During the lattice relaxation, an energy equal to the thermal depth $E_{th}{}^e$ of an electron is converted to phonons. These transient phonons can enhance a succeeding majority hole capture. During the lattice

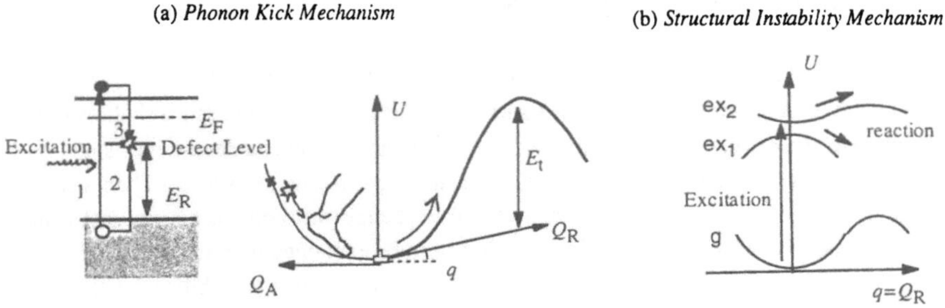

Figure 3. (a) Phonon kick mechanism [15] and (b) structural instability mechanism for defect reactions in semiconductors.

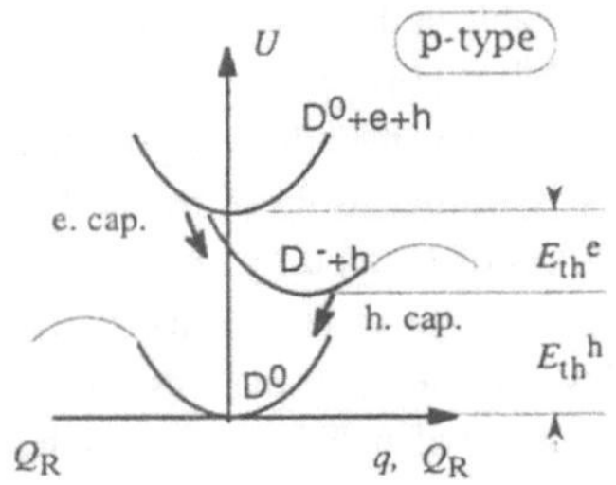

	e. cap.	h. cap.	$E_{th}^e \sim E_g$	$E_{th}^h \sim E_g$
p-type	thermally activation	enhanced	if $q=Q_R$, phonon kick by e. cap. **struct. instability**	if $q=-Q_R$, phonon kick by h. cap. **recomb. enhanced**
n-type	enhanced	thermally activation	if $q=Q_R$, phonon kick by e. cap. **recomb. enhanced**	if $q=-Q_R$, phonon kick by h. cap. **struct. instability**

Figure 4. An example of the configuration coordinate diagram for a defect in p-type semiconductors and the table of probable defect reaction path. Enhanced defect reaction occurs in the hatched cases.

relaxation after a hole capture, an energy equal to the thermal depth E_{th}^h of a hole is converted to phonons.

When we introduce the CCD for defect reactions, we should mind that there are many situations depending on the variety of charged states, the magnitude of E_{th}^e and E_{th}^h, the relation between the relaxation mode q and the reaction coordinate Q_R, the curvature of adiabatic potentials at the equilibrium of another potentials, and so on. If the signs of the curvatures are opposite and $q=Q_R$, the structural instability mechanism takes place (Fig. 3(b): the excitation to ex_1). If E_{th}^e is large and $q=Q_R$ for the defect in Fig. 3, the phonon kick mechanism takes place by an electron capture, and this corresponds to the structural instability mechanism shown in Fig. 3(b) (the excitation to ex_2). If E_{th}^h is large and $q=-Q_R$ for the defect in Fig. 3, the phonon kick mechanism takes place by a hole capture which is enhanced by an electron capture. This could be the most effective path to the defect reaction and nothing but so called the recombination enhanced mechanism. A detailed discussion is given elsewhere [21].

6. Summary

So far we have discussed the off-center instability of a substitutional impurity in covalent semiconductors and the off-center type self-trapping of an exciton in alkalihalides. In the ground state, two materials are different both in the chemical bonding and the lattice structure. Therefore the characters of the elementary excitations are different and have been discussed in different sections of the solid state physics (see the left part of Fig. 1). However, a local electronic excitation, which is realized by introducing of an impurity or self-trapping, can induces another possibility of the bonding, which is spontaneously accompanied by a symmetry breaking. In excited states of these materials, electronic states are remixed among s- and p-orbitals of the constituents in cooperation with the atomic rearrangement. As discussed in section 3, when a large atomic displacement occurs, the elementary excitations (electron, hole and the lattice) which are defined in the starting "solid state structure", lose their own meaning. There we must do with a more basic scheme (all valence electrons and all ions), in which we can discuss various electronic-atomic processes in all-types of s-p bonding materials (from covalent IV to ionic I-VII): see the right part of Fig. 1. Intensive study on the atomic displacements induced by electronic excitations will present us possibility of novel structures (both local bonding and bulk structure) of the materials as well as on a deep understanding of the ground state structure.

Acknowledgments

The author would like to thank Professor Y. Kayanuma and Professor K. Maeda for valuable discussions. The present work is supported by the Proposal-Based Advanced Industrial Technology R&D Program of NEDO Japan.

References

[1] S. Nakajima, Y. Toyozawa and R. Abe, *The Physics of Elementary Excitations* (Springer-Verlag, Berlin, 1983)

[2] D. V. Lang and R. A. Logan, Phys. Rev. Lett. **39**, 635 (1977).

[3] D. J. Chadi and K. J. Chang, Phys. Rev. Lett. **61**, 873 (1988), Phys. Rev. B**39**, 10063 (1989).

[4] M. Saito, A. Oshiyama and O. Sugino, Phys. Rev. B**47**, 13205 (1993).

[5] J. Dabrowski and M. Scheffler, Phys. Rev. Lett. **60**, 2183 (1988).

[6] D. J. Chadi and K. J. Chang, Phys. Rev. Lett. **60**, 2187 (1988).

[7] K. L. Brower, Phys. Rev. B**26**, 6040 (1982).

[8] K. Murakami, H. Kuribayashi and K. Masuda, Phys. Rev. B**38**, 1589 (1988).

[9] H. A. Jahn and E. Teller, Proc. Roy. Soc. A**161**, 220 (1937), and H. A. Jahn, Proc. Roy. Soc. A**164**, 117 (1938).

[10] Y. Shinozuka, Materials Science Forum **83-87**, 527 (1992).

[11] K. S. Song and R. T. Williams, *Self-Trapped Excitons* (Springer-Verlag, Berlin, 1993)

[12] K. Kan'no, K. Tanaka and T. Hayashi, Rev. Solid State Sci. **4**, 383 (1990).

[13] Y. Kayanuma, *Defect Processes Induced by Electronic Excitation in Insulators*, ed. N. Itoh (World Scientific, 1989) Chapter 2.
[14] Y. Kamiura, M. Tsutsune, M. Hayashi, Y. Yamashita and F. Hashimoto, Materials Science Forum **196-201**, 903 (1995).
[15] by the courtesy of Prof. K. Maeda.
[16] J. D. Weeks, J. C. Tully and L. C. Kimerling, Phys. Rev. B**12**, 3286 (1975).
[17] H. Sumi, Phys. Rev. B**29**, 4616 (1985), J. Phys. C**17**, 6071 (1984).
[18] J. C. Bourgoin and J. M. Corbett, Phys. Lett. **38A**, 135 (1972), and Radiation Effects **36**, 157 (1978).
[19] M. K. Sheinkman, JETP Lett. **38**, 330 (1983), and M. K. Sheinkman and L. C. Kimerling, *Defect Control in Semiconductors*, ed. K. Sumino (North-Holland, 1990) p. 97.
[20] Y. Shinozuka, Jpn. J. Appl. Phys. **32**, 4560 (1993).
[21] Y. Shinozuka, in preparation.

Part VIII

Optical Properties of Fullerenes and Photo-Induced Polymerization

Lattice Relaxations Theory of Localized Excitations and Applications to Fullerenes

C.L. Wang[1], *W.Z. Wang*[2], *Z.B. Su*[2], *and L. Yu*[2,3]

[1]China Center of Advanced Science and Technology (World Laboratory), Beijing 100080, China

[2]Institute of Theoretical Physics, CAS, Beijing 100080, China

[3]International Center of Theoretical Physics, I-34100 Trieste, Italy

C_{60} fullerenes have the highest point group symmetry, I_h, which is subject to Jahn-Teller distortions due to the electron-phonon coupling with localized excitations. These localized excitations, including polarons and self-trapped excitons (STE), are accompanied by strong lattice distortions which lead to a dynamical symmetry breaking from I_h to D_{5d} symmetry for both polaron and STE cases. Hence the lattice relaxation effects are essential in describiig their dynamics. We present the lattice relaxation theory generalized to include the multi-electron background and the multi-phonon process, using the fullerene system as an illustrative example. Moreover, the self-consistency of the electron states with lattice configurations is taken into account within this framework. Making use of the formalism developed from the lattice relaxation theory, we calculate the non-radiative decay rate from an electron-hole pair state to a D_{5d} STE, the photo-luminescense and photo-absorption spectra related to a D_{5d} STE, which is derived from the dipole forbidden transition, as well as the splitting of the six-fold degenerate D_{5d} polaron state due to dynamical Jahn-Teller effects. The obtained results are discussed in comparison with available experiments and possible observation consequences are predicted.

1. Introduction

The lattice relaxation theory was developed in the early 50s by Huang and Rhys[1], Pekar[2], Lax[3], Kubo and Toyozawa[4], and many others in connection with the study of colour centers in solids. Two of the present authors have generalized this theory to multi-electron systems in studying conducting polymers[5]. Since then, we have successfully applied this theory to calculate non-radiative transition probabilities and radiation transition rates in polyacetylene and M-X chains[6], as well as photo-luminescence spectrum[7] and Raman spectrum[8] in fullerenes.

In a many-electron system, different electron states correspond to different lattice configurations, and these features are demonstrated more clearly in low-dimensional systems than in three-dimensional ionic crystals or semiconductors, since in the former systems the lattice configurations are very sensitive to the electron motion. It should be emphasized that the lattice configurations of the initial state and the final state in an optical transition are different. In modern languages, different configurations, *i. e.* different dynamical symmetry breakings, are referred to as different phonon vacua, which give rise to multi-phonon

Springer Series in Solid-State Sciences, Volume 124
Relaxations of Excited States and Photo-Induced Structural Phase Transitions
Ed.: K. Nasu

processes. Moreover, in a many-body system, not only different lattice configurations create different electron states, but also different electron states are in turn to affect the corresponding lattice configurations through the Bogoliubov-De Gennes (BdeG) formalism, which make solutions for both electron wave functions and lattice configurations self-consistent. The self-consistency guarantees the total system energy taking its functional minimum with respect to electron wave functions and lattice configurations in the adiabatic approximation. The self-consistency can also be understood as follows: When we solve the eigenvalue problem for the electrons, the effective potential of electrons depends on the lattice configuration, and when the electron distribution changes, the lattice configuration should follow these changes so that the effective potential will be re-configured in accordance with the new electron distribution. The localized excitations in such systems are accompanied by strong lattice distortions, hence it is essential to use the lattice relaxation theory to study the dynamics of these localized excitations.

In this paper, we outline the main features of the lattice relaxation theory of multi-electron systems using the fullerene system as an illustrative example. We first calculate the non-radiative decay rate from an electron-hole pair state to an STE. Then we calculate the photo-luminescenes and photo-absorption spectra in the presence of a parity-forbidden transition related to STE. Finally we discuss the dynamical Jahn-Teller effects of six-fold degenerate D_{5d} polarons.

2. The non-radiative decay process

After an electron is excited from the highest occupied molecular orbital (HOMO) state h_u into the lowest unoccupied molecular orbital (LUMO) state t_{1u} in C_{60} fullerene with I_h symmetry, the system will experience a non-radiative decay process into an STE state with D_{5d} symmetry. In this section, we explicitly formulate and calculate its non-radiative decay rate.

The model Hamiltonian[11, 12] adopted to describe the fullerenes is a revised SSH Hamiltonian [9], which has grasped the main physics of dynamical symmetry breaking due to 1D Peierls instability and the nonlinearity of excitations in conducting polymers [10]. In order to consider the optical transitions between symmetry broken states and the effect of *e-e* correlations of C_{60}, this Hamiltonian with the Hubbard interactions included can be expressed as [5, 7]: $H = H_e + H_{ph} + H_{int}$ with

$$\begin{aligned} H_e = & - \sum_{<i,j>,s} t_0(c^\dagger_{i\,s}c_{j\,s} + h.c.) + U\sum_i n_{i\uparrow}n_{i\downarrow} + V \sum_{<i,j>,s,s'} n_{i\,s}n_{j\,s'} \\ & + \sum_\nu \alpha(< b^\dagger_\nu > + < b_\nu >)F^\nu + \sum_\nu \hbar\omega_\nu < b^\dagger_\nu >< b_\nu > \quad , \\ H_{ph} = & \sum_\nu \hbar\omega_\nu[(b^\dagger_\nu - < b^\dagger_\nu >)(b_\nu - < b_\nu >) + \frac{1}{2}] \quad , \\ H_{int} = & \sum_\nu (\alpha F^\nu + \hbar\omega_\nu < b_\nu >)(b^\dagger_\nu - < b^\dagger_\nu >) + h.c. \quad , \end{aligned} \tag{1}$$

where

$$F^\nu = \frac{\hbar}{\sqrt{2\hbar\omega_\nu M}} \sum_{<i,j>,s} (\vec{\xi}_i^\nu - \vec{\xi}_j^\nu) \cdot \frac{\vec{l}_1(ij)}{l_1(ij)} (c_{i,s}^\dagger c_{j,s} + h.c.) \ . \tag{2}$$

Here α is the e-ph coupling constant, $\vec{\xi}_i^\nu$ is the amplitude of normal mode ν in the homogeneous ground state [12], and $\vec{l}_1(ij)/l_1(ij)$ is the unit vector along the unrenormalized bond $< i, j >$. $b_\nu^\dagger$ (b_ν) is the phonon creation (annihilation) operator for mode ν, while $< b_\nu^\dagger >$, $< b_\nu >$ are the order parameters of dynamic symmetry breaking. They represent the equilibrium phonon positions with respect to the unrenormalized origin, and can be determined self-consistently within the BdeG formalism [10] through the gap equations $\hbar\omega_\nu < b_\nu > = -\alpha < e|F^\nu|e >$, where $|e>$ is the many-electron state found from the corresponding BdeG equations. Here the unrestricted Hartree-Fork approximation has been used to deal with the e-e interactions in H_e, which will affect the electronic and lattice structures of the ground state and the exciton formation. The set of parameters is chosen as: $t_0 = 1.75eV$, $\alpha = 4.75eV/\mathring{A}$, $U = 4.0eV$, $V = 0.15eV$. For the homogeneous ground state, the electronic HOMO-LUMO gap is 2.00eV, while the difference of the bond lengths is $0.04\mathring{A}$. It is noted that only the totally symmetric A_g modes can contribute to $< b_\nu >$, *i.e.* only these modes can condense to induce the bond length alternation in the I_h-symmetric ground state of the fullerenes. In this non-radiative decay process, the initial state is an electron-hole pair, whereas the final state is a D_{5d} STE state.

We can choose either the initial or the final state lattice configuration as Q_ν. The electron state $|e>$ and the phonon state $|\tilde{n}>$, each of which depends on the lattice configuration and diagonalizes the non-interacting H_e and H_{ph}, respectively, should be chosen such that the following equations are valid

$$<e|H_{int}|e>= 0, \quad <\tilde{n}|H_{int}|\tilde{n}>= 0. \tag{3}$$

The former equation leads to the gap equation, whereas the latter one provides a definition of $Q_\nu =< \tilde{n}|\hat{Q}_\nu|\tilde{n} >= \sqrt{\frac{\hbar}{2\omega_\nu M}} < \tilde{n}|b_\nu^\dagger + b_\nu|\tilde{n} >$, representing the dynamical symmetry breaking. The initial state $|i>$ is represented as $|i>= |e_i> \bigotimes |\tilde{n}_i>$, whereas the final state as $|f>= |e_f> \bigotimes |\tilde{n}_f>$.

Within the lattice relaxation formalism applied to the fullerenes, the non-radiative transition probability in the Born approximation is given by [5]

$$W = \frac{2\pi}{\hbar} \underset{i}{\mathcal{A}v} \sum_f |<f|H_{int}|i>|^2 \, \delta(W_{if} + \sum_q (n_q^i + \frac{1}{2})\hbar\omega_q - \sum_q (n_q^f + \frac{1}{2})\hbar\omega_q), \tag{4}$$

where $\underset{i}{\mathcal{A}v}$ means statistical average over the initial states and a summation over the required final states is understood. The energy difference between the initial and final states W_{if} is expressed as

$$W_{if} = \sum_{k_i} \epsilon_{k_i} - \sum_{k_f} \epsilon_{k_f} + \frac{1}{2} \sum_\nu M\omega_\nu^2 {Q_\nu^i}^2 - \frac{1}{2} \sum_\nu M\omega_\nu^2 {Q_\nu^f}^2, \tag{5}$$

where the summations $\sum_{k_i}$ and $\sum_{k_f}$ are carried over all occupied single electron states. The Hamiltonian (1) can be split in two ways, one using the initial lattice configuration, the other using the final lattice configuration. Either H^i_{int} for the initial state or H^f_{int} for the final state can be used here to calculate the matrix element in (4) and the results should be equivalent[5]. Hereafter, we use H^i_{int} in the numerical calculations.

Making use of the steepest descent method and taking the low temperature limit, the non-radiative transition probability can be explicitly evaluated to yield

$$W = \sqrt{\frac{2\pi}{\hbar\omega_b W_{if}}} \left(\frac{S}{p}\right)^p e^{(p-S)} \left(\frac{1}{4\hbar} G_1 (1-\frac{p}{S})^2 + \frac{1}{2\omega_b} G_2 \frac{p}{S}\right), \tag{6}$$

where $p = W_{if}/\hbar\omega_b$, $\omega_b^2 = K/M$; G_1 and G_2 are electronic matrix elements,

$$\begin{aligned} G_1 = & \left| \sum_\nu (\alpha \sqrt{\frac{2M\omega_\nu}{\hbar}} < e_i|F_\nu|e_f > + M\omega_\nu^2 < e_i|e_f > Q^i_\nu)(Q^f_\nu - Q^i_\nu) \right|^2 , \\ G_2 = & \frac{1}{M} \sum_\nu \left| \alpha \sqrt{\frac{2M\omega_\nu}{\hbar}} < e_i|F_\nu|e_f > + M\omega_\nu^2 < e_i|e_f > Q^i_\nu \right|^2 , \end{aligned} \tag{7}$$

where the Huang-Rhys factor S, equal to $\frac{E_r}{\hbar\omega_b}$, is the average number of phonons (with energy $\hbar\omega_b$) emitted during the transition to bring the initial configuration into the final one [5], and the lattice relaxation energy

$$E_r = \sum_\nu \frac{M\omega_\nu^2}{2} |Q^f_\nu - O^i_\nu|^2 \tag{8}$$

is a key quantity in this theory. The overlap integral $< e_i|e_f >$ is a Slater determinant, which can be decomposed into a product of $< e_i|e_f >_\downarrow$ and $< e_i|e_f >_\uparrow$. The element of $< e_i|e_f >_\sigma$ can be expressed in terms of single electron wavefunctions of the initial and final states,

$$< \bar{k}_i|k_f >_\sigma = \sum_{j=1}^N \bar{\Psi}^*_{k_i\sigma}(j) \Psi_{k_f\sigma}(j) \tag{9}$$

where k_i and k_f are occupied single electron levels of the initial and final state, respectively. N is the total number of sites in the fullerene. To compute the matrix element $< e_i|F^\nu|e_f >$, one resolves to calculate $< e_i|c^\dagger_n c_m|e_f >_\sigma$, which is a summation of J slater determinants (J is the total number of occupied state with spin σ). The k_fth Slater determinant is a determinant $< e_i|e_f >_\sigma$ with its k_fth column $\begin{pmatrix} < \bar{k}_{i1}|k_f >_\sigma \\ < \bar{k}_{i2}|k_f >_\sigma \\ \cdots \\ < \bar{k}_{iJ}|k_f >_\sigma \end{pmatrix}$ replaced by $\begin{pmatrix} \bar{\Psi}^*_{k_{i1}\sigma}(n)\Psi_{k_f\sigma}(m) \\ \bar{\Psi}^*_{k_{i2}\sigma}(n)\Psi_{k_f\sigma}(m) \\ \cdots \\ \bar{\Psi}^*_{k_{iJ}\sigma}(n)\Psi_{k_f\sigma}(m) \end{pmatrix}$, where k_{i1}, k_{i2}, $\cdots$, k_{iJ} are labels of the occupied initial states[6]. The numerical calculations of the non-radiative decay rate for the present set of parameters is performed by using the above formulas. We got $W = 2.60 \times 10^{11}$ sec^{-1}, which is comparable with a numerical simulation of dynamical process from an electron-hole pair state to an STE D_{5d} state, whose formation time is equal to 2.11×10^{-13} sec.

3. The parity forbidden transition related to STE

For a neutral fullerene C_{60}, the electronic HOMO state is h_u, while the LUMO state is t_{1u} which has the same parity, so the pure electronic transition from HOMO to LUMO is dipole forbidden. In some experimental reports, the observed sub-gap vibronic structures are attributed to the phonon-assisted STE, and this vibronic transition is assigned to the singlet sector of the lowest STE [13, 14]. Hence when we calculate the photo-luminescence and the photo-absorption spectra, the optical intensities from higher electronic states should be lent to trigger the actual transition with the assistance of the corresponding phonon process [15]. We introduce intermediate multi-electron states through the first order perturbation of e-ph interactions H_{int} representing the effect of quantum lattice fluctuations on electrons at the equilibrium lattice configuration,

$$\begin{aligned} |e_i n_i> &= |e_i^0 n_i^0> + \sum_{s(\neq i)} \frac{<e_s^0 n_s^0|H_{int}|e_i^0 n_i^0>}{\mathcal{E}_i^0 - \mathcal{E}_s^0} |e_s^0 n_s^0> , \\ |e_f n_f> &= |e_f^0 n_f^0> + \sum_{t(\neq f)} \frac{<e_t^0 n_t^0|H_{int}|e_f^0 n_f^0>}{\mathcal{E}_f^0 - \mathcal{E}_t^0} |e_t^0 n_t^0> . \end{aligned} \quad (10)$$

Here we should emphasize that the unperturbed intermediate states $|e_s^0 n_s^0>$ and $|e_t^0 n_t^0>$ are defined in different phonon vacua, $|e_i^0 n_i^0>$ and $|e_f^0 n_f^0>$, respectively. The transition between $|n_i^0>$ and $|n_s^0>$ (or between $|n_f^0>$ and $|n_t^0>$) through interaction H_{int} is a single phonon process, whereas the transition between $|n_f^0>$ and $|n_s^0>$ (or between $|n_i^0>$ and $|n_t^0>$) is a multi-phonon process because they belong to different phonon vacua. The dipole moment of this transition can be given in the low temperature limit (*i.e.* $n_i^{0\nu} = 0$) as

$$\begin{aligned} <e_f n_f|\hat{r}|e_i n_i> &= \sum_{s(\neq i)} \frac{<e_s^0 n_s^0|H_{int}|e_i^0 n_i^0>}{\mathcal{E}_i^0 - \mathcal{E}_s^0} <e_f^0|\hat{r}|e_s^0><n_f^0|n_s^0> \\ &+ \sum_{t(\neq i)} \frac{<e_f^0 n_f^0|H_{int}|e_t^0 n_t^0>}{\mathcal{E}_f^0 - \mathcal{E}_t^0} <e_t^0|\hat{r}|e_i^0><n_t^0|n_i^0> \\ &= \sum_l \Big[M_{fi}^l <n_f^{0l}|1> + M_{fi}^{l+} \sqrt{n_f^{0l}} <(n_f^{0l}-1)|0> \\ &+ M_{fi}^{l-} \sqrt{n_f^{0l}+1} <(n_f^{0l}+1)|0> \Big] \prod_{l\neq l'} <n_f^{0l'}|0>, \end{aligned} \quad (11)$$

where

$$\begin{aligned} M_{fi}^l &= \sum_{s(\neq i)} \frac{<e_s^0|\alpha F^l|e_i^0>}{\epsilon_i^0 - \epsilon_s^0 - \hbar\omega_l} <e_f^0|\hat{r}|e_s^0> , \\ M_{fi}^{l\pm} &= \sum_{t(\neq f)} \frac{<e_f^0|\alpha F^l|e_t^0>}{\epsilon_f^0 - \epsilon_t^0 \pm \hbar\omega_l} <e_t^0|\hat{r}|e_i^0> . \end{aligned} \quad (12)$$

Here ϵ_s^0, ϵ_i^0 are the unperturbed electronic energies of the states $|e_s^0>$ and $|e_i^0>$, while the total energy difference is $\mathcal{E}_s^0 - \mathcal{E}_i^0 = \epsilon_s^0 - \epsilon_i^0 + \sum_l \hbar\omega_l n_s^l$. We can see from Eq. (11) that the actual transition can be interpreted as going through two channels. One is via a direct radiative transition from state $|e_i^0 n_i^0>$ to state $|e_t^0 n_t^0>$, then via a non-radiative transition to state $|e_f^0 n_f^0>$. The other one is first via a non-radiative transition from state $|e_i^0 n_i^0>$ to state $|e_s^0 n_s^0>$, then via a direct radiative transition to state $|e_f^0 n_f^0>$. The actual optical intensity is effectively provided by the transition from the first-order perturbed mixed state $|e_i n_i>$ to the final state $|e_f n_f>$ in the multi-phonon processes. Thus the optical spectral function of this phonon-assisted transition in the low temperature limit can be obtained through the standard formulae,

$$I(\hbar\omega) \sim \underset{i}{Av} \sum_f \left|< e_f^0|\hat{r}|e_s^0 >\right|^2 \delta(\hbar\omega - \mathcal{E}_{if}) . \tag{13}$$

The Franck-Condon overlap integrals $< n_f^l|1 >$ and $< n_f^l|0 >$ in Eq. (11) are related to the lth component of the Huang-Rhys factors S_l via $< n_f^l|n_i^l >= e^{-\frac{S_l}{2}} \left(\frac{n_i^l!}{n_f^l!}\right)^{\frac{1}{2}} (-S_l)^{\frac{n_f^l - n_i^l}{2}} L_{n_i^l}^{n_f^l - n_i^l}(S_l)$, where $L_n^m(z)$ is the associated Laguerre polynomial. Making use of the above formulae, we have obtained the luminescence and absorption spectra due to the singlet STE (see Figure 1 of reference [7]), which agree very well with the observation by Wang for luminescence [13] and by Whetten *et al.* for absorption [16].

4. Quantum fluctuations between six D_{5d} polarons

We studied the multi-mode Jahn-Tellar (JT) problem of the anion C_{60} within BdeG formalism, and obtained a D_{5d} symmetry structure of C_{60}^{1-} [12]. The self-trapped excitons in neutral and charged C_{60} molecules were also studied using this framework [7]. We believe this self-consistent formalism is very efficient in describing the static JT distortions of this type of multi-electron, multi-mode systems, since the traditional treatment for the Jahn-Teller effect is a single electron scheme in nature while our discussion is a many electron scheme consistent with the ion displacement. However, such a quasi-static approach is not sufficient. In fact, there are six equivalent D_{5d} polarons corresponding to six five-fold axes in a C_{60} molecule. These six polaron states are degenerate in energy and can be connected by quantum tunneling, which may give rise to splitting of the degenerate states. Therefore, it is important to consider the dynamic JT effects. By using the lattice relaxation theory, we are able to calculate this splitting and find that the quantum tunneling indeed leads to a further increase of the polaron binding energy by $\sim$ 15 meV and the full I_h symmetry is recovered. To describe these six polaron states, we construct an origin-shifting operator $\hat{U}^\beta$, which is used to translate the unrenormalized phonon vacuum of lattice vibration to one of the six phonon vacua given by D_{5d} polaronic configurations $\{Q_\mu^\beta\}$, $\beta = 1, \ldots, 6$,

$$\hat{U}^\beta = \exp\{-\frac{1}{\hbar}\sum_\mu \sqrt{\frac{M\hbar\omega_\mu}{2}}(\hat{b}_\mu - \hat{b}_\mu^\dagger)Q_\mu^\beta\} \quad . \tag{14}$$

The shifted phonon operator of βth D_{5d} polaron has the form of $\hat{b}_\mu^\beta = \hat{U}^\beta \hat{b}_\mu \hat{U}^{\beta\dagger}$, and the corresponding phonon vacuum is defined as $\hat{b}_\mu^\beta|\tilde{o}_\beta>= 0$, where $|\tilde{o}^\beta>= \hat{U}^\beta|0>$, and

$$<\tilde{o}_{\beta'}|\tilde{o}_\beta>= \exp\{-\sum_\mu \frac{M\omega_\mu}{4\hbar}(Q_\mu^{\beta'} - Q_\mu^\beta)^2\}. \tag{15}$$

The quantum tunneling between these six minima is a tunneling process of the self-consistent lattice plus multi π-electron states through quantum fluctuations. Taking the states at the six equivalent D_{5d} polaronic minima of the adiabatic potential as the zero-order degenerate states, the tunneling state should be a combination of the six non-orthogonal D_{5d} polaron multi-electron states with different phonon vacua, $|\Phi_i>= \sum_{\beta=1}^6 C_{\beta i}|\Psi_\beta>$, where $|\Psi_\beta>= |e_\beta> \bigotimes |\tilde{o}_\beta>$, and can be solved by the degenerate perturbation theory. The secular equation is given by:

$$\|<\Psi_{\beta'}|H|\Psi_\beta> -E <\Psi_{\beta'}|\Psi_\beta>\| = 0\ . \tag{16}$$

To explicitly obtain the matrix elements, we need again to calculate Slater determinants $< e_{\beta'}|(c_{i\,s}^\dagger c_{j\,s} + c_{j\,s}^\dagger c_{i\,s})|e_\beta >$ and $< e_{\beta'}|e_\beta >$. Due to symmetry consideration, we only need to calculate three kinds of matrix elements. The diagonal matrix elements $< \Psi_{\beta'}|H|\Psi_{\beta'} >$ $(\beta = 1, 2, \cdots, 6)$ are all equal, being the potential energy of D_{5d} polarons. For the C_{60}^{1-} problem, the off-diagonal matrix elements only have two different values depending on the angle between the two axes, $< \Psi_1|H|\Psi_2 >$ and $< \Psi_2|H|\Psi_4 >$. Moreover, since the angles of these two pairs of axes are supplementary to each other, we have $< \Psi_2|H|\Psi_4 >= - < \Psi_1|H|\Psi_2 >$. The above analysis can also be applied to evaluate $< \Psi_{\beta'}|\Psi_\beta >$. The secular equation (16), now becomes

$$\left\|\begin{matrix} \sigma & \tau & \tau & \tau & \tau & \tau \\ \tau & \sigma & \tau & -\tau & -\tau & \tau \\ \tau & \tau & \sigma & \tau & -\tau & -\tau \\ \tau & -\tau & \tau & \sigma & \tau & -\tau \\ \tau & -\tau & -\tau & \tau & \sigma & \tau \\ \tau & \tau & -\tau & -\tau & \tau & \sigma \end{matrix}\right\| = 0, \tag{17}$$

where $\sigma = H_{11} - E$, $\tau = U - ES$, and $H_{11} =< \Psi_1|H|\Psi_1 >$, $U =< \Psi_1|H|\Psi_2 >$ and $S =< \Psi_1|\Psi_2 >$. It is easy to obtain two three-fold degenerate eigenvalues from Eq. (17),

$$E_\mp = \frac{H_{11} \mp \sqrt{5}U}{1 \mp \sqrt{5}S}. \tag{18}$$

The six degenerate D_{5d} polaron states span a six-dimensional space which will be reduced to two three-fold irreducible representations, $T_{2u} \bigoplus T_{1u}$, where the higher E_- is identified to be T_{2u} levels, while the lower E_+ corresponds to T_{1u} levels. The eigenvectors can also be determined analytically. From Eq. (18), we find the energy splitting, $E_- - E_+ = 2(SH_{11} - U)\sqrt{5}/(1 - 5S^2)$. Analyzing this expression, we find that the factor $2(SH_{11} - U)$ is actually equal to $-2 < \Psi_2|H_{int}|\Psi_1 >$ which is the energy splitting for the two D_{5d} polaron problem, and $\sqrt{5}/(1 - 5S^2)$ is an enhancement factor due to the presence of 5 neighbours. Numerical estimation indicates $E_- - E_+ = 0.105\ eV$ For C_{60}^{1-}. We have also studied the C_{60}^{2-} polaron problem. The dynamical JT effect might be observed in isolated molecules or liquid solutions, whereas the static JT effect is most likely to be observable in solids or solid matrices, since in the latter there exists an external influence on the molecules which tends to effectively fix the direction of symmetry breaking[17]. We believe that the tunneling split states may be observable in the electronic resonant Raman scattering experiments[8].

5. Concluding remarks

We find the lattice relaxation approach for the many-electron systems has many application on systems, such as polyacetylenes, M-X chains, and fullerenes, where the many-body effects are important. This theory, combined with the self-consistent BdeG formalism can correctly describe lattice relaxation effects and is able to catch the essential physics on multi-electron background and multi-phonon process during optical transitions and non-radiative transitions. We believe it will find more applications in the future.

Acknowledgment

Two of the authors (clw and zbs) appreciate valuable collaboration with Yasha Yi and Lin Tian. This work is partly supported by the National Natural Science Foundation of China.

References

[1] K. Huang and A. Rhys, Proc. Roy. Soc. **A204**, 406 (1950).
[2] S. I. Pekar, Zh. Eks. Teo. Fis. **20**, 510 (1950).
[3] M. Lax, J. Chem. Phys. **20** 1752 (1952).
[4] R. Kubo and Y. Toyozawa, Progr. Theor. Phys. (Kyoto), **13**, 160 (1955).
[5] Z. B. Su and L. Yu, Phys. Rev. bf B27, 5199 (1983); (E)**B29**, 2309 (1984); Commun. in Theor. Phys. (Beijing) **2**, 1203; 1323; 1341 (1983).
[6] C. L. Wang, W. Z. Wang, G. L. Gu, Z. B. Su and L. Yu, Phys. Rev. **B48**, 10788 (1993).
[7] W. Z. Wang, C. L. Wang, Z. B. Su and L. Yu, Phys. Rev. Lett. **72**, 3550 (1994).
[8] W. Z. Wang, C. L. Wang, A. R. Bishop, L. Yu and Z. B. Su, Phys. Rev. **B51**, 10209 (1995).
[9] W. P. Su, J. R. Schrieffer and A. J. Heeger, Phys. Rev. Lett. **42**, 1698 (1979); Phys. Rev. **B 22**, 2209 (1980); **28**, 1138(E) (1983).
[10] See , *e.g.*, L. Yu, *Solitons and Polarons in Conducting Polymers*, (World Scientific publishing Company, Singapore, 1988); and *Conjugated Conducting Polymers*, H. Kiess, ed., (Springer, Berlin, 1992).

[11] G. W. Hayden, E. J. Mele, Phys. Rev. **B36**, 5010 (1987); K. Harigaya, J. Phys. Soc. **B60**, 4001 (1991); B. Friedman, Phys. Rev. **B45**, 1454 (1992).
[12] W. M. You, C. L. Wang, F. C. Zhang and Z. B. Su, Phys. Rev. **B47**, 4765 (1993).
[13] Y. Wang, J. Phys. Chem. **96**, 764(1992).
[14] C. Reber *et al.*, J. Phys. Chem. **95**, 2127 (1991); A. Skumanich, Chem. Phys. Lett. **182**,486(1991); S. C. Graham *et al.*, Synth. Met. **49-50**, 531(1992); X. Wei *et al. ibid*, **49-50**, 549 (1992); B.C. Hess et al., to appear in Mol. Cryst. & Liq. Cryst.
[15] See *e.g.*, G. F. Imbusch, in *Luminescence Spectroscopy*, ed. by M. D. Lumb, (Academic Press, London, 1978); K. K. Rebane, *Impurity Spectra of Solids*, (Plenum, N. Y.-London, 1970).
[16] R. L. Whetten *et al.*, Mater. Res. Soc. Symp. Proc. **206**, 639 (1991).
[17] W. Z. Wang, C. L. Wang, A. R. Bishop, L. Yu, Z. B. Su and J. T. Gammel, in *FULLERENES* edited by R. S. Ruoff and K. M. Kadish, (The Electrochemical Society, Inc., Pennington, 1995), p1224.

Exciton Relaxation and Electron Pairing in Fullerene

X. Sun[1], G.P. Zhang[1], K.H. Lee[2], T.Y. Park[2], R.T. Fu[1], and D.L. Lin[3]

[1]National Laboratory of Infrared Physics, Academia Sinica, and Department of Physics, Fudan University, Shanghai 200433, China
[2]Departments of Chemistry and Physics Education, Won Kwang University, Iksan City 570-749, Korea
[3]Department of Physics, State University of New York, Buffalo, NY 14260, USA

Abstract. The dynamical processes of photoexcitation and charge transfer in C_{60} are studied by calculating the time dependences of the bond structure and electronic spectrum. The evolutions of the bond distortion and gap states indicate that the relaxation time of the self-trapping exciton is 80 fs, but the relaxation time of the charge transfer is 300 fs. Meanwhile the calculation shows that the electron-lattice interaction in C_{60} can induce the electron pairing with the binding energy 0.05 eV.

1. Introduction

The photoexcitation (PE) and charge transfer (CT) in C_{60} are two fundamental processes manipulating the electrical, optical and chemical properties of the Fullerene. Both experimental and theoretical studies have shown that the bond structure and electronic states of C_{60} are changed by PE or CT [1-5], and its symmetry is reduced from I_h to D_{5d}. At present, the consequences produced by PE or CT in C_{60} have been well investigated; PE leads to a self-trapped exciton [3] and CT results in a polaron [4]. However, this is merely the static aspect of PE and CT, their dynamic aspect is only known a little. In this paper the dynamics of PE and CT is studied by revealing the evolution of forming the STE and polaron, then the relaxation times of STE and polaron can be determined. It is interesting to find that, although both PE and CT are ultrafast processes, the PE is four times faster than the CT.

In order to demonstrate the dynamical processes of the CT and PE, it is needed to describe the static features of the polaron and STE in C_{60}.

For the bond structure, it is easier to depict its distortion if the Buckyball is oriented in such a way that its top and bottom faces are pentagons, then the ball consists of 13 layers of bonds [3]. After the

Springer Series in Solid-State Sciences, Volume 124
Relaxations of Excited States and Photo-Induced Structural Phase Transitions
Ed.: K. Nasu

CT or PE, the bond structure is distorted, but the distortion possesses a laminar struture, i. e., the bonds in the same layer have the same change in their lengths. It is notable that the distortion is concentrated in the equatorial area near the 7th layer [3, 4], and the distortion for the PE is about twice larger as that of the CT.

For the electronic states, the HOMO and LUMO of neutral C_{60} have five- and three-fold degeneracy respectively. After the CT or PE, both the HOMO and LUMO are split, the state A_{1u} is lifted up from the HOMO, and the state A_{2u} is pulled down from the LUMO, the split in PE is about 0.1 eV, but in CT it is only 0.05 eV.

Therefore the effects of the CT and PE on the C_{60} are displayed in the changes of the bond lengths and the emergence of the gap states A_{1u} and A_{2u}. Then the dynamical processes of the CT and PE can be demonstrated by showing the time-dependences of the bond lengths and the levels of A_{1u} and A_{2u}. These will be done in the sections 2 and 3. In section 4, we study another topic, the electron pairing in C_{60}, which is related to the charge transfer with two electrons.

2. Formalism

The previous works [3,4] have shown that the C_{60} can be modeled by a tight-binding Hamiltonian:

$$\hat{H} = -\sum_{i,j,s} t_{ij}(\hat{C}^{\dagger}_{i,s}\hat{C}_{j,s} + h.c.) + \frac{K}{2}\sum_{i,j}(|\mathbf{r}_i - \mathbf{r}_j| - D_0)^2, \quad (1)$$

The hopping integral t_{ij} only depends on the distance $D_{ij} = |\mathbf{r}_i - \mathbf{r}_j|$. Since the atom displacement $\mathbf{Q}_i = \mathbf{r}_i - \mathbf{r}_i^0$, where $\mathbf{r}_i^0$ is the equilibrium position of the atom i in neutral C_{60}, is much smaller than the bond length, the change of the distance can be expressed as,

$$\Delta D_{ij} = |\mathbf{r}_i - \mathbf{r}_j| - |\mathbf{r}_i^0 - \mathbf{r}_j^0| = \mathbf{n}_{ij} \cdot (\mathbf{Q}_i - \mathbf{Q}_j), \quad (2)$$

where $\mathbf{n_{ij}}$ is an unit vector in the direction from the site i to site j.

Then

$$t_{ij}(D_{ij}) = t_{ij}(|\mathbf{r}_i^0 - \mathbf{r}_j^0| + \Delta D_{ij}) = t_0 + \alpha \mathbf{n}_{ij} \cdot (\mathbf{Q}_i - \mathbf{Q}_j), \quad (3)$$

where $t_0 = t_{ij}(|\mathbf{r}_i^0 - \mathbf{r}_j^0|), \qquad \alpha = dt_{ij}/dD_{ij}.$

Substituting the Eq.(3) into the Eq.(1), the Hamiltonian reads,

$$\hat{H} = -\sum_{i,j,s} t_0(\hat{C}^{\dagger}_{i,s}\hat{C}_{j,s} + h.c.) - \sum_{i,j,s} \alpha \mathbf{n}_{ij} \cdot (\mathbf{Q}_i - \mathbf{Q}_j)$$

$$\cdot(\hat{C}_{i,s}^{\dagger}\hat{C}_{j,s}+h.c.)+\frac{K}{2}\sum_{i,j}(|\mathbf{r}_i-\mathbf{r}_j|-D_0)^2, \qquad (4)$$

the second term is the electron-phonon interaction, and α the coupling constant. The Ref. [3] has given the parameters t_0=1.8eV, α=3.5eV/$\AA$, D_0 = 1.54$\AA$ and K=15.0eV/$\AA^2$. From this Hamiltonian, the equilibrium positions of sixty carbon atoms $\{\mathbf{r}_i\}$ (i =1, 2, 3, ..., 60) can be determined by minimizing the total energy $E(\{\mathbf{r}_i\})$,

$$\frac{\delta E(\{\mathbf{r}_i\})}{\delta \mathbf{r}_i}=0. \qquad (5)$$

In the case of neutral C_{60}, it gives a dimerized bond structure with symmetry I_h, the lengths of the single and double bonds are 1.43 and 1.39 Å respectively and the energy difference between LUMO and HOMO is 1.83 eV. These results agree with the experimental data [6].

Now we turn to consider the dynamics of CT and PE. As one extra electron transfers to the neutral C_{60}, or an electron is excited from HOMO to excited states, the original bond structure with symmetry I_h can not be held and all atoms will adjust their positions. The force $\mathbf{f}_i$, which exerts on the atom i, can be calculated from,

$$\mathbf{f}_i=-\frac{\delta\ E(\mathbf{r}_i)}{\delta\ \mathbf{r}_i}. \qquad (6)$$

Since the carbon atom is much heavier than the electron, the classical dynamical equations can be applied to the motions of the carbon atoms. As one may note that all the carbon atoms sit in the surface of a ball with radius r_0, and the radial relaxation is much smaller than the tangential ones, therefore the carbon atoms move mainly on the surface, and the dynamical equation for each atom reads,

$$\ddot{\theta}_i-\dot{\phi}_i^{\,2}\ sin\theta_i\ cos\theta_i+\lambda\ \dot{\theta}_i\ =\ -\frac{1}{m}\frac{1}{{r_0}^2}\frac{\delta\ E(\{\theta_j,\phi_j\})}{\delta\theta_i}, \qquad (7)$$

$$\ddot{\phi}_i+2\ \dot{\theta}_i\ \dot{\phi}_i\ cot\theta_i+\lambda\ sin\theta_i\ \dot{\phi}_i\ =\ -\frac{1}{m}\frac{1}{{r_0}^2\ sin^2\ \theta_i}\ \frac{\delta\ E(\{\theta_j,\phi_j\})}{\delta\phi_i} \qquad (8)$$

In the case of CT, one extra electron is in LUMO; In the case of PE, one electron is in LUMO and one hole in HOMO. In the beginning, the atoms sit in the original positions $\{\theta_i^0,\phi_i^0\}$ of the neutral C_{60}. The combined dynamical equations (7) and (8) can be solved numerically step by step. The positions $\{\theta_i,\phi_i\}$ and velocities $\{\dot{\theta}_i,\dot{\phi}_i\}$ at the end of each step serve as the initial conditions of the next step. Consequently the positions of all atoms at any time can be obtained.

Since the vibration period τ_0 of the atoms is 10^{-14} sec, the interval τ of the step and damping λ in Eqs. (7) and (8) should satisfy,

$$\tau << \tau_0 << \frac{1}{\lambda}. \tag{9}$$

In our calculation, we take $\tau = 2.11 \times 10^{-15} sec$, $\lambda = 10^{13}/sec$.

3. Relaxation of CT and PE

The numerical results of the evolution of the bond distortion in the case of CT are showed in Fig. 1, the series of figures represent the changes of the bond lengths at the moments t=0, 60, 240, 300 and 420 fs. It can be seen that, in the early stages, the changes are rather diverse, even the bonds in the same layer get very different changes. After 300fs the changes in the same layer are close to each other, it means that the laminar structure of the bond distortion has formed. And we see that the configuration at 420 fs has been very close to that of the static

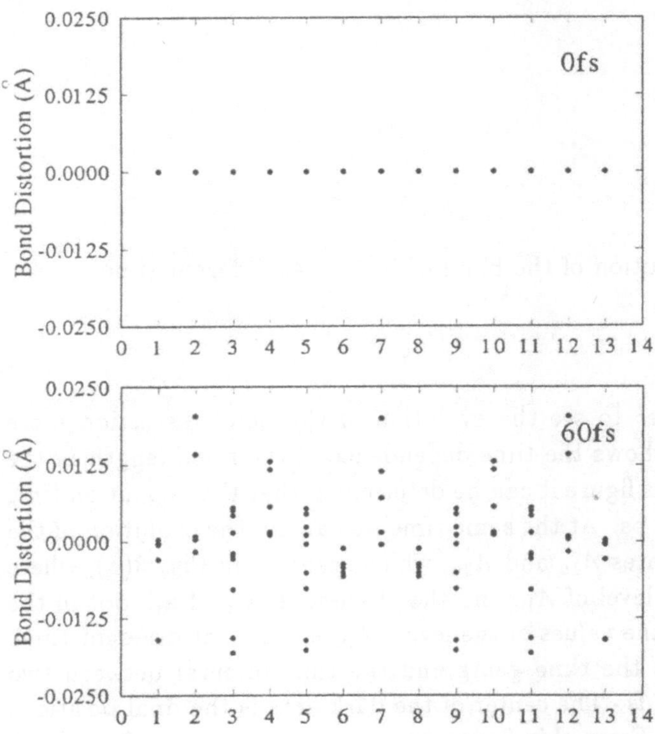

Figure 1

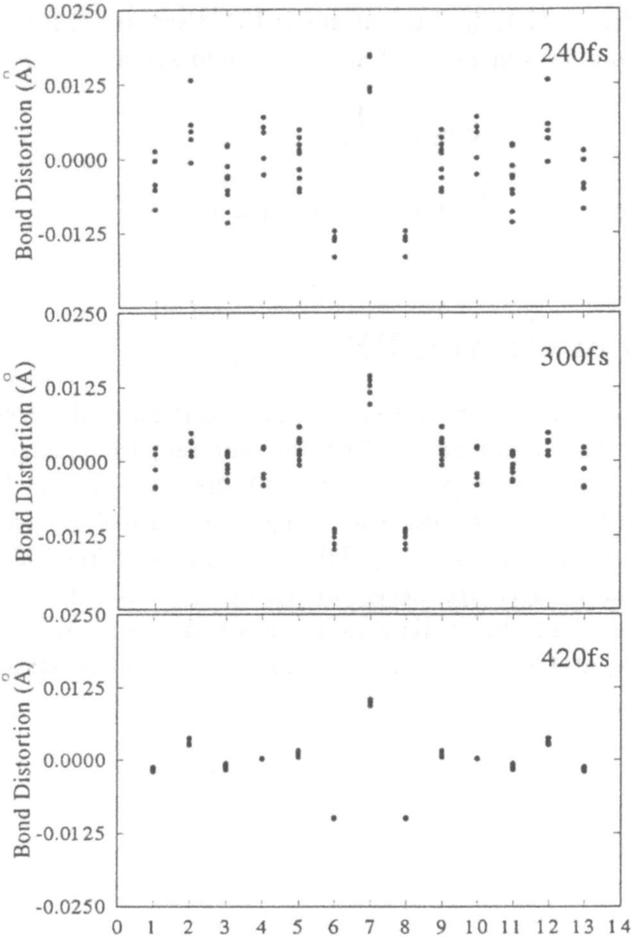

Fig. 1 The evolution of the bond distortion at different time.

distortion. In order to see the evolution of the bond distortion more clearly. Fig. 2(b) shows the time dependence of the bond length in the 7th layer. From this figure it can be determined that the relaxation time of CT is $T_{CT} = 0.3$ ps. At the same time we can get the evolution of the levels of the gap states A_{1u} and A_{2u}, which is shown in Fig. 3(b), where the ordinate is the level of A_{1u} and the abscissa is A_{2u}. Each dot in the picture represents the values of the levels A_{1u} and A_{2u} at different time. The arrow denotes the time going and the time interval between two successive dots is 2 fs. The center of the dark area is the final positions of the gap states. From this figure one can observe that, after about

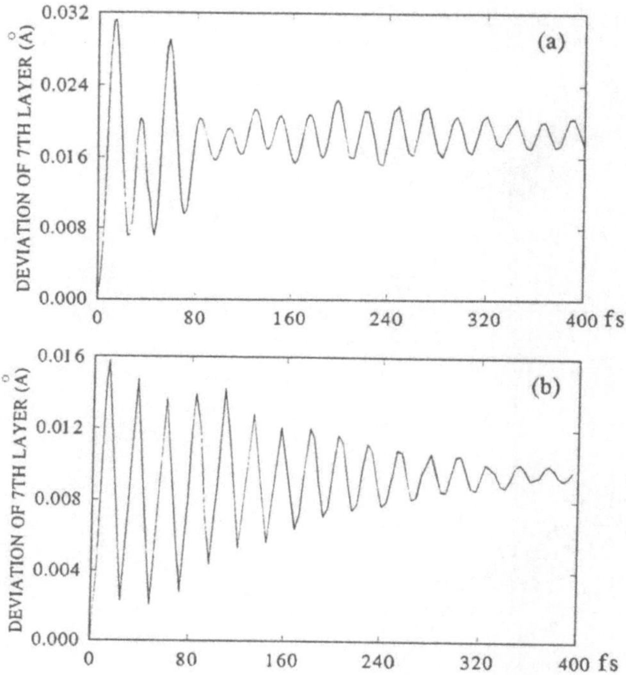

Fig. 2 The time dependence of the change of the bond length in the 7th layer (a) STE, (b) CT.

10 turns, the dot reaches the dark area, and it also indicates that the relaxation time of the CT is 0.3 ps.

For the STE, the time-dependent change of the bond length in the 7th layer is shown in Fig. 2(a), and the evolution of the gap states A_{1u} and A_{2u} is shown in Figure 3(a). These two figures indicate that the relaxtion time T_{STE} of STE is about 80 fs.

Comparing the STE with the CT, it can be seen that the dynamical process of STE is four time faster than that of CT. The physical reason for such a big difference is understandable: In the case of STE, an electron sits in the LUMO while a hole in HOMO, then both the electron and hole exert the forces on the atoms; In the case of CT, there is only one electron in LUMO, then the driven force is weaker than that in STE. This difference can also explain why the bond distortion and level splits in STE are twice bigger as that in CT.

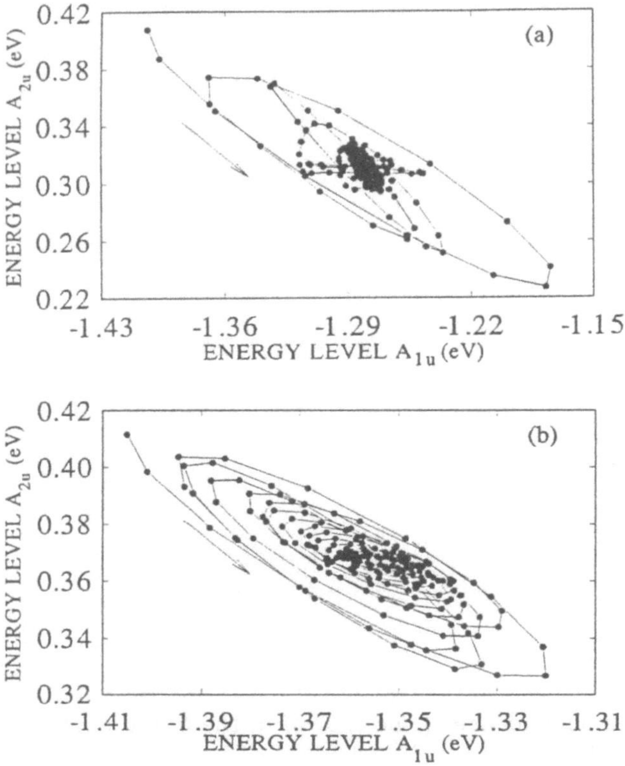

Fig. 3 The evolution of the gap states A_{1u} and A_{2u}. (a) STE, (b) CT.

4. Electron Pairing

One of the most prominent properties of C_{60} is that the alkali-metal doped Fullerenes M_3C_{60} (M=K, Rb, Cs) are superconductors with fairly high T_c about 30K. The essential issue to understand the mechanism of its superconductivity is to reveal the origin of the electron pairing. In this section the electron pairing is studied on the basis of the charge transfer.

Due to the electron-lattice interaction, the transferred electrons are trapped by the bond distortion in C_{60}. The charge transfer with one electron leads to the single polaron $C_{60}{}^{-}$, and the charge transfer with two electrons leads to the bipolaron $C_{60}{}^{2-}$. When two electrons are transferred, there are two possibilities: (1) Two electrons go to the same C_{60} to form a bipolaron $C_{60}{}^{2-}$; (2) Two electrons go to different C_{60} to form two single polarons $C_{60}{}^{-}$. Denote E_0, E_1 and E_2 as the energies of the neutral C_{60}, single polaron C_{60}^{-} and bipolaron C_{60}^{2-}, then the energy diffrerence,

$$\mathcal{E}_{pair} = (E_2 + E_0) - 2E_1 \quad (10)$$

is the binding energy of the electron pairing. If $\mathcal{E}_{pair}$ is negative, two electrons will prefer to enter the same C_{60} rather than go to two different C_{60} separately, this is the effect of the electron pairing. Our calculation shows that $\mathcal{E}_{pair} = -0.05$ eV. It indicates that the electron pair can be induced by the electron-lattice interaction in C_{60}.

Acknowledgements

The authors are grateful to Dr. Z. G. Yu for the helpful discussions. This work was supported by the National Science Foundation and National Education Commission of China and the Grant BSRI 95-3484 Korea.

References

1 R. J. Sension, A. Z. Szarka, G. R. Smith, and R. M. Hochstrasser, Chem. Phys. Lett. **185**, 179 (1991).

2 T. Kato, T. Kadama, and T. Shida, Chem. Phys. Lett. **205**, 405 (1993).

3 R. L.Fu, R. T. Fu, and X. Sun, Phys. Rev. B **45**, 17615 (1993).

4 B. Friedman, Phys. Rev. B **45**, 1454 (1992); K. Harigaya, Phys. Rev. B **45**, 13676 (1992); R. T. Fu, Z. Chen, and X. Sun, Chin. Phys. Lett. **9**, 541 (1992).

5 N. S. Sariciftrci, L. Smilomtz, A. J. Heeger, and F. Wudl, Science **258**, 1474 (1992).

6 F. Li and J. Lannin, Phys. Rev. Lett. **65**, 1905 (1990).

Photochemical Transformations and Polymerization in Solid C_{60}

P.C. Eklund, J.M. Holden, and A.M. Rao

Department of Physics and Astronomy and Center for Applied Energy Research, University of Kentucky, Lexington, KY 40506, USA

1. Introduction

In this paper we review theory and experiment on the photopolymerization of solid C_{60} and related C_{60} polymers. In addition, we summarize experimental results on a second interesting phototransformation in solid C_{60}, viz., the photo-enhanced doping, or diffusion, of molecular oxygen into the interstitial voids in the host lattice. Both of these phototransformations render the transformed solids insoluble in organic solvents, and this property has been patented for applications as photoresists[1]. Using vibrational spectroscopy as a tool, we discuss recent infrared and Raman studies of the photopolymer and other C_{60} polymers produced at high pressure and high temperature (HPHT). Finally, recent results on the identification of charge transfer excitons in solid C_{60} will be given. These optical excitations have been proposed as the "trigger mechanism" for the photopolymerization process[2].

Solid C_{60} is a molecular crystal in which the molecules are bound together by weak van der Waals forces. Suitably doped, the resulting solid can either be an organic ferromagnet or moderately high T_c superconductor[3]. In the pristine solid, the molecules are positioned in a close packed array on a face centered cubic (fcc) lattice (a_0 = 14.7 Å). Above the molecular orientation ordering temperature T_{01} = 260 K, the molecules spin freely and rapidly about their lattice positions[4-8]. It is now well known that the electronic and vibrational properties of the pristine solid are strongly connected to the properties of the molecule itself. Consistent with the "soccer ball" structure proposed first by Smalley, Kroto and co-workers, the C_{60} molecule exhibits a single line NMR spectrum, indicating 60 equivalently bonded carbon atoms[9]. Symmetry then requires that the molecular shell be comprised of 12 isolated pentagonal rings, needed for the curvature of the molecular surface, and 20 hexagonal carbon rings (Fig. 1a). The general chemical requirement of four bonds per carbon atom is then realized with double bonds located at the fusion between hexagonal rings, in agreement with theoretical calculations. These 30 double bonds provide numerous sites for intermolecular attachment to form new polymeric C_{60} phases or for the attachment of chemical side groups[10].

Besides the application of light or HPHT, polymerization in solid C60 has also been reported to occur by electron impact[11], in a plasma discharge[12], and in the compound M_1C_{60}, where M=alkali metal [13-16]. In the M_1C_{60} system, the electron transfer from the M atoms to the C_{60} lattice, under specific synthesis conditions, has been found to lead to negatively charged polymer chains. The various routes to polymerization in C_{60} solids has sparked interest in a general mechanism. It seems that a contraction of the intermolecular separation is a reasonable first step for all of these processes. Of course, the details of how the intermolecular bonds then form is the key theoretical issue. A sufficient contraction can be induced by pressure, but high temperature also appears necessary to obtain certain C_{60} polymer phases[17-22]. In the case of M_1C_{60}, the Coulomb interaction between the alkali metal ions and the C_{60} anions has been proposed to drive the lattice distortion leading to the chain polymerization[23]. Recently, Tsubo and Nasu have calculated the optical absorption of solid C_{60} and have found "charge transfer" excitons between ~1.6 and 2.9 eV[2]. These excitations are molecular in character but the weak intermolecular coupling allows the hole and electron in the photo excited state to reside on different, adjacent molecules. The Coulomb attraction between electron and hole is then postulated to lead to a reduction in the intermolecular separation and "trigger" the formation of intermolecular bonds. In this paper, we present clear experimental evidence for these CT excitons by quantitative analysis of gas and solid phase optical absorption. The results are compared with theory.

In earlier work, we reported that the irradiation of C_{60} films with visible or UV light in the *absence* of oxygen (this is important and is discussed below), leads to a photo-induced

Springer Series in Solid-State Sciences, Volume 124
Relaxations of Excited States and Photo-Induced Structural Phase Transitions
Ed.: K. Nasu

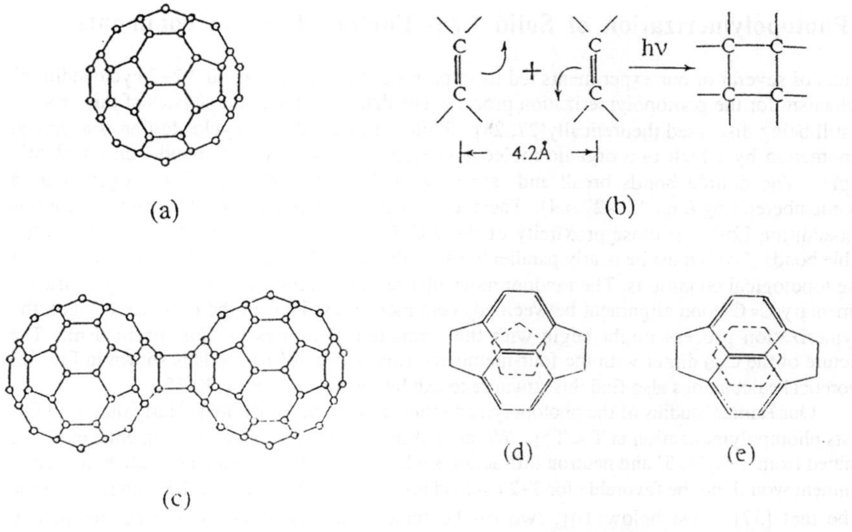

Fig. 1: (a) Schematic view of a C_{60} molecule showing the arrangement of pentagons and hexagons. The C=C are formed at the fusion of two hexagons; (b) The "2 + 2" cycloaddition reaction between parallel carbon double bonds on adjacent molecules is shown schematically; (c) A C_{60} dimer produced by the photochemical "2 + 2" cycloaddition reaction; (d) and (e) : Two possible orientations of adjacent C_{60} molecules in the merohedral disordered glass phase (see text).

transformation of the fcc solid to a new contracted fcc solid phase in which the C_{60} molecules were proposed to be covalently linked[24]. On the basis of vibrational spectra[24], it was concluded that in the phototransformed phase the molecular cage remained in tact but the I_h symmetry was lost due to the formation of intermolecular bonds. A photochemical "2 + 2" cycloaddition reaction mechanism was proposed in which double bonds on neighboring molecules are broken and reform as a four-membered ring joining the molecular cages[24].

Recently, several high pressure-high temperature (HPHT)-induced C_{60} polymers have been reported[17-22]. To date, based on Rietveld refinement of powder x-ray diffraction (XRD) data, four HPHT polymeric phases of solid C_{60} have been proposed: rhombohedral phase (R) [17-19, 22], tetragonal phase (T) [19], orthorhombic phase (O) [19], and a contracted fcc (F) phase [18]. In these HPHT polymers, the covalent linking between the C_{60} molecules was also proposed to be via the "2 + 2" mechanism. The O phase represents a one dimensional (1D) polymer while the R and T phases correspond to 2D polymers [19]. Based on ^{13}C NMR study of the C_{60}-benzyene adduct, Hoke et al. reported direct experimental evidence for attachment of the benzyene ring to C_{60} via the four membered ring[25]. From a recent nuclear magnetic resonance (NMR) study[26] on the R and O phases of the pressure induced C_{60} polymers, the proposed four membered ring linking C_{60} molecules was experimentally verified. Furthermore, similar to the photo-induced C_{60} polymer, the R, T and O polymer phases were also found insoluble in toluene, and upon annealing them in an inert atmosphere at temperatures ~200 °C and ambient pressure, they also were found to revert back to pristine C_{60} [18, 19].

During the study of the effect of O_2 on the polymerization rate, we discovered that the diffusion of O_2 molecules into interstitial positions in the host fcc C_{60} lattice was dramatically accelerated by the presence of visible or UV light[27, 28]. From α-particle back scattering, infrared and Raman experiments on the same C_{60} films exposed to 488 nm light and 1 atm oxygen, a photochemical enhancement of the oxygen diffusion rate by a factor of ~10 was observed[27, 28]. This large photochemical effect has never been explained.

2. Photopolymerization of Solid C_{60}- Further General comments

Results of several of our experiments led us to propose the photochemical "2+2 cycloaddition" mechanism for the photopolymerization process. The details of the photophysics of this process are still being discussed theoretically[27, 28]. Photochemical "2 + 2" cycloaddition is a general phenomenon by which two organic molecules, each containing a C=C bond, can covalently couple. The double bonds break and reform with the two molecules fused together by a four-membered ring (i.e., "2 + 2" = 4). There are two general topochemical requirements for this cycloaddition [29]: (i) close proximity of these C=C bonds (separation <4.2 Å), and (ii) the double bonds (C=C) must be nearly parallel to each other (Fig. 1b). Above T_{01}, solid C_{60} satisfies these topological conditions. The random molecular reorientations, in effect, virtually guarantee momentary C=C bond alignment between adjacent molecules. Thus in the presence of light, the polymerization process might begin with the formation of dimers by this mechanism. The structure of the C_{60} dimer with the four-membered ring between C_{60} shells is shown in Fig. 1c. Theoretical calculations also find this structure to exhibit the lowest energy[30-35].

Our Raman studies of the photopolymerization process demonstrated clearly that solid C_{60} resists photopolymerization at $T < T_{01}$. We argued that below T_{01}, other experimental evidence obtained from x-ray[4, 5] and neutron diffraction studies[6, 36] indicate that the relative molecular alignment would not be favorable for 2+2 cycloaddition, i.e., the topochemical requirements could not be met [37]. Just below T_{01}, two of the three rotational degrees of freedom in the orientationally disordered phase are lost, and the molecules undergo a "ratchet-like" orientational motion about four specific (111) directions (instead of random directions as in the high-T phase). At lower temperatures $T_{02} < T_{01}$, a second transition to a merohedrally disordered (MD) phase occurs in which two distinct relative molecular alignements are "frozen in" such that either an electron-rich double bond on one molecule faces the electron-deficient opening in a pentagon (~83%) or hexagon (~17%) of the adjacent molecule (Figs. 1d and 1e). Thus in both the "ratcheting " and in the MD phase, the number of parallel bond alignments made per unit time is considerably reduced and the system should not dimerize efficiently based on topochemical arguments.

Above T_{01},the rapid spinning of the molecules guarantees the general 2+2 topochemical requirements can be met. However, as larger oligomers form, they are orientationally trapped in the host lattice, and only hindered rotations are possible. Thus, many C=C bond misalignments then exist. Nevertheless, the C_{60} oligomers appear to grow in size in spite of this rotational hindrance. For example, in the HPHT phases, large volumes of one, two or three dimensional, covalently bonded C_{60} networks can be produced.

The presence of interstitial O_2 in the C_{60} lattice has been shown to inhibit strongly the photoinduced polymerization in solid C_{60}[24, 38]. We interpret this observation as indirect evidence for placing the first excited spin triplet state (T_1) in the photochemical reaction pathway[37]. Photophysical experiments on C_{60} molecules in solution indicate that C_{60} in the singlet ground state is first excited to higher energy spin singlet states and then a virtually a 100% effective "intersystem crossing" to excited triplet states occurs[39, 40]. Finally, via "internal conversions" or radiationless transitions within the triplet manifold, the molecule relaxes to the T_1 state, which is relatively long lived[39, 41]. Similar photophysical experiments report that the introduction of O_2 in the solution dramatically reduces the C_{60} T_1 lifetime[39]. Thus, the observation that interstitial O_2 quenches the photopolymerization process suggests that the T_1 state is somehow involved.

Little is yet known about the structure of the C_{60} photopolymer. X-ray diffraction studies as a function of photon dose reveal that the fcc lattice undergoes a continuous contraction from the initial value a_0=14.2 Å to 13.80 Å[42]. It should be mentioned that it has been proposed recently, based on UPS studies of phototransformed C_{60} films, that the phototransformed system is really a collection of dimers[43]. This conclusion is based on a fit of their data to calculated electronic levels[35] for the dimer including phenomenological broadening. It is not clear whether broadened states for trimers and/or higher oligomers would also fit the data just as well. Furthermore, laser desorption mass spectra (LDMS) studies of phototransformed C_{60} films show clear evidence for clusters of C_{60} molecules $(C_{60})_N$ for N as high as 8 or 10, while the pristine film showed only the mass corresponding to N=1 in the same experiment[44]. The LDMS data is therefore inconsistent with a photo-dimer model for phototransformed C_{60}. Further x-ray diffraction studies are underway to resolve the structural issues.

3. Vibrational Spectroscopy as a probe of C60 polymers

The very high icosahedral (I_h) symmetry for an isolated C_{60} molecule yields only 46 distinct intramolecular vibrational mode frequencies[45] although 180 - 6 = 174 normal modes are expected for a sixty atom molecule. Because of the weak intermolecular interactions, the high frequency ($\omega > 200$ cm^{-1}) spectrum of the fcc C_{60} solid is essentially that of isolated molecules in solution or in the gas phase. Calculations have shown that the intramolecular vibrational modes with $\omega < 800$ cm^{-1} are expected to exhibit approximate radial character and those above 800 cm^{-1} take on a more tangential character. Below 200 cm^{-1}, Raman- and infrared-active *inter* molecular modes are observed[46]. Group theory for an isolated C_{60} molecule predicts that only ten ($8H_g\{5\} + 2A_g\ \{1\}$) of the 46 intramolecular frequencies are Raman-active, and only four ($4F_{1u}$ $\{3\}$) are infrared active[45], where the symmetry labels of the icosahedral group (I_h) are given and

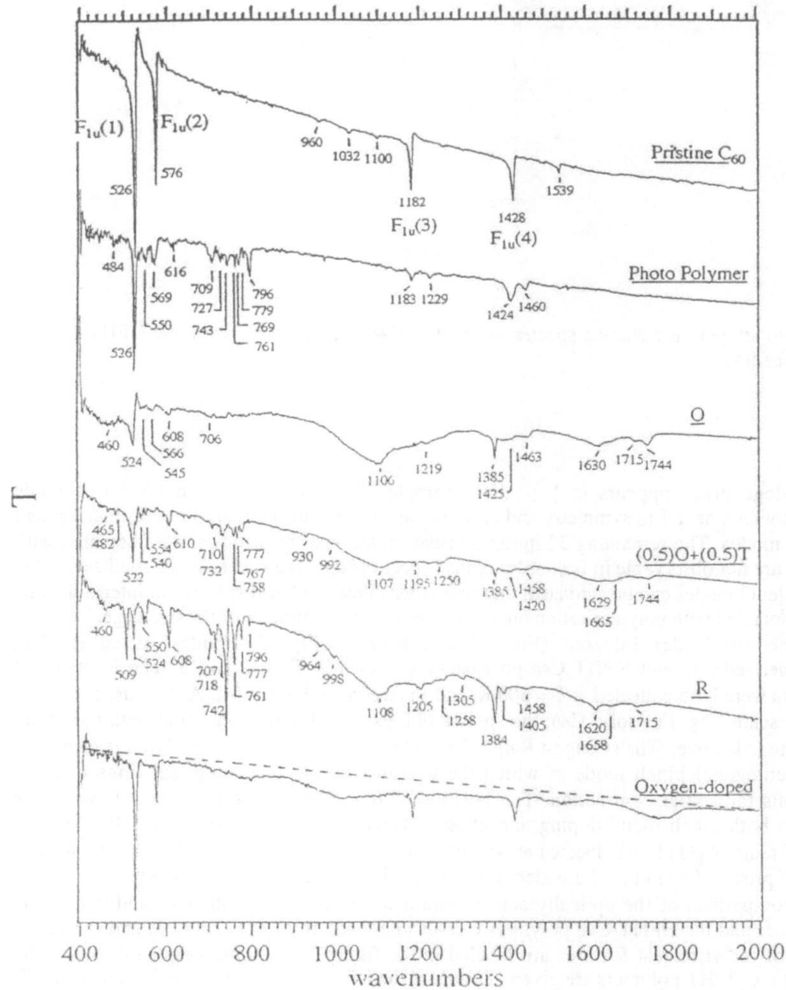

Fig. 2: Room temperature FTIR transmission spectra for pristine C_{60}, photopolymerized and HPHT C_{60} polymers, and oxygen doped C_{60} (see text).

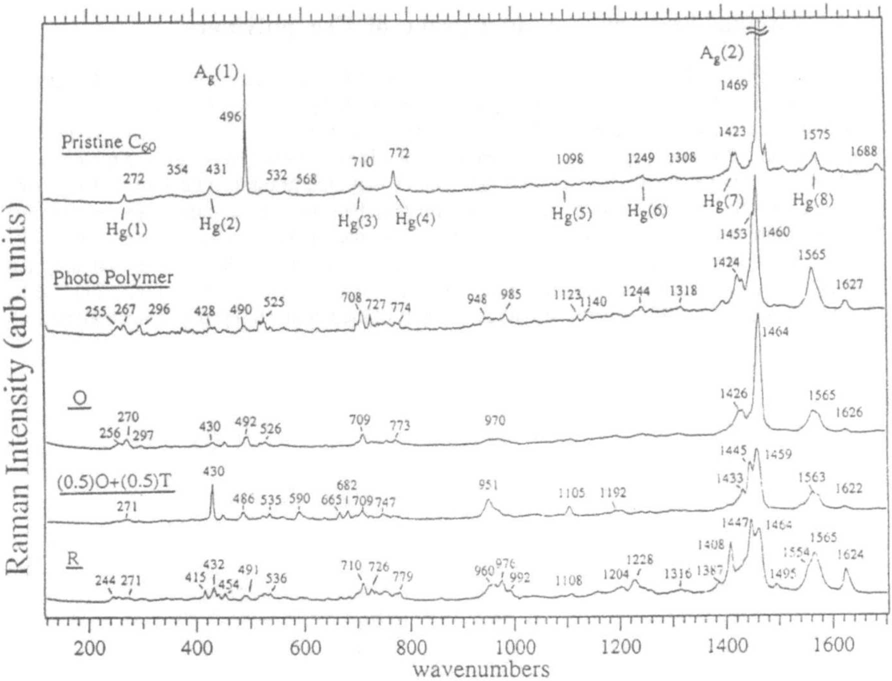

Fig. 3: Room temperature Raman spectra for pristine C_{60}, photopolymerized and HPHT C_{60} polymers (see text).

the mode degeneracy appears in { }. For example, there are four distinct infrared mode frequencies which have F_{1u} symmetry and each frequency is identified with three distinct, related vibrational modes. The remaining 32 intramolecular vibrational frequencies are associated with modes that are not observable in first order optical spectroscopy. However, as we shall see, some of these "silent" modes can be activated, and first order-allowed modes split when intermolecular bonding associated with polymerization breaks the icosahedral symmetry of the C_{60} cage.

The first-order infrared (Fig. 2) and Raman (Fig. 3) modes of solid C_{60}, photopolymerized C_{60} and HPHT C_{60} polymers are shown in Figs. 2 and 3. The Raman and infrared data were both collected at T = 300 K, and an Argon ion laser (5145 Å) was used to excite the Raman scattering. For solid C_{60}, the spectra in Figs. 2 and 3 are consistent with the group theory discussed above. The strongest Raman line is the $A_g(2)$ line at 1469 cm^{-1} and is identified with the pentagonal pinch mode in which the C-atoms in the 12 pentagonal rings undergo simultaneous tangential contraction. The position of this Raman line has been shown to be sensitive to both alkali-metal doping and photopolymerization in the solid state[46-50]. The "breathing" radial $A_g(1)$ line is located at 493 cm^{-1} in Fig. 3; the remaining 8 lines in the Raman spectrum of pristine C_{60} in Fig. 3 are identified with the Raman active H_g modes[50].

A comparison of the optically active vibrational modes of pristine C_{60} and that of the photopolymer and the HPHT C_{60} polymers can also be made from Fig. 2 and 3, and the center frequency of the strongest features are labeled in the figure. First some comments about the structure of the HPHT polymers are given. Details of the HPHT synthesis carried out in a "belt", hydrostatic pressure apparatus by Nunez-Regueiro and coworkers have appeared elsewhere[19]. Depending on the temperature and the pressure in the "belt", one of three polymerized phases can be obtained, as determined by a Rietveld refinement of the x-ray powder diffraction data: a one dimensional (orthorhombic phase (O), a two-dimensional rhombohedral phase (R) and a second

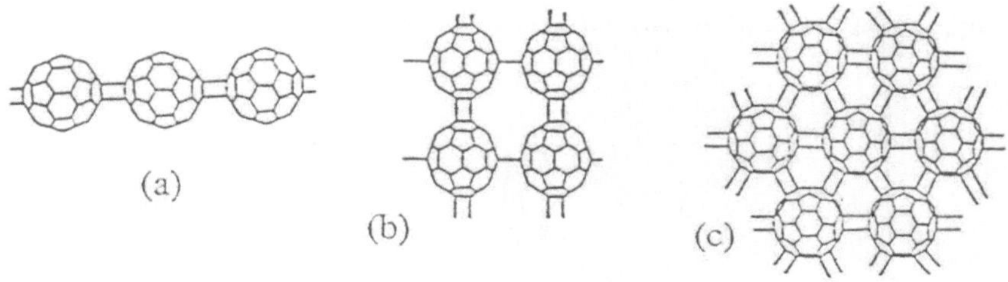

Fig. 4: Schematic arrangement of C_{60} molecules in (a) 1D orthorhombic (O), (b) 2D tetragonal (T), and (c) 2D rhombohedral (R) C_{60} polymer phases. After Ref. 19.

two-dimensional tetragonal phase (T). The structure for these C_{60} HPHT polymers is shown schematically in Fig. 4. Similar to the photodimer and photopolymer, a four-membered ring is proposed as the link between the cages. Recent NMR data has confirmed the proposal of four-membered links for the O and R phases[26]. One of the samples was a 50%:50% mixture of the O and T phases, as the T phase material is difficult to isolate[51]. Because so many atoms reside in the respective unit cells of these HPHT and photopolymers, the vibrational spectra are too complex to analyze in detail without theoretical calculations. However, some general comments will be made briefly. A more detailed account is forthcoming[51]. By comparison to the infrared and Raman spectra of the pristine van der Waals bonded solid, it can be seen that the interfullerene bonds provide a strong symmetry breaking mechanism for the C_{60} cage which activate many new infrared (Fig. 2) and Raman (Fig. 3) modes. Sharp lines are evident in the polymer spectra indicating the cages remain in tact. Many of these sharp lines appear to be derived from A_g-, H_g- and F_{1u}- C_{60} modes. Several of these lines appear to downshift, consistent with the loss of C=C double bonds from the molecular cage, which were broken to form intermolecular bonds. These lines also split in the polymer as the I_h symmetry is broken. Similar to the HPHT polymers, the photopolymer also is seen to exhibit numerous infrared and Raman-active lines relative to the pristine solid, and for the same reasons discussed above for the HPHT polymers.

At the bottom of Fig. 2, we also show for comparison, the infrared transmission spectrum of oxygen-doped C_{60} film (d ~5800 Å) prepared by simultaneous exposure to light (488 nm, 15 mW/mm^2) and 1 atm oxygen for 25 hours. At shorter exposure, very little change in the spectrum over that for C_{60} was observed[27]. During this period, O_2 simply diffuses into interstitial positions of the lattice, albeit the diffusion rate is accelerated dramatically by the presence of the light[27, 28]. After long exposure to the light, a chemical attack of the C_{60} cages is eventually observed (C=O stretch at 1750 cm^{-1}) and broad bands form[27]. This has been interpreted as evidence for multiple attachment of the O atoms to the cage. Similar, broad infrared bands are also observed to co-exist with sharp lines in the HPHT polymer spectra. They need not signal the presence of broken C_{60} cages, but may simply arise from multiple radial bonds, as in the case of the oxygenated film discussed above.

The strongest line ($A_g(2)$, pentagonal pinch mode) in the Raman spectrum of pristine C_{60} downshifts from 1469 cm^{-1} to 1460 cm^{-1} in the photopolymerized material, loses intensity and becomes depolarized[50]. This downshift of the most intense Raman line has been observed in molecular dynamics calculations of a C_{60} dimer, trimer and higher oligomers[31, 33]. A downshift of ~10 cm^{-1} for the $A_g(2)$ derived-mode frequency was predicted for the C_{60} dimer and trimer; and another peak downshifted by ~20 cm^{-1} was predicted for longer C_{60} polymers[33]. It has therefore been suggested that the photopolymerized C_{60} solid predominantly consists of C_{60} dimers and trimers. As discussed above, this view has been used to interpret UPS data and is inconsistent with LDMS data.

Detailed Lorentzian lineshape analysis of the Raman data in the vicinity of the pentagonal pinch mode frequency for the data revealed six distinct Raman lines at very nearly the same frequency in each polymeric phase[51]. Only the relative intensities of these lines were found sensitive to the phase of the fullerene polymers [51]. Assuming that the strongest peak in the Raman spectra for the polymerized phases is always the $A_g(2)$- derived mode, the degree of frequency softening for this mode as a function of the intermolecular separation (a_I) deduced from

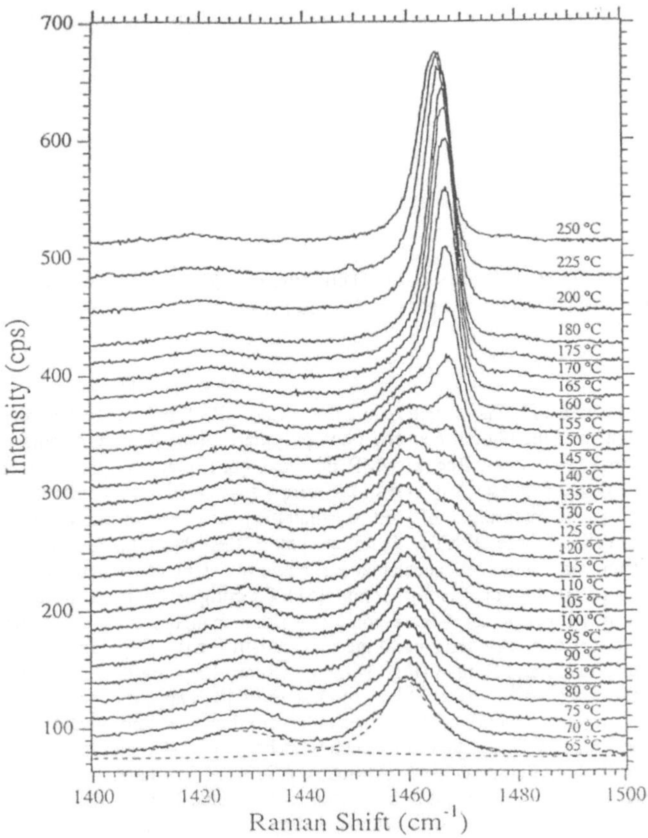

Fig. 5: Raman spectra obtained at various temperatures in the vicinity of the pentagonal pinch mode frequency for a photopolymerized C_{60} film on a suprasil substrate. The dashed curves in the bottom spectrum represent a Lorentzian lineshape fit to the data.

XRD data can be plotted. A frequency softening of $\Delta\omega/\Delta a_I$ ~58 cm^{-1}/Å and ~19 cm^{-1}/Å was obtained, respectively, for the infrared-active $F_{1u}(4)$- derived mode and the Raman-active $A_g(2)$-derived mode[51]. For comparison, the $F_{1u}(4)$- derived mode and the $A_g(2)$- derived mode frequencies in the alkali metal (M) doped C_{60} (M_xC_{60}) exhibit, respectively, a softening of only 14 and 6 cm^{-1} per electron transferred from the alkali metal to the C_{60} molecule[46-49].

Using the sensitivity of the pentagonal pinch mode ($A_g(2)$) to study the thermal stability of photopolymerized C_{60} films[52]. Raman spectra taken at low laser intensity was recorded as a function of temperature in the interval 65 °C < T < 260 °C (Fig. 5). The thermal production of the monomer was monitored by the strength of the 1469 cm-1 line. At ~65 °C, the Raman spectrum of the phototransformed film exhibited a broad band at 1458 cm^{-1} characteristic of the photopolymer. As the temperature was increased, a narrow weak line at 1469 cm^{-1} (characteristic of C_{60} monomer) appears and grows with increasing temperature at the expense of the diminishing 1458 cm^{-1} line. This was taken as a clear evidence for a back transformation to the van der Waals (monomeric) solid by the thermal detachment of C_{60} monomers from oligomers[52]. An Arrhenius plot of the Raman intensity ratio (I_{1459}/I_{1469}) was found to yield an activation energy of 1.25 eV which was identified as the thermal barrier to break the four membered rings[52]. A theoretical estimate of 2.4 eV for this energy barrier has been obtained recently by Adams *et al.* [31]. Porezag *et al.*[33] determined the value ~1.6 eV for this thermal energy barrier using the density functional tight binding (DF-TB) approach.

We have already discussed that the covalent attachment of two C_{60} monomers to form a dimer breaks the icosahedral symmetry of an isolated C_{60} molecule and therefore activates numerous new Raman- and infrared-allowed bands (Figs. 2 and 3). However, new *interball* or intermolecular modes are anticipated which we have not discussed. They involve the stretching or twisting of the four membered ring joining adjacent C_{60} molecules. A new Raman line at 118 cm^{-1} in the room temperature spectrum of the photopolymerized film was observed; and this mode was absent in the room temperature spectrum of pristine C_{60}[52]. The Raman line at 118 cm^{-1} was identified with an interball stretching mode[24, 52]. This mode frequency in the C_{60} dimer was calculated subsequently for the "2 + 2" dimer (Fig. 1c) by several groups: 104 cm^{-1} [30], 81 cm^{-1} [34], 69 cm^{-1} [31], and 89 cm^{-1}[33]. An infrared-active, torsional interball mode for the "2 + 2" dimer has also been predicted by Menon and Subbaswamy at 227 cm^{-1} [30] but has not yet been observed experimentally. It should be mentioned that using the same calculational scheme, Adams *et al.* [31] finds the interball stretching mode in the C_{60} dimer to upshift from 69 to 119 cm^{-1} if the dimers are coupled together by similar four-membered rings to form the neutral linear C_{60} polymer chain[31]. This brings their theoretical results into closer agreement with experiment.

We have studied the temperature dependence of the Raman-active "interball" mode at 118 cm^{-1} for a C_{60} film initially polymerized with a Hg arc source[52]. As anticipated, the 118 cm^{-1} line was observed to disappear with increasing temperature in concert with the appearance of the 1469 cm^{-1} Raman line for C_{60} monomers (Fig. 5). A fit to the temperature dependence of the experimental integrated intensity (area under the 118 cm^{-1} line) was obtained and found to exhibit the same exponential form and activation energy as that for the 1459 cm^{-1} mode but with a different Raman cross section parameter[52].

4. Charge Transfer Excitons: a Possible Trigger Mechanism for Photopolymerization

The optical absorption of solid C_{60} has been calculated by several authors from a one-electron or a many-electron perspective[53]. Since solid C_{60} is a nearly ideal molecular crystal, it is important to treat the molecular electronic states properly, including the Coulomb interaction between electron and hole in the photoexcited (or excitonic) state. If the problem is solved allowing the possibility for the electron-hole pair to be delocalized over neighboring molecules, so called "charge-transfer" (CT) excitons may be found. The degree of delocalization is variable, depending on the particular excitation, some of the excitons are found to be essentially localized on one molecule. Both "localized" (L) or Frenkel excitons and delocalized charge-transfer (CT) excitons are therefore possible and have been calculated[54-57]. Near the absorption edge, the calculations find that the largest oscillator strength in the spectrum lies primarily with localized (Frenkel) excitons[54-57]. However, the CT excitons become important a few hundred meV above the absorption threshold[56, 57].

To experimentally identify CT excitons we have carried out quantitative analysis of the gas phase and solid C_{60} optical absorption. Polycrystalline C_{60} films of known mass were transferred under inert atmosphere to quartz optical cells (path length 355 mm) and subsequently sealed at $\leq 10^{-8}$ torr vacuum. The C_{60} in the film was vaporized by placing the cells in a 1 meter long furnace where gas phase absorption spectra in the range of 1.25 to 6 eV were obtained. For sufficiently small gas concentrations (N_g), the absorption coefficient (a_g) for gas phase C_{60}, can be related to the imaginary part of the molecular polarizability (α_2) as $\alpha_2 \simeq a_g\lambda_0 / 8\rho^2 N_g$, where λ_0 is the wavelength of light. The real part of the molecular polarizability, α_1, is obtained from a parametric Lorentzian oscillator fit to the absorption data,

$$\alpha_M^{Lor}(\omega) = \alpha_0 + \Sigma_j (f_j / (\Omega_j^2 - \omega^2 - i\,\omega\Gamma_j)) \quad , \quad (1)$$

where the f_j, Ω_j and Γ_j are the oscillator strengths, positions and widths, respectively, and α_0 accounts for oscillator strength from transitions above the data range (< 6 eV). The fit to absorption data is not sensitive to α_0, so α_0 must be adjusted to match the dielectric constant obtained previously by ellipsometry on thin, solid films[58]. Although the Lorentzian lineshape

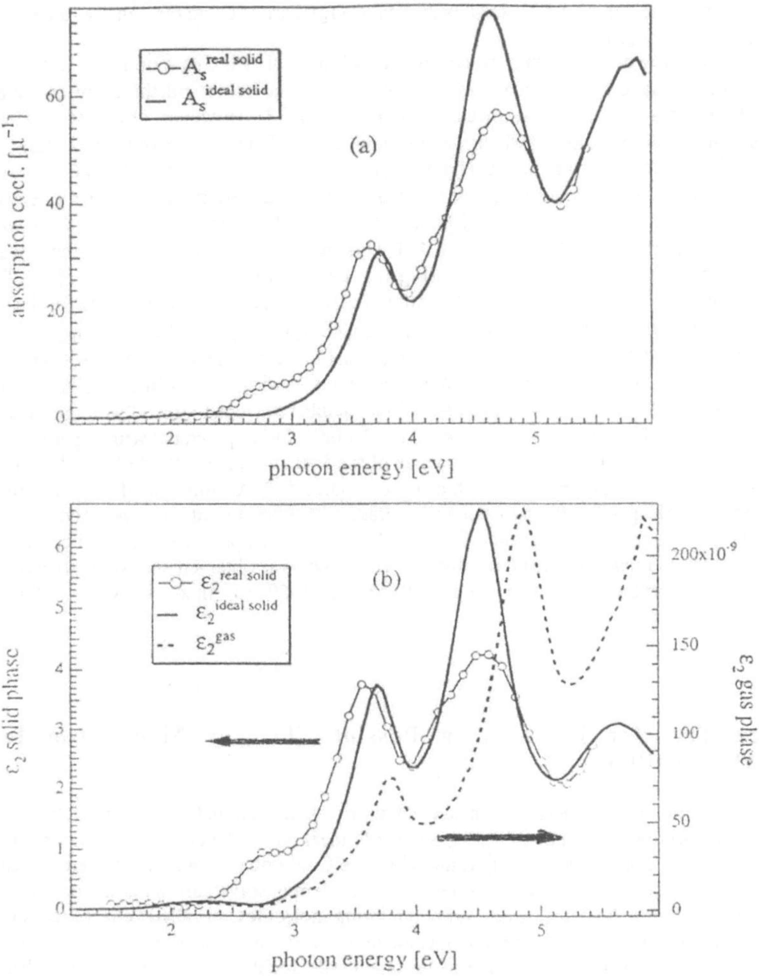

Fig. 6: (a) Comparison of the optical absorption coefficient for an ideal C_{60} molecular solid constructed using Claussius-Mossotti and for a thin C_{60} film obtained by ellipsometry and near normal incidence reflection/transmission spectroscopy; (b) Comparison of $\varepsilon_2^{real\ solid}$, $\varepsilon_2^{ideal\ solid}$ and ε_2^{gas} (see text).

has first principles validity, it must be noted that theory shows there are many more transitions in the spectra than there are oscillators in the fit. Therefore, our Lorentzian analysis should be regarded as a convenient method of obtaining quantitative dispersive information from absorptive information consistent with causality.

Using $\alpha_M \simeq \alpha_1^{Lor} + i\alpha_2$, we can construct the dielectric function, $\varepsilon(\omega) = \varepsilon_1 + i\,\varepsilon_2$, of the "ideal molecular solid", where there is no intermolecular interaction except the dipole screening from neighboring molecules. This is carried out by appealing to the Claussius-Mossotti relationship,

$$(\varepsilon(\omega) - 1) / (\varepsilon(\omega) + 2) = (4\pi/3)\, N_S\, \alpha_M(\omega), \qquad (2)$$

where N_S is the molecule density for solid C_{60}. The real solid dielectric function has been obtained by ellipsometry[58] and R, T spectroscopy. Figure 6b shows $\varepsilon_2^{real\ solid}$, $\varepsilon_2^{ideal\ solid}$ and ε_2^{gas}. Comparing $\varepsilon_2^{ideal\ solid}$ to ε_2^{gas}, it can be seen that the effect of Claussius-Mossotti "renormalization" manifests itself primarily as a strong redshift and redistribution of apparent oscillator strengths. For simple cases, the degree of "renormalization" can be shown to be proportional to the oscillator strength of the transition and so the effect is most prominent above 3 eV in C_{60}, where there are intense dipole-allowed transitions.

Comparing $\varepsilon_2^{ideal\ solid}$ to $\varepsilon_2^{real\ solid}$ it is evident that most of the strong features above 3 eV in the real solid are well described by dipole allowed transitions in the noninteracting molecular solid. Therefore, these excitations must be L excitons. Below 3 eV, the real solid exhibits additional transitions, which after broadening, give rise to a wide band centered at ~2.8 eV. It is important to note that their 2.8 eV band is not seen in the "ideal solid". (This is demonstrated in Fig. 6a where we plot the absorption coefficient of the ideal molecular solid, $A_S^{ideal\ solid}$ along with the absorption coefficient for the real solid, $A_S^{real\ solid}$. Since C_{60} molecules must be in close proximity (as in a solid) for appreciable orbital overlap, we attribute this peak in the real solid not anticipated from gas phase data to the CT exciton predicted by Nasu et. al[2, 57]. The CT exciton has been forwarded as the mechanism by which photopolymerization takes place. If this is indeed the case, the quantum yield for phototransformation ought to be peaked in the region of this band (2.5 to 3.2 eV).

Acknowledgement: The authors are grateful for support of this research from NSF #OSR - 9452895. PCE acknowledges many discussions with Prof. M.S. Dresselhaus and Dr. G. Dresselhaus.

5. References

1. US Patents: US 08-037056 A; US 08-073533 A
2. T. Tsubo and K. Nasu, Jpn. J. Appl. Phys., **34**, 261 (1995).
3. M.S.Dresselhaus, G. Dresselhaus, and P.C. Eklund, *Science of Fullerenes and Carbon Nanotubes.*, Academic Press (1996).
4. P.A. Heiney *et al.*, Phys. Rev. Lett., **66**, 2911 (1991).
5. P.A. Heiney , J. Phys. Chem. Solids, **53**, 1333 (1992).
6. W.I.F. David *et al.*, Europhys. Lett., **18**, 219 (1992).
7. R.D. Johnson, D.S. Bethune. and C.S. Yannoni, Acc. Chem. Res., **25**. 169 (1992).
8. R. Tycko *et al.*, Phys. Rev. Lett., **67**, 1886 (1991).
9. R. Taylor *et al.*, J. Chem. Soc. Chem. Commun., **20**, 1423 (1990).
10. M.S.Dresselhaus, G. Dresselhaus, and P.C. Eklund, *Science of Fullerenes and Carbon Nanotubes.*, Chapter 10, Academic Press (1996).
11. Y. B. Zhao, *et al.*, Appl. Phys. Lett., **64**, 577 (1994).
12. N. Takahashi, *et al.*, J. Appl. Phys., **74**, 5790 (1993).
13. J. Winter and H. Kuzmany, Solid State Commun., **83**, 1321 (1992).
14. S. Pekker, *et al.*, Science, **265**, 1077 (1994).
15. P. W. Stephens, *et al.*, Nature, **370**, 636 (1994).
16. O. Chauvet, *et al.*, Phys. Rev. Lett., **72**, 2721 (1994).
17. O. Bethoux, *et al.*, Proceedings of the Materials Research Society, Abstract No. G2.9, p. 202 (1993).
18. Y. Iwasa *et al.*, Science, **264**, 1570 (1994).
19. M. Nunez-Regueiro *et al.*, Phys. Rev. Lett., **74**, 278 (1995).
20. M. Nunez-Regueiro *et al.*, (in preparation).
21. I. O. Bashkin *et al.*, J. Phys. Condens. Matter, **6**, 7491 (1994).
22. G. Oszlanyi and L. Forro, Solid State Commun., **93**, 265 (1995).
23. Y. Iwasa in *Optical Properties of Low Dimensional Materials.*, Eds. T. Ogawa and Y. Kanemitsu. 1995, World Scientific: Singapore.
24. A. M. Rao *et al.*, Science, **259**, 955 (1993).

25. S. J. Hoke *et al.*, J. Org. Chem., **57**, 5069 (1992).
26. C. Goze *et al.*, Phys. Rev. B, (in press).
27. A. M. Rao *et al.*, J. Mat. Res., **8**, 2277 (1993).
28. C. C. Eloi *et al.*, J. Mat. Res., **8**, 3085 (1994).
29. K. Venkatesan and V. Ramamurthy, "Photochemistry in organized and Constrained Media" Ch. 4. 1991, New York: VCH. p. 133.
30. M. Menon, K.R. Subbaswamy and M. Sawtarie, Phys. Rev. B, **49**, 13966 (1994).
31. G. B. Adams *et al.*, Phys. Rev., **B50**, 17471 (1994).
32. M. R. Pederson and A.A. Quong, Phys. Rev. Lett., **74**, 2319 (1995).
33. D. Porezag *et al.*, Phys. Rev. B, **52**, 14963 (1995).
34. D. L. Strout *et al.*, Chem. Phys. Lett., **214**, 576 (1993).
35. G. B. Adams and J.B. Page,Chapter 5 *in*. Fullerene Polymers and Fullerene-polymer Composites, Eds. P.C. Eklund and A.M. Rao. (to be published), Springer-Verlag.
36. W.I.F David *et al.*, Nature, **353**, 147 (1991).
37. P. Zhou *et al.*, Chem. Phys. Lett., **211**, 337 (1993).
38. P. Zhou *et al.*, Appl. Phys. Lett., **60**, 2871 (1992).
39. J. W. Arbogast *et al.*, J. Phys. Chem, **95**, 11 (1991).
40. R. R. Hung and J.J. Grabowski, J. Phys. Chem., **95**, 6073 (1991).
41. R. E. Haufler *et al.*, Chem. Phys. Lett., **179**, 449 (1991).
42. P. W. Stephens (personal comm.)
43. G. P. Lopinski, J.R. Fox, and J.S. Lannin, Chem. Phys. Lett., **239**, 107 (1995).
44. D. S. Cornett *et al.*, J. Phys. Chem., **97**, 5036 (1993).
45. G. Dresselhaus, M.S. Dresselhaus and P.C. Eklund, Phys. Rev. B, **45**, 6923 (1992).
46. M.S.Dresselhaus, G. Dresselhaus, and P.C. Eklund, *Science of Fullerenes and Carbon Nanotubes.*, Chapter 11, Academic Press (1996).
47. P. C. Eklund *et al.*, Thin Solid Films, **257**, 185 (1995).
48. P. C. Eklund *et al.*, Thin Solid Films, **257**, 211 (1995).
49. M. S. Dresselhaus, G. Dresselhaus and P.C. Eklund, J. Raman Spect., **27**, 351 (1996).
50. P. C. Eklund *et al.*, J. Phys. Chem. Solids, **53**, 1391 (1992).
51. A. M. Rao (in preparation).
52. Y. Wang*et al.*, Chem. Phys. Lett., **217**, 413 (1994).
53. M.S.Dresselhaus, G. Dresselhaus, and P.C. Eklund, *Science of Fullerenes and Carbon Nanotubes.*, Chapter 13, Academic Press (1996).
54. S. G. Louie and E.L. Shirley, J. Phys. Chem. Solids, **54**, 1767 (1993).
55. E.L. Shirley and S.G. Louie, Phys. Rev. Lett.,**71**, 133 (1993).
56. K. Harigaya and S. Abe, Mol. Cryst. Liq. Crist., **256**, 825 (1994).
57. T. Tsubo and K. Nasu, Solid State Commun., **91**, 907 (1994).
58. S. L. Ren *et al.*, Appl. Phys. Lett., **59**, 2678 (1991).

Ultrafast Relaxation of Excitons and Photopolymerization Process in Fullerenes

A. Nakamura, M. Ichida, M. Sakai, and T. Yajima

Center for Integrated Research in Science and Engineering, and Department of Applied Physics, Nagoya University, Chikusa-ku, Nagoya, 464-01 Japan

Abstract. We have studied relaxation dynamics of photoexcitation and initial processes of the photopolymerization in C_{60} molecules and solids by means of time-resolved spectroscopies. The femtosecond pump and probe spectroscopy shows that the photoexcitations in C_{60} solution are relaxed into the triplet state in 900 ps, while for C_{60} solid the fast relaxation occurs in 1 ps. The fast relaxation of charge-transfer (CT) excitons being characteristic of the C_{60} solid is interpreted in terms of the self-trapping of CT excitons in the simple cubic lattice. In the luminescence spectrum of the pristine crystal, we observe luminescence due to the S_1-S_0 transition which is partially allowed by the Hertzberg-Teller mechanism and luminescence originating from the self-trapped CT excitons. From the comparison of the similar spectral features observed for the pristine and photopolymerized crystals we discuss configurations of the photoinduced dimer and self-trapped CT exciton, and propose that the self-trapped CT exciton is a precursor to the photopolymerization.

1. Introduction

Since the discovery of the fullerene family of carbon cluster molecules, a great deal of optical study has been carried out to investigate its electronic structures using conventional spectroscopic methods [1-5]. The fullerene molecule and solid show interesting electronic and optical properties because of highly symmetric structure and rigidity of molecule itself. In fullerene crystals, there are two very different bond strengths. The one is intermolecular bond and the other is the intramolecular bond. The former is largely van der Waals in character, while the latter is covalent. When the fullerene crystals are optically excited, the excitation and de-excitation occur in two different ways; the intramolecular relaxation in the isolated molecule and the intermolecular relaxation in the crystal lattice [6].

Recently, it has been reported that exposure to visible or ultraviolet light causes solid C_{60} to polymerize, where the molecules cross link via covalent bonds [7,8]. The measurements of the laser desorption spectrum of phototransformed C_{60} film revealed that the formation of C_{120}, C_{180} and so on in solid state. A 2+2 cyclo addition photochemical reaction has been proposed for the photopolymerization mechanism [7]. It is of great interest to investigate how such a permanent lattice distortion is induced and stabilized after the optical excitation.

In this paper we review the ultrafast relaxation dynamics of photoexcitations and initial processes of photopolymerization which were studied by the pump and probe

Springer Series in Solid-State Sciences, Volume 124
Relaxations of Excited States and Photo-Induced Structural Phase Transitions
Ed.: K. Nasu

spectroscopy and the time-resolved luminescence spectroscopy. In C_{60} solid, we observe a characteristic absorption band due to the intermolecular charge-transfer exciton (CT exciton) which is composed of an electron and a hole being separated at the adjacent molecules at the lattice site. Relaxation of CT excitons takes place accompanying with distortion of the simple cubic lattice by the strong interaction between CT exciton and intermolecular vibration. This self-trapped CT exciton state is suggested to be a precursor to the photoinduced dimer.

2. Experimental

C_{60} and C_{70} fullerenes were synthesized and purified using a Kratchmer-Huffman method[9]. Samples of C_{60} and C_{70} in solution were prepared by dissolving C_{60} and C_{70} powders in spectroquality toluene. Thin films were evaporated on quartz substrates from Mo boats at 700 C with the pressure of 2×10^{-6} Torr. A thickness of the film was about 150 nm. Single crystals of C_{60} were prepared from pure (>99.9%) powders by a sublimation method. The typical size of the crystal is ~0.5 mm. The experiments were performed using a conventional pump and probe setup. The femtosecond pumping pulses were generated by the second harmonic of an amplified Ti:sapphire laser system at 10 Hz. The duration and the energy of the pulse are 200 fs and ~10 μJ, respectively. Transient absorption spectra were measured by white continuum pulses in the spectral range of 400-800 nm. The luminescence spectra were measured by using a 1 m grating monochromator with a cooled photomultiplier or a CCD camera. The time-resolved luminescence spectra were measured by a time-correlated single photon counting method using second harmonic pulses of a mode-locked Ti:sapphire laser with a wavelength of 752 nm.

3. Results and discussion

3.1 Absorption spectra

Figure 1 shows absorption spectra of C_{60} in solution and C_{60} film in the visible to ultraviolet region. In the spectrum of C_{60} in toluene we observe a weak absorption band due to the HOMO-LUMO forbidden transition around 2 eV [10]. The strong absorption band peaked at 3.7 eV is assigned to the transition from h_u to t_{1g} states. In the solid state, the allowed transition band is broadened and shifted to the lower energy side owing to the condensation. In the energy region where the very weak absorption is observed for isolated C_{60} molecules, we see a strong absorption band which is characteristic of the solid C_{60}. Recently, Tsubo and Nasu have calculated excitonic transition energies and oscillator strengths for the fcc C_{60} crystal, and demonstrated that there appears an excitonic absorption band around 2.5 eV [11]. As the probability that the electron and hole forming the exciton are found on the same molecule is almost zero, this exciton is composed of the electron and hole which are separated at the adjacent molecules. Therefore, we can attribute the absorption band at 2.7 eV to the intermolecular charge transfer exciton (CT exciton).

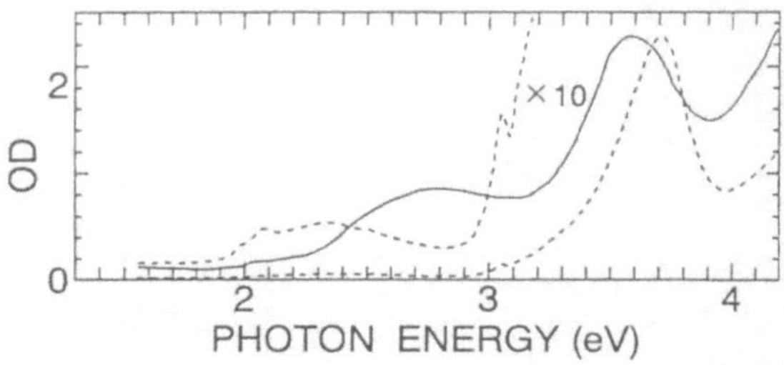

Fig.1: Absorption spectra of C_{60} in toluene (dashed curve) and C_{60} thin film (solid curve).

3.2 Relaxation dynamics of photoexcitations in C_{60}

Figure 2(a) and (b) show the absorption spectra of C_{60} thin film measured at 80 K [6]. The weak absorption band corresponding to the HOMO-LUMO gap transition is observed around 2 eV and the absorption band due to CT excitons is observed between 2.3 and 3.2 eV. Shown in Fig.2(c) are differential absorption spectra for t_d=0, 100 and 490 ps in C_{60} thin film at 80 K. The photon energy of pumping pulses corresponds to the higher energy side of the CT exciton band. For t_d=0 ps we observe an induced absorption band below 2.4 eV, while the decrease in the absorption is observed in the energy region of 2.4-3.0 eV. The similar results have been also observed by the excitation with the photon energy of 4.1 eV [14]. Comparing the induced absorption band with the absorption band in 1.8-3.0 eV observed for the C_{60} solution, the absorption increase is ascribed to the transition from the LUMO t1u band to the higher conduction bands. The absorption decrease near 2.6 eV in Fig.2(c) indicates the bleaching of the CT exciton band and is composed of two bleaching peaks at 2.5 and 2.7 eV. For t_d=100 ps both the induced absorption and the bleaching are partially recovered but the bleaching appears again for t_d=490 ps. These behaviors imply that there are several components in the decay kinetics and the component of induced absorption is superimposed on the bleaching component in the spectral region of 2.3-2.8 eV. The spectral and temporal behaviors observed for C_{60} film are quite different from those of molecules in solution 6,13]. In solution, only the induced absorption is observed and its decay time is about 900 ps, which is governed by the formation of the triplet state. It is worthy to note that the triplet absorption cannot be observed for solid C_{60} even in the late stage of the delay time. This is because the decay time of the CT excitons is very fast compared to the intersystem crossing time and the triplet state is not sufficiently formed.

The time variations of absorption change measured at 2.3 and 2.7 eV are illustrated in Fig.3. Although we see the oscillatory behavior for 1-5 ps, it seems to be the scattering of experimental data points. The decay behavior monitored at the photon energy of 2.3 eV corresponds to the decay of the induced absorption, which exhibits two-component decay. Assuming a two-component exponential decay, the decay times of fast- and slow-component are analyzed to be 1 ps and ~400 ps, respectively. From the argument of the spectral behavior, these decay times are found

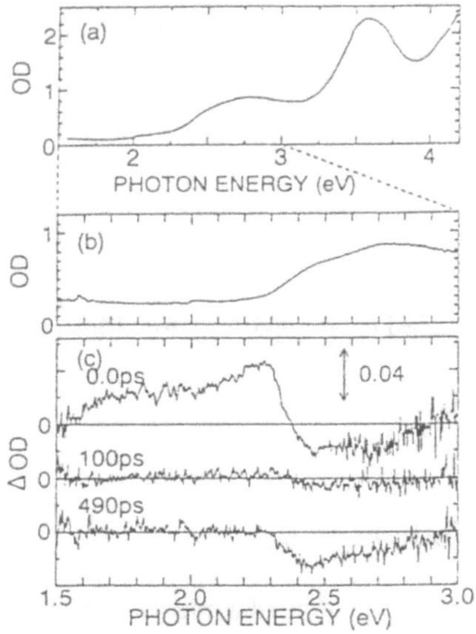

Fig.2. Absorption spectra (a and b), and the differential absorption spectra (c) for C_{60} thin film at 80 K. The delay times are 0, 100, and 490 ps.

to be the lifetimes of the electrons in the t_{1u} conduction band. The decay behavior measured at 2.7 eV indicates the recovery of the bleaching of the CT exciton band. The fast decay component with the decay time of 1 ps corresponds to the relaxation of the CT excitons and the slow one with the decay time of >1 ns corresponds to the recovery of population in the ground state. The fast decay behavior of the photoexcitation in solid C_{60} is contrast with the slow decay behavior observed for the isolated molecules in solution [6]. If we estimate periods of the translational modes using the values of 40.9 and 54.7 cm^{-1} measured by the far-infrared spectroscopy[15], periods of 0.8 and 0.6 ps are obtained. These values are in agreement with the fast decay time. We have also performed measurements of time evolution at 4.2 K and found the temperature independent behavior of the relaxation dynamics.

We propose the following scenario for the relaxation process of the CT excitons [6,14]. The good coincidence of the fast decay time with the period of the translational mode suggests that the lattice relaxation of the sc geometry plays an important role in the relaxation of CT excitons excited on the adjacent molecules in the sc lattice. The lattice relaxation in the sphere-like molecule has been proposed both theoretically and experimentally [16,17]. However, the different behaviors of decay kinetics for C_{60} solution and solid C_{60} lead us to consider the relaxation of the sc lattice. Figure 4 shows the schematic representation of configurational coordinate diagram which explains the relaxation dynamics of CT excitons. The coordinate Q represents the displacement of the intermolecular distance. The strong coupling between the intermolecular vibration and the CT exciton gives rise to the distortion of the equilibrium distance between C_{60} molecules during the course of the exciton

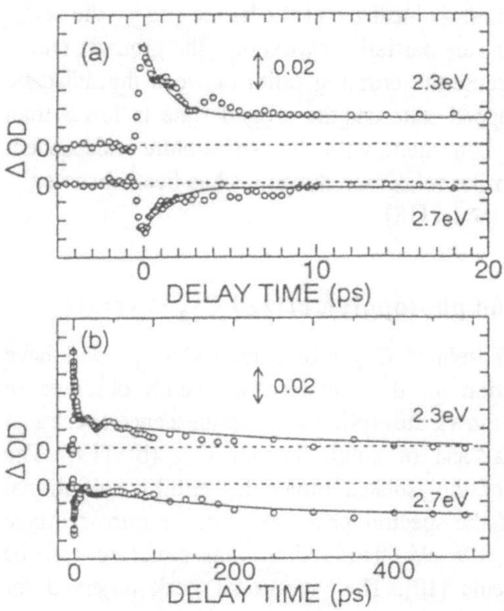

Fig.3. Time variations of absorption change measured at 2.3 and 2.7 eV for C_{60} thin film at 80 K.

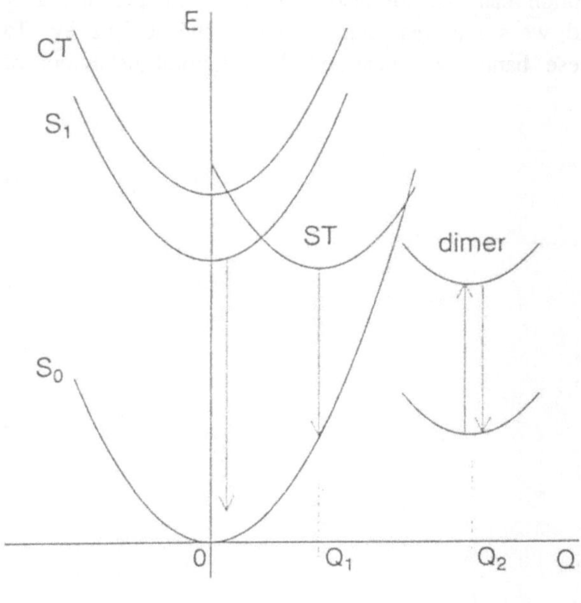

Fig.4. Schematic illustration of adiabatic potential energy surfaces of C_{60} solid. Q: displacement of intermolecular distance, CT: charge transfer exciton, ST: self-trapped CT exciton.

relaxation. Consequently, the CT exciton is localized at the lattice site by the self-trapping. These self-trapped CT excitons are partially relaxed into the ground state by the dynamic relaxation process, because the crossing point between the adiabatic potential surfaces (APS) of the self-trapped state and the ground state is lower than the optical excitation energy. This is consistent with the temperature independent behaviors of the decay kinetics. The similar relaxation dynamics has been observed for C_{70}, and is interpreted by the same model [18].

3.3 Luminescence in pristine and photopolymerized C_{60} crystals

Although the luminescence spectra of isolated C_{60} molecules and C_{60} solid have been intensively studied, the explanation for the luminescence bands observed in solid is still controversial. Figure 5 shows absorption and luminescence spectra of isolated C_{60} molecules in toluene (a) and of single crystal C_{60} (b) [19]. The absorption and luminescence spectra of the isolated molecules exhibit vibrational structure, and the energy positions of the spectral peaks exhibit the mirror image taking the 0-0 transition energy as 1.9 eV. This vibrational structure can be explained by the Hertzberg-Teller scheme [10]. The vibrational peak observed for solid C_{60} is shifted to the lower energy side by 42 meV, although the absorption spectrum for solid C_{60} shows the structure similar to that for isolated molecules. Therefore, the spectral feature observed near the edge for solid C_{60} can be also attributed to the partially allowed S_1-S_0 transition. However, the spectral feature of luminescence spectrum for solid C_{60} is quite different; the main luminescence band below 1.8 eV and the absorption band do not display a mirror image. In higher energy side of the main band, we see a weak luminescence band at 1.82 eV. To investigate the origin of these bands, we measured the temporal behaviors of

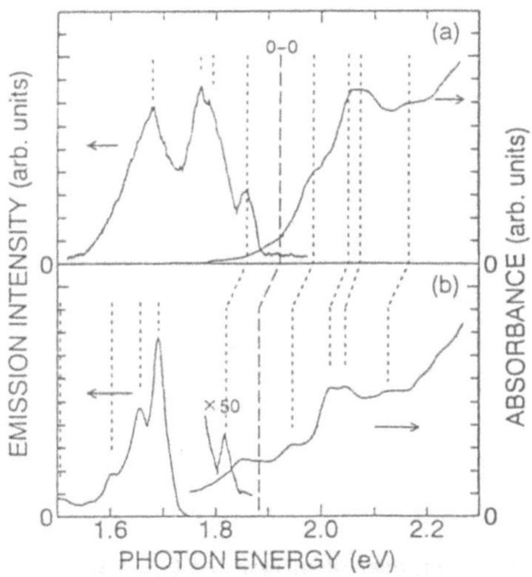

Fig.5. Absorption and luminescence spectra of isolated C_{60} molecules in toluene solution (a) and of C_{60} single crystal at 80 K (b).

luminescence, and found that the the decay times of the main band and the weak band are 1.2 ns and 500 ps, respectively. The decay time of the weak band is in good agreement with the lifetime of the LUMO state. Therefore, we can assign this band to the S_1-S_0 transition with emission of intramolecular t1u mode, i.e. Frenkel exciton luminescence.

If the band at 1.82 eV is interpreted by the S_1-S_0 transition assisted with the t_{1u} mode, the energy positions of the emission peaks cannot be assigned to any vibrational mode. Therefore, we can conclude that the main band is not related with the S_1-S_0 transition. Using the configurational coordinate diagram which explains the relaxation dynamics of CT excitons, the main band is suggested to originate from the minimum of the APS of self-trapped CT excitons. Although the most of excitons are relaxed into the ground state nonradiatively, the excitons which could not pass through the barrier stay on the minimum and recombine radiatively.

We have investigated luminescence spectra of the C_{60} crystal exposed to the Ar laser light at the power density of about 100 W/cm^2. Measuring Raman spectrum, we confirmed that the photo-induced polymerization occurs at this power density. Shown in Fig.6 are luminescence spectra of the pristine and photopolymerized crystals. In the polymerized crystal, luminescence band due to the polymers, probably dimers is broadened and shifted to the lower energy side by 35 meV, compared to that of self-trapped excitons. The broadening of the spectrum might be due to the simultaneous formation of dimers, trimers and so on. The similarity of the overall spectral features suggests that displacement beyond the self-trapped CT exciton configuration corresponds to the dimer configuration. In the configurational coordinate diagram shown in Fig.4, the minimum of the APS of the dimer state lies on the larger displacement of the coordinate.

From these results we can propose that the self-trapping of CT excitons accompanied by the lattice relaxation is a precursor to the photopolymerization. At

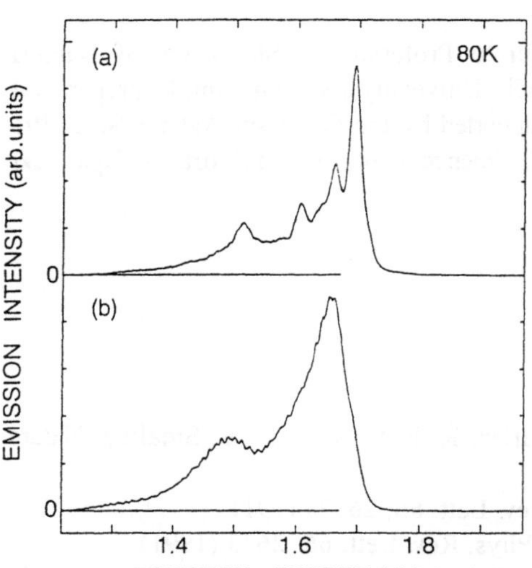

Fig.6. Luminescence spectra of pristine (a) and photopolymerized (b) C_{60} crystals.

room temperature which is higher than the spinning transition temperature [20], it is possible to achieve the parallel alignment between reactive carbon double bonds on adjacent molecules. To form a C_{60} dimer, the optical excitation should initially cause the displacement of adjacent molecules and as a result the double bonds are broken and reformed as a four-sided ring. The formation process proposed here is quite analogous to that of the F center (a vacancy trapping an electron) and H center (an interstitial halogen atom) in alkali halides where the self-trapped exciton is the precursor to the formation of the these defects [21].

4. Conclusions

We have investigated the ultrafast relaxation dynamics and initial processes of the photopolymerization. Comparing the results of spectral and temporal behaviors observed for C_{60} molecules both in solution and in solid, we discussed the relaxation processes being characteristic of the isolated molecule and the solid state. With the excitation of the intermolecular CT excitons in C_{60} solid, the fast relaxation occurs in 1ps, which is interpreted in terms of the self-trapping of CT excitons. In C_{60} solution, however, the much slower relaxation has been observed and we have found that the decay of the singlet state is governed by the formation of the triplet state with the characteristic time of 900 ps. The self-trapped state which induces the transient distortion of the sc lattice seems to be a precursor to the photopolymerization.

In luminescence spectrum of the pristine crystal, the luminescence bands originating from both the Frenkel excitons and the self-trapped CT excitons are observed. The luminescence spectrum of the photopolymerized crystal exhibits the spectral feature similar to that of self-trapped CT excitons, which suggests that the photoinduced dimer has the configuration corresponding to the successive relaxation of self-trapped CT excitons.

Acknowledgement. We are very grateful to Professor H. Shinohara of Nagoya University and Professor Y. Saitoh of Mie University for their sample preparation and useful discussion. This work was supported by the Grants-in Aid for Scientific Research from the Ministry of Education, Science, Culture and Sports of Japan and Tatematsu Foundation.

References

[1] H. W. Kroto, J. R. Heath, S. C. O'Brien, R. F. Curl and R. E. Smalley, Nature 318, 162 (1985).
[2] S. Saitoh and A. Oshiyama, Phys. Rev. Lett. 66, 2637 (1991).
[3] Q.-M. Zhang, J.-Y. Yi, J. Bernholc, Phys. Rev. Lett. 66, 2633 (1991).
[4] C. Reber, L. Yee, J. Mckiernan, J. I. Zink, R. S. Williams, W. M. Tang, D. A. A. Ohlberg, R. L. Whetten and F. Diederich, J. Phys. Chem., p5, 2127 (1991).

[5] A. F. Hebard, R. C. Haddon, R. M. Fleming and A. R. Kotan, Appl. Phys. Lett. 59, 2109 (1991).
[6] A. Nakamura, M. Ichida, T. Yajima, H. Shinohara and Y. Saitoh, J. Lumin. 66&67, 383 (1996).
[7] A. M. Rao, P. Z. Zhou, Kai-An Wang, G. T. Hager, J. M. Holden, Ying Wang, W-T. Lee, Xiang-Xie Bi, P. C. Eklund, D. S. Cornett, M. A. Duncan and I. J. Amster, Science 259, 955 (1993).
[8] P. Z. Zhou, Z-H. Dong, A. M. Rao and P. C. Eklund, Chem. Phys. Lett. 211, 337 (1993).
[9] W. Kratschmer, L. D. Lamb, K. Forstiropoulos and D. R. Huffman, Nature 347, 354 (1990).
[10] S. Leach, M. Vervloet, A. Despre, E. Breheret, J. P. Hare, T. J. Dennis, H. W. Kroto, R. Taylor and D. R. M. Walton, Chem. Phys. 160, 451 (1992).
[11] T. Tsubo and K. Nasu, J. Phys. Soc. Jpn. 63, 2401 (1994).
[12] R. V. Besasson, T. Hill, C. Lambert, E. J. Land, S. Leach and T. G. Truscott, , Chem. Phys. Lett. 201, 326 (1993).
[13] R. j. Sension, C. M. Philiips, A. Z. Szarka, W. J. Ramanow, A. R. McGhie, J. P. McCauley, Jr., A. B. Smith III and R. M. Hochstrasser, J. Phys. Chem. 95, 6075 (1991).
[14] A. Nakamura, T. Tokizaki, S. Iwai, Y. L. Lee, Y. Saitoh and H. Shinohara, Proc. Int. Conf. on the Physics of Semiconductors, Beijin, 1992, (World Scientific, 1992) p.1828.
[15] S. Huant, J. B. Robert, G. Chouteau, P. Bernier, C. Fabre and A. Rassat, Phys. Rev. Lett. 69, 2666 (1992). and S. A. FitzGerald and A. J. Sievers, Phys. Rev. Lett. 70, 3175 (1993).
[16] M. Matus, H. Kuzmany, E. Sohmen, Phys. Rev. Lett. 68, 2822 (1992).
[17] W. Z. Wang, C. L. Wang, Z. B. Su and L. Yu, Phys. Rev. Lett. 3550 (1994).
[18] M. Ichida, M. Sakai, T. Yajima and A. Nakamura, Progress in Crystal Growth & Characterization on Materials, in press.
[19] A. Nakamura, M. Ichida and T. Yajima, Progress in Crystal Growth & Characterization on Materials, in press.
[20] P. A. Heiney, J. E. Fischer, A. R. McGhie, W. J. Romanow, A. M. Denenstein, J. P. McCanley,Jr., and A. B. Smith,III, Phys. Rev. Lett. 66, 2911 (1991).
[21] S. Iwai, T. Tokizaki, A. Nakamura, T. Shibata, K. Tanimura, N. Itoh and A. Shluger, Phy. Rev. Lett. 76, 1691 (1996).

Index of Contributors

Springer Series in Solid-State Sciences

Editors: M. Cardona P. Fulde K. von Klitzing H.-J. Queisser

1 **Principles of Magnetic Resonance** 3rd Edition By C. P. Slichter
2 **Introduction to Solid-State Theory** By O. Madelung
3 **Dynamical Scattering of X-Rays in Crystals** By Z. G. Pinsker
4 **Inelastic Electron Tunneling Spectroscopy** Editor: T. Wolfram
5 **Fundamentals of Crystal Growth I** Macroscopic Equilibrium and Transport Concepts By F. E. Rosenberger
6 **Magnetic Flux Structures in Superconductors** By R. P. Huebener
7 **Green's Functions in Quantum Physics** 2nd Edition By E. N. Economou
8 **Solitons and Condensed Matter Physics** Editors: A. R. Bishop and T. Schneider
9 **Photoferroelectrics** By V. M. Fridkin
10 **Phonon Dispersion Relations in Insulators** By H. Bilz and W. Kress
11 **Electron Transport in Compound Semiconductors** By B. R. Nag
12 **The Physics of Elementary Excitations** By S. Nakajima, Y. Toyozawa, and R. Abe
13 **The Physics of Selenium and Tellurium** Editors: E. Gerlach and P. Grosse
14 **Magnetic Bubble Technology** 2nd Edition By A. H. Eschenfelder
15 **Modern Crystallography I** Fundamentals of Crystals, Symmetry, and Methods of Structural Crystallography 2nd Edition By B. K. Vainshtein
16 **Organic Molecular Crystals** Their Electronic States By E. A. Silinsh
17 **The Theory of Magnetism I** Statics and Dynamics By D. C. Mattis
18 **Relaxation of Elementary Excitations** Editors: R. Kubo and E. Hanamura
19 **Solitons** Mathematical Methods for Physicists By. G. Eilenberger
20 **Theory of Nonlinear Lattices** 2nd Edition By M. Toda
21 **Modern Crystallography II** Structure of Crystals 2nd Edition By B. K. Vainshtein, V. L. Indenbom, and V. M. Fridkin
22 **Point Defects in Semiconductors I** Theoretical Aspects By M. Lannoo and J. Bourgoin
23 **Physics in One Dimension** Editors: J. Bernasconi and T. Schneider
24 **Physics in High Magnetics Fields** Editors: S. Chikazumi and N. Miura
25 **Fundamental Physics of Amorphous Semiconductors** Editor: F. Yonezawa
26 **Elastic Media with Microstructure I** One-Dimensional Models By I. A. Kunin
27 **Superconductivity of Transition Metals** Their Alloys and Compounds By S. V. Vonsovsky, Yu. A. Izyumov, and E. Z. Kurmaev
28 **The Structure and Properties of Matter** Editor: T. Matsubara
29 **Electron Correlation and Magnetism in Narrow-Band Systems** Editor: T. Moriya
30 **Statistical Physics I** Equilibrium Statistical Mechanics 2nd Edition By M. Toda, R. Kubo, N. Saito
31 **Statistical Physics II** Nonequilibrium Statistical Mechanics 2nd Edition By R. Kubo, M. Toda, N. Hashitsume
32 **Quantum Theory of Magnetism** 2nd Edition By R. M. White
33 **Mixed Crystals** By A. I. Kitaigorodsky
34 **Phonons: Theory and Experiments I** Lattice Dynamics and Models of Interatomic Forces By P. Brüesch
35 **Point Defects in Semiconductors II** Experimental Aspects By J. Bourgoin and M. Lannoo
36 **Modern Crystallography III** Crystal Growth By A. A. Chernov
37 **Modern Chrystallography IV** Physical Properties of Crystals Editor: L. A. Shuvalov
38 **Physics of Intercalation Compounds** Editors: L. Pietronero and E. Tosatti
39 **Anderson Localization** Editors: Y. Nagaoka and H. Fukuyama
40 **Semiconductor Physics** An Introduction 6th Edition By K. Seeger
41 **The LMTO Method** Muffin-Tin Orbitals and Electronic Structure By H. L. Skriver
42 **Crystal Optics with Spatial Dispersion, and Excitons** 2nd Edition By V. M. Agranovich and V. L. Ginzburg
43 **Structure Analysis of Point Defects in Solids** An Introduction to Multiple Magnetic Resonance Spectroscopy By J.-M. Spaeth, J. R. Niklas, and R. H. Bartram
44 **Elastic Media with Microstructure II** Three-Dimensional Models By I. A. Kunin
45 **Electronic Properties of Doped Semiconductors** By B. I. Shklovskii and A. L. Efros
46 **Topological Disorder in Condensed Matter** Editors: F. Yonezawa and T. Ninomiya

Springer Series in Solid-State Sciences

Editors: M. Cardona P. Fulde K. von Klitzing H.-J. Queisser

47 **Statics and Dynamics of Nonlinear Systems** Editors: G. Benedek, H. Bilz, and R. Zeyher

48 **Magnetic Phase Transitions** Editors: M. Ausloos and R. J. Elliott

49 **Organic Molecular Aggregates** Electronic Excitation and Interaction Processes Editors: P. Reineker, H. Haken, and H. C. Wolf

50 **Multiple Diffraction of X-Rays in Crystals** By Shih-Lin Chang

51 **Phonon Scattering in Condensed Matter** Editors: W. Eisenmenger, K. Laßmann, and S. Döttinger

52 **Superconductivity in Magnetic and Exotic Materials** Editors: T. Matsubara and A. Kotani

53 **Two-Dimensional Systems, Heterostructures, and Superlattices** Editors: G. Bauer, F. Kuchar, and H. Heinrich

54 **Magnetic Excitations and Fluctuations** Editors: S. W. Lovesey, U. Balucani, F. Borsa, and V. Tognetti

55 **The Theory of Magnetism II** Thermodynamics and Statistical Mechanics By D. C. Mattis

56 **Spin Fluctuations in Itinerant Electron Magnetism** By T. Moriya

57 **Polycrystalline Semiconductors** Physical Properties and Applications Editor: G. Harbeke

58 **The Recursion Method and Its Applications** Editors: D. G. Pettifor and D. L. Weaire

59 **Dynamical Processes and Ordering on Solid Surfaces** Editors: A. Yoshimori and M. Tsukada

60 **Excitonic Processes in Solids** By M. Ueta, H. Kanzaki, K. Kobayashi, Y. Toyozawa, and E. Hanamura

61 **Localization, Interaction, and Transport Phenomena** Editors: B. Kramer, G. Bergmann, and Y. Bruynseraede

62 **Theory of Heavy Fermions and Valence Fluctuations** Editors: T. Kasuya and T. Saso

63 **Electronic Properties of Polymers and Related Compounds** Editors: H. Kuzmany, M. Mehring, and S. Roth

64 **Symmetries in Physics** Group Theory Applied to Physical Problems 2nd Edition By W. Ludwig and C. Falter

65 **Phonons: Theory and Experiments II** Experiments and Interpretation of Experimental Results By P. Brüesch

66 **Phonons: Theory and Experiments III** Phenomena Related to Phonons By P. Brüesch

67 **Two-Dimensional Systems: Physics and New Devices** Editors: G. Bauer, F. Kuchar, and H. Heinrich

68 **Phonon Scattering in Condensed Matter V** Editors: A. C. Anderson and J. P. Wolfe

69 **Nonlinearity in Condensed Matter** Editors: A. R. Bishop, D. K. Campbell, P. Kumar, and S. E. Trullinger

70 **From Hamiltonians to Phase Diagrams** The Electronic and Statistical-Mechanical Theory of sp-Bonded Metals and Alloys By J. Hafner

71 **High Magnetic Fields in Semiconductor Physics** Editor: G. Landwehr

72 **One-Dimensional Conductors** By S. Kagoshima, H. Nagasawa, and T. Sambongi

73 **Quantum Solid-State Physics** Editors: S. V. Vonsovsky and M. I. Katsnelson

74 **Quantum Monte Carlo Methods in Equilibrium and Nonequilibrium Systems** Editor: M. Suzuki

75 **Electronic Structure and Optical Properties of Semiconductors** 2nd Edition By M. L. Cohen and J. R. Chelikowsky

76 **Electronic Properties of Conjugated Polymers** Editors: H. Kuzmany, M. Mehring, and S. Roth

77 **Fermi Surface Effects** Editors: J. Kondo and A. Yoshimori

78 **Group Theory and Its Applications in Physics** 2nd Edition By T. Inui, Y. Tanabe, and Y. Onodera

79 **Elementary Excitations in Quantum Fluids** Editors: K. Ohbayashi and M. Watabe

80 **Monte Carlo Simulation in Statistical Physics** An Introduction 2nd Edition By K. Binder and D. W. Heermann

81 **Core-Level Spectroscopy in Condensed Systems** Editors: J. Kanamori and A. Kotani

82 **Photoelectron Spectroscopy** Principle and Applications 2nd Edition By S. Hüfner

83 **Physics and Technology of Submicron Structures** Editors: H. Heinrich, G. Bauer, and F. Kuchar

84 **Beyond the Crystalline State** An Emerging Perspective By G. Venkataraman, D. Sahoo, and V. Balakrishnan

85 **The Quantum Hall Effects** Fractional and Integral 2nd Edition By T. Chakraborty and P. Pietiläinen

86 **The Quantum Statistics of Dynamic Processes** By E. Fick and G. Sauermann

87 **High Magnetic Fields in Semiconductor Physics II** Transport and Optics Editor: G. Landwehr

88 **Organic Superconductors** By T. Ishiguro and K. Yamaji

89 **Strong Correlation and Superconductivity** Editors: H. Fukuyama, S. Maekawa, and A. P. Malozemoff

MIX
Papier aus verantwortungsvollen Quellen
Paper from responsible sources
FSC® C105338

If you have any concerns about our products,
you can contact us on
ProductSafety@springernature.com

In case Publisher is established outside the EU,
the EU authorized representative is:
Springer Nature Customer Service Center GmbH
Europaplatz 3, 69115 Heidelberg, Germany

Printed by Libri Plureos GmbH
in Hamburg, Germany